# X-Ray Microscopy and Spectromicroscopy

Springer-Verlag Berlin Heidelberg GmbH

Jürgen Thieme
Günter Schmahl
Dietbert Rudolph
Eberhard Umbach
(Eds.)

# X-Ray
## Microscopy
## and
## Spectromicroscopy

Status Report
from the Fifth International Conference,
Würzburg, August 19–23, 1996

The book contains the invited papers
(With 249 Figures and 14 Tables)
The CD-ROM contains all contributions to the conference
(With 593 Figures and 35 Tables, many in colour)

 Springer

Dr. Jürgen Thieme

Professor Dr. Günter Schmahl

Dr. Dietbert Rudolph

Forschungseinrichtung Röntgenphysik
Georg-August-Universität Göttingen
Geiststrasse 11, D-37073 Göttingen, Germany

E-mail: jthieme@gwdg.de
        gschmah@gwdg.de
        drudolp@gwdg.de

Professor Dr. Eberhard Umbach

Experimentelle Physik II
Universität Würzburg
Am Hubland, D-97074 Würzburg, Germany

E-mail: umbach@physik.uni-wuerzburg.de

---

*Cover picture:* X-ray microscopic image of the microstructure formed by colloidal particles within a dystric cambisol. (Thieme et al., Aggregation of Colloids Observed by X-Ray Microscopy. Page II-15).

Library of Congress Cataloging-in-Publication Data
X-ray microscopy and spectromicroscopy / Jürgen Thieme ... [et al.], (eds.). p. cm. Based on presentation to the International Conference of X-Ray Microscopy and Spectromicroscopy, XRM 96, which took place in Würzburg, August 19-23, 1996.
ISBN 978-3-642-72108-3    ISBN 978-3-642-72106-9 (eBook)
DOI 10.1007/978-3-642-72106-9

1. X-ray microscopy–Congresses. I. Thieme, Jürgen. II. International Conference of X-Ray Microscopy and Spectromicroscopy (1996: Würzburg, Germany)
QH212.X2X25 1998 502'.8'2–dc21 98-15577

Additional material to this book can be downloaded from http://extras.springer.com

© Springer-Verlag Berlin Heidelberg New York 1998
Originally published by Springer-Verlag Berlin Heidelberg 1998
Softcover reprint of the hardcover 1st edition 1998

Typesetting: Camera ready by the first editor
Cover design: *design & production GmbH*, Heidelberg

SPIN: 10646159        57/3144 - 5 4 3 2 1 0 - Printed on acid-free paper

# Preface

This book is based on presentations to the International Conference of X-Ray Microscopy and Spectromicroscopy, XRM 96, which took place in Würzburg, August 19–23, 1996. The conference also celebrated the 100th anniversary of the discovery of X-rays by Wilhelm Conrad Röntgen on November 8, 1895, in Würzburg.

This book contains state-of-the-art reviews and up-to-date progress reports in the field of X-ray microscopy and spectromicroscopy, including related new X-ray optics and X-ray sources. It reflects the lively activities within a relatively new field of science which combines the development of new instruments and methods with their applications to numerous topical scientific questions. The applications range from biological and medical topics, colloid physics, and soil sciences to solid-state physics, material sciences, and surface sciences. Their variety demonstrates the interdisciplinary and cooperative character of this field and the growing demand for microscopic and spectromicroscopic information on the nanometer scale and under specific sample conditions, for example in wet (natural) surroundings or on a solid surface. We emphasize that the enormous progress in this field that has been made recently and will be made in the future, concerning the resolution, contrast, and versatility, results from the availability of high-brilliance soft X-ray radiation provided especially by the third-generation X-ray synchrotron sources such as ALS in Berkeley, ELETTRA in Trieste, or BESSY II in Berlin. Of similar interest is the further development of improved X-ray optics and X-ray condensers with high numerical apertures and high-resolution X-ray objectives, both with high efficiencies.

In a transmission X-ray microscope (TXM) the X-ray objective generates an enlarged image of the specimen. In a scanning transmission X-ray microscope (STXM) the X-ray objective forms a microprobe used to examine the specimen one pixel at a time. Experiments performed with both types of instrument are presented in this book, including the new developments of cryo-X-ray microscopy and X-ray microscopic tomography. A variety of scientific applications have been addressed in fields of biological and medical research (e.g., cells, sperm, algae, bone), in colloid physics and soil sciences (e.g., interactions of soil colloids with surfactants and micro-organisms, aggregation phenomena, lipid membranes), and in solid-state physics (magnetic domains).

Spectromicroscopy and microspectroscopy have added a new, spectroscopic, dimension to the microscopy of surfaces. The microscopic information can be obtained either by imaging photoexcited electrons by scanning focused X-ray beam across a sample surface, whereas the spectroscopic information is derived from the variation of the photon energy and/or from the energy analysis of the emitted electrons. Both approaches have several advantages and are realized in different instruments, some of which are presented and discussed in this book. The availability of such instruments has, of course, lead to numerous applications, some of which are described here. They cover a wide range from medical research to material sciences and applications (e.g., polymers, thin-film solar cells, magnetic layers).

To keep the book to a reasonable size, only some of the contributions have been printed. However, all contributions listed in the table of contents can be found on the attached compact disk.

In our capacity as organizers we would like to thank the University of Würzburg for its hospitality and assistance during the conference. We are grateful to the Deutsche Forschungsgemeinschaft and the Balzers Hochvakuum AG as the main sponsors.

We would like to thank all authors for their cooperation and all of our coworkers who have contributed to the organization of the conference, especially Dr. Rainer Fink, Dr. Kirsten Thieme, Lieselotte Reichert, and Britta Leinemann.

Göttingen and Würzburg,
Mai 1998

*Jürgen Thieme*
*Günter Schmahl*
*Dietbert Rudolph*
*Eberhard Umbach*

# Table of Contents

---

All contributions listed here can be found on the compact disk attached to the book, the
contributions without a * can also be found as a black & white printed version in the book.

## Part II    X-Ray Microscopy Applications

## Part III    Microspectroscopy and Spectromicroscopy

## Part IV    X-Ray Optics

## Part V    X-Ray Sources

# Part I

# X-Ray Microscopy Projects

# X-Ray Microscopy in Berkeley

W. Meyer-Ilse[1], H. Medecki[1], J. T. Brown[1], J. M. Heck[1], E. H. Anderson[1],
A. Stead[2], T. Ford[2], R. Balhorn[3], C. Petersen[4], C. Magowan[5], D. T. Attwood[1]

[1] Center for X-ray Optics, Lawrence Berkeley National Laboratory,
Berkeley, CA 94720, USA, E-Mail: W_Meyer-Ilse@LBL.GOV
[2] School of Biological Sciences, Royal Holloway,University of London,
Egham, Surrey TW20 0EX, UK
[3] Lawrence Livermore National Laboratory, Livermore, CA 94550, USA
[4] San Francisco General Hospital – University of California San Francisco,
San Francisco CA 94110, USA
[5] Life Sciences Division, Lawrence Berkeley National Laboratory, USA

**Abstract.** A new high resolution soft X-ray microscope (XM-1) has been used in a variety of applications. It is a conventional transmission microscope with a zone plate condenser and objective. A mutual indexing system incorporates state-of-the-art visible light microscopy and precise positioning of samples. XM-1 has a spatial resolution of 43 nm, as measured with a knife edge object, using the 10% to 90% intensities. It is used in collaboration with other groups to investigate variety of mostly biological samples. In our most extensive study, the life cycle of malaria parasites (*Plasmodium falciparum*) in intact human red blood cells was mapped. Abnormalities in the parasites development with protease inhibitor treatments and membrane protein deficiencies have been investigated and were linked to parasite mortality. New structures in green alga (*Chlamydomonas*), uniquely visible with soft X-rays, have been confirmed and analyzed in unfixed samples. In addition XM-1 is used to map the morphological variation of genetically altered sperm cells. We also give a brief introduction of the history of X-ray microscopy

## 1 Development of X-Ray Microscopy

The Center for X-ray Optics built and operates a new high resolution soft X-ray microscope, called XM-1 [1]. This instrument provides transmission images from samples up to ten microns thick. It uses bending magnet radiation from the Advanced Light Source, a third generation synchrotron facility at the Berkeley National Laboratory. XM-1 is a new design, that follows the optical setup of the microscope pioneered by the University of Göttingen [2]. A condenser zone plate, fabricated by the University of Göttingen [3], is used to illuminate the sample and acts as a linear monochromator. The radiation transmitted through the sample is then enlarged by an objective zone lens plate fabricated by CXRO in collaboration with IBM/Yorktown Heights [4]. Especially new with XM-1 are the full incorporation of state-of-the-art visible light microscopes and its ease of operation guaranteed by precision mechanics and computer control.

**X-Ray Microscopy and Spectromicroscopy**
Eds.: J. Thieme, G. Schmahl, D. Rudolph, E. Umbach
© Springer-Verlag Berlin Heidelberg 1998

New microscopy techniques have always opened up new scientific activities. That was the case with the early instruments made by Anton van Leeuwenhoek and is still true for more recent techniques like scanning probe and near field optical microscopy. In return the development of new instruments is motivated by scientific demands. The resolution of modern visible light microscopes is limited by the wavelength. Use of shorter wavelength radiation, like electrons or X-rays, provides significantly better spatial resolution, but also imposes limitations to the samples observed. Electrons for instance require vacuum and can penetrate only thinly sectioned samples. Other optical techniques, including near field scanning optical (NSOM) and single molecule fluorescence are useful for membrane studies, but not to study thick material.

X-ray microscopy primarily uses a natural contrast mechanism for biological specimen in water. The development of instruments, that take advantage of this contrast mechanism, is ongoing. Improved spatial resolution, as compared to visible light microscopy, can only be achieved if X-ray lenses with sufficient numerical aperture are used. Since the index of refraction for soft X-rays is near unity and absorption is not negligible, refractive lenses have impracticable high absorption losses. Early X-ray microscopes by Kirkpatrick and Baez used reflective optics [5]. Zone plate lenses for X-ray microscopes were discussed by Baez [6] and are now used for highest resolution instruments. Present X-ray microscopes exceed the resolution of visible light microscopes by more than a factor of five. The developments in X-ray microscopy over the last fifteen years are documented in these proceedings, from conferences held with a period of three years [7–10].

## 1.1  Contrast Mechanisms in Transmission X-Ray Microscopy

Contrast in X-ray microscopy is from the interaction of X-ray photons with the electrons in the sample molecules. Most work uses contrast primarily from photoelectric absorption, but phase contrast is used as well [11]. The atomic cross sections for photon energies away from ionization thresholds are well characterized and published in the Henke-Tables [12]. Near those thresholds, the photoelectric interactions are more complex because the chemical states of the electrons affect the cross sections as well. Spectromicroscopy uses these structures to determine chemical states in the samples. Today the majority of X-ray microscopy studies are done in the water window, which is for photon energies between the K shell absorption edges of Oxygen (543 eV, 2.3 nm) and Carbon (284 eV, 4.4 nm). For X-ray energies just below the Oxygen edge (e.g., 517 eV, 2.4 nm), the absorption of mostly carbon containing organic material is about an order of magnitude less than the absorption of water, permitting a natural contrast. The penetration depth of these soft X-rays is also ideally suited to image intact cells with a thickness of a few microns. This is probably the mostly stressed advantage of X-ray microscopy as compared to other methods like electron microscopy, where transmission is only feasible with much thinner samples. Single cells often have to be sectioned for electron microscopy.

The photoelectric absorption, which provides contrast in soft X-ray micro-scopy, also damages the sample. The hope of several early proponents of X-ray microscopy to investigate living samples like in visible light microscopy is therefore not possible. However, it is possible to prepare samples in different stages of development and then make conclusions based on statistical methods; a common method used with electron microscopy.

The dose required for an image with given signal to noise ratio can be calculated from the Henke-Tables and instrument parameters, but it is more difficult to determine the damage caused by such a dose. The damage depends on the sample. Chromatin in sperm for instance is resistant enough to allow imaging with soft X-rays at high spatial resolution and good signal to noise without structural damage. In general living cells do not survive imaging at high spatial resolution without intolerable damage. In some cases however, X-ray microscopy provides unique information from living cells at reduced resolution. The dense spheres in *Chlamydomonas* described below are such an example. There are two methods to overcome the limitations from radiation dose. The first one is to build an instrument that records the image before any damage becomes apparent. This requires fast X-ray sources with sufficient brightness and available photon flux for highest quality X-ray microscopy. Such sources are not yet available, but might be in the future. The second method introduces sample fixation to make it more stable against structural changes from radiation. Chemical fixation, for instance with glutaraldehyde, is frequently used, but cryogenic sample fixation is of great interest [13, 14] too.

## 1.2  Optical Setups

Different instrumental setups are used to take advantage of the contrast mechanisms provided with soft X-rays. These include direct imaging, scanning, contact, projection, and holographic methods. In scanning microscopes the sample is scanned through a small focal spot [15]. The transmitted photons are detected and used to build up a digital image in a computer. The smallest focal spot requires a spatially coherent illumination of the zone plate lens and nanometer-precision scanning capabilities. In a conventional microscope setup, like our XM-1, the sample is illuminated with incoherent soft X-rays. The transmitted photons are enlarged by a zone plate lens to form an image on an X-ray area detector. Usually this is an X-ray sensitive CCD detector, which is connected to a computer to generate a digital image. The image formation, which means the relation between the transmission of the sample and the image detected, is partially coherent for both scanning and conventional X-ray microscope [16] and therefore nonlinear. The non-linearity is especially visible for feature sizes approaching the diffraction limit.

## 1.3  The X-Ray Microscope XM-1

The conventional soft X-ray microscope XM-1 used for these studies was designed with highest performance and ease of use in mind. It incorporates an external Zeiss Axioplan visible light microscope for sample location and analysis, before X-ray imaging of the sample. The visible light image positions and X-ray image positions are mutually indexed to each other. The precision of the indexing system is better than 2 µm over a field of 3 mm and better than 1 µm in focus, which allows easy adjustment and focusing of specific cells. The X-ray optical setup is analogous to bright field visible light microscopes. The sample is illuminated with a condenser zone plate lens. An enlarged image of the transmitted radiation is formed by an objective zone plate lens. The X-ray wavelength (photon energy) of the microscope can be adjusted allowing spectromicroscopy techniques without compromising spatial resolution.

Samples, 5-10 micron thick, are imaged at atmospheric pressure and can be in a liquid environment. Wet samples are mounted between two silicon nitride films, each approximately 100 nm thick. The images are recorded directly on an X-ray sensitive CCD camera, using a magnification of typically 2400x, and stored digitally with other parameters of the microscope. The field of view is about 10-15 microns. Exposure times vary from fractions of a second to several tens of seconds.

The spatial resolution in our X-ray microscope is primarily determined by the width of the outermost zone of the objective lens, which is presently 35 nm. We achieved a resolution of 43 nm, measured between 10% and 90% intensity in the image of a knife edge test sample. Theoretically the outermost zone width of 35 nm could provide a resolution 31 nm [17] in our microscope. In biological application the spatial resolution alone is not sufficient to attain new information; contrast and signal to noise and sample preparation are equally important [18].

## 2  Biological Applications

Soft X-ray microscopy is ideally suited to investigate intracellular structures in intact cells. We studied the development of the intraerythrocytic stages of the malaria parasite *Plasmodium falciparum*. The repetitive 48 hour life cycle of intraerythrocytic infection is responsible for the morbidity and mortality of malaria. Better understanding of the process of intracellular parasite maturation and the interactions between the parasite and its host erythrocyte can contribute to efforts to devise novel approaches to the control of this deadly disease. We studied hundreds of images of infected cells collected and glutaraldehyde fixed at intervals throughout the 48 hour life cycle (Fig. 1a-b) in normal and in pathologic red blood cells and have used these to evaluate normal structural development of the malaria parasites. We then investigated development in two different unfavorable environments; protease inhibitor treated erythrocytes and genetically abnormal elliptocytes [19]. Cysteine protease inhibitors block globin hydrolysis [20], which causes the digestive vacuole of the parasite to expand and fill with un-degraded hemoglobin [21, 20]. Treatment *in vitro* with leupeptin or benzyloxycarbonyl, [(Z)-Phe-Arg-CH$_2$F] (ZFR), a peptide fluoromethyl ketone inhibitor of cathepsin L and other cysteine proteinases, has been demonstrated to result in death of parasites [20]. By bright field and electron microscopy, digestive vacuoles appear enlarged and filled with material that does not detectably differ from the contents of the red blood cell cytosol [22].

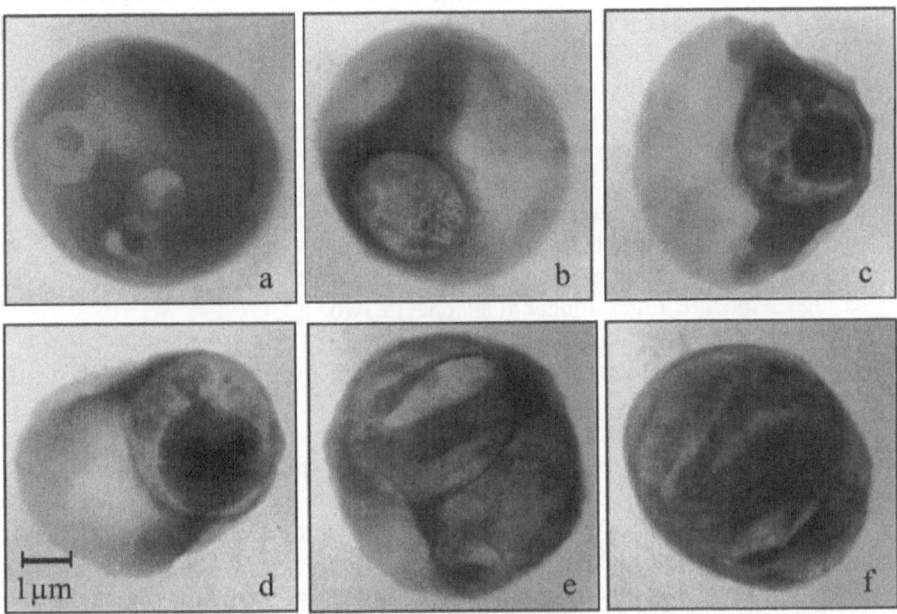

**Fig. 1.** *Plasmodium falciparum* malaria parasites in human red blood cells: (a) young ring stage parasites, (b) 12 hours old parasite, (c-d) parasites with unusually dense digestive vacuoles from protease inhibitor treatment, (e-f) cleft-like structures in ZFR treated parasites.

Two previously undetected anomalies can be recognized in protease inhibitor treated parasites. The majority of trophozoite stage parasites examined after being cultured for 12 hours or more in the presence of leupeptin or ZFR provide unique and intriguing images of aberrant digestive vacuoles that occupy approximately 25% of the parasite volume (Fig. 1c–d).

The second anomaly is detected in the structure of some ZFR, but not leupeptin, treated parasites. These trophozoites lack obvious digestive vacuoles. Parasites show more partitioning and have unusual cleft-like structures with redistributed mass (Fig. 1e–f) compared with untreated parasites.

## 2.1 Chlamydomonas

*Chlamydomonas*, a unicellular flagellated green alga has been investigated in an unfixed vegetative stage, and extensive studies were done to compare findings with contact X-ray methods [23]. These algae are important to the ecological balance of both the food chain and as producers of oxygen.

Images from initially living *Chlamydomonas* could be obtained with exposure times of approximately one second (Fig. 2). These images confirm the existence of X-ray dense spheres reported elsewhere with contact X-ray microscopy and visible light differential interference contrast (DIC) [24]. To our knowledge, these spheres have never been reported in electron microscopy studies. For these alga exposure times longer than one to two seconds causes severe radiation damage. We were not able to detect the spheres after chemically fixing the algae with glutaraldehyde. Images taken after

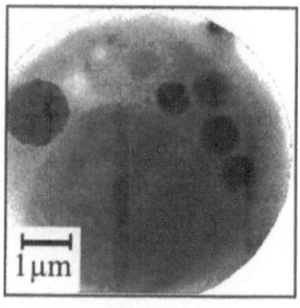

**Fig. 2.** X-ray dense spheres in *Chlamydomonas.*

cryo-fixation by the Göttingen group [13] however appear to preserve the structures seen here with living cells. We can only speculate on the content and function of these spheres. It has been suggested that the dark appearance of the spheres is from large amounts of Ca, which has a stronger absorption than C. We tested that hypothesis by imaging the algae with X-rays below the Ca L-edge (346 eV, 3.6 nm). The images did not change appreciatively, meaning the increased absorption of the spheres relative to the cell is not from Ca.

To investigate the elemental composition of those spheres further, we analyzed images taken above and below the absorption edges of oxygen, manganese, cobalt, and iron. For these studies the algae were dried because the absorption from the water would have prohibited the oxygen studies, and the long expo-

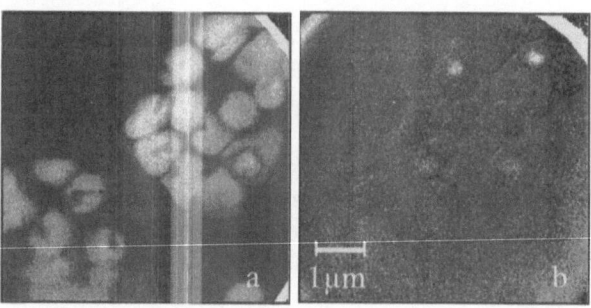

**Fig. 3.** Elemental distribution in *Chlamydomonas*, calculated from images above and below the binding energies of oxygen (a) and iron (b).

sure times needed for the detection of the metal elements would have not been possible with living cells. It is therefore expected that the morphology has been changed as compared to the living cells, but the elements found are expected to be the same as in the original cell. Fig. 3 shows the results for iron and oxygen. In a dried cell the spheres are not visible at 2.4 nm wavelength (inside the water window), but when imaged at 2.3 nm (absorption edge of oxygen), they are clearly visible again. This means that they contain significantly more oxygen then the rest of the cell. Some of these spheres contain also iron, manganese, cobalt, or a combination thereof (Fig. 3).

## 2.2 Sperm Cells

Soft X-ray microscope images have been used to examine the uniformity of chromatin organization within the heads of sperm from several mammals [25] (Fig. 4). Sperm chromatin is particularly well suited for imaging with X-rays. Since the DNA is packaged in a highly compacted state, X-ray images of the sperm heads show structural details that cannot be observed using other techniques. These images are providing new insight into the importance of the timely synthesis of protamine 1, one of the two nuclear proteins that package DNA in spermatids and sperm.

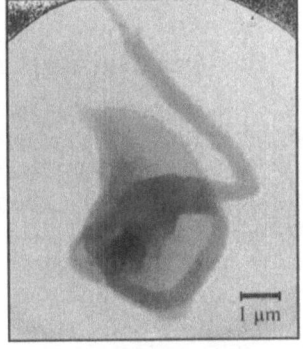

Sperm nuclei are extremely resistance to radiation damage from exposures to X-rays at the intensities required to form images in an unfixed state. This appears to be due, at least in part, to the high thiol content of the protamines that envelope the DNA. Observations were made of sperm in the fixed state as well and showed

**Fig. 4.** Transgenic sperm cell from mouse.

excellent spatial resolution. This stability of sperm cells against radiation allows for multiple images of unfixed material.

## 2.3 Cryptosporidium

*Cryptosporidium* is a parasite commonly found in lakes and rivers, particularly in those contaminated with animal waste and sewage. Occasionally it finds its way into drinking water supplies. The parasite is of about 4 to 6 micron in size and resistant to chlorination. Recent outbreaks in Las Vegas (1994) and Milwaukee (1993) caused about 140 deaths and approximately 400,000 cases of severe diarrhea and vomiting. *Cryptosporidium* is important to human health because it can be fatal to immunocompromised individuals. We recorded a first series of X-ray microscopy images of *Cryptosporidium*. These images were made from formalin fixed, wet samples. The images show a sporozoite emerging from the oocyst (Fig. 5a) an empty oocyst with the associated residuum (Fig. 5b).

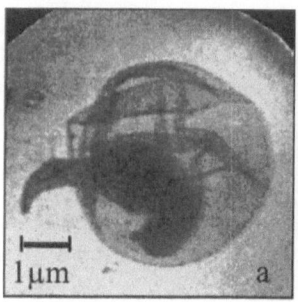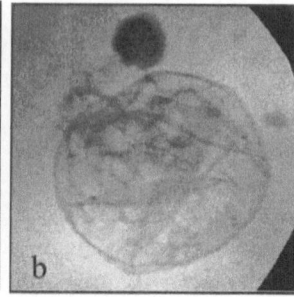

**Fig. 5.** *Cryptosporidium:*
(a) sporozoite emerging oocyst,
(b) empty oocyst with residium.

## 3  Non-Biological Applications

Soft X-ray microscopy is advantageous wherever high spatial resolution images from thick samples are required. If the sample is in a liquid environment, X-rays might be the only possible method. In addition to that, the possibility to gain elemental or chemical information through spectromicroscopy, makes X-ray transmission to the method of choice in a variety of applications in materials sciences and environmental re-search.

Fig. 6 shows images of silicate ($SiO_x$) micro-spheres suspended in toluene ($C_6H_5CH_3$-liquid at room temperature) imaged at different wavelengths near the absorption edge of the oxygen K-shell. At a photon energy of 539 eV (2.30 nm) the absorption of oxygen is at a maximum, whereas at the slightly lower photon energy of 533 eV (2.33 nm) the absorption

**Fig. 6.**
*(see text)*

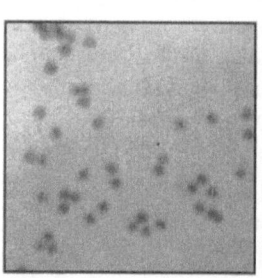

of oxygen is negligible. As only the silicate micro-spheres contain oxygen, they are clearly visible at 2.30 nm and almost invisible (actually slightly brighter than the medium) at 2.33 nm.

## 4  Future Plans

We are currently developing a cryogenic sample holder to be used in our soft X-ray microscope. The holder will allow us to take longer exposure times without increasing sample damage and eventually multiple view imaging to reveal tomographic information. For multiple view imaging a rotation stage is planned to be added. A new electron beam writing tool (NanoWriter) is being commissioned which will provide us with higher resolution condenser and objective zone plate lenses. In addition the NanoWriter will allow us to choose optimized lens parameters (focal length, number of zones, outermost zone width) for different applications. For instance we plan to improve the spectral resolution of the instrument by using a small monochromator pinhole with a low aberration condenser zone plate. The reduced  field of view will be compensated by electronically stitching together individual images like tiles. The stitching accuracy has already been demonstrated with our present sample stage to be 60 nm RMS.

For a greater flexibility in spectromicroscopy applications we hope to obtain funding to construct a scanning X-ray microscope that combines the brightness from an ALS undulator source with the instrumental features build into our present microscope.

## 5  Conclusion

We have built a conventional soft X-ray microscope and successfully used it in a variety of applications. We imaged the 48 hour life cycle of the malaria parasite *Plasmodium falciparum* in human erythrocytes and studied the influence of protease inhibitor treatments. We studied the green algae *Chlamydomonas* and investigated the content of dense spheres of unknown function and composition. In other studies [26] we imaged *Cryptosporidium*, sperm cells, bacterial spores [27] and non-biological samples.

## Acknowledgments

We gratefully acknowledge the staff of Berkeley Lab's Center for X-ray Optics, which contributed in multiple ways to the success of the microscope and beamline. Especially E. Hall, W. Low, and J. Smithwick contributed significantly to the construction of XM-1. The objective zone plates were fabricated in a collaboration with IBM/Yorktown Heights, N.Y. We thank our collaborators Dr. D. Rudolph and Prof. G. Schmahl, University of Göttingen, who built the condenser zone plate. We are grateful to Ms Linda Geniesse for help with the figure preparation, and the staff at the Advanced Light Source.

This research is supported by the United States Department of Energy, Office of Basic Energy Sciences, and the Office of Health and Environmental Research under contract DE-AC 03-76SF00098. C. M. received support from NIH grant #DK32094-10 for the malaria project. The operation of the microscope for biological applications is supported through the Laboratory Directed Research and Development (LDRD) grants awarded to us by the Berkeley Laboratory, which is highly appreciated.

# References

1   Meyer-Ilse, W., Medecki H., Jochum L., Anderson, E., Attwood, D., Magowan, C., Balhorn, R., Moronne, M., Rudolph, D., Schmahl G., *Synchrotron Radiation News* **8** (1995) 29-33

2   Schmahl, G., Rudolph, D., Niemann, B., Christ, O., *Quarterly Reviews of Biophysics* **13**, 3 (1980) 297-315

3   Hettwer, M., This volume.

4   Anderson, E. H., Kern, D., In [9] 75-78

5   Kirkpatrick, A., Baez, V., J., *J. Opt. Soc. Am.* **38**, (1948) 766-774

6   Baez, V., *J. Opt. Soc. Am.* **42**, (1952) 756

7   Schmahl, G., Rudolph, D. (Eds.): *X-Ray Microscopy* (Springer 1984)

8   Sayre, D., Howells, M., Kirz, J., Rarback, H. (Eds.): *X-Ray Microscopy II* (Springer 1987)

9   Michette, A., Morrison, G. R., Buckley, C. J. (Eds.): *X-Ray Microscopy III* (Springer 1990)

10  Aristov, V. V., Erko, A. I. (Eds.): *X-Ray Microscopy IV* (Chernogolovka, Russia 1993)

11  Schmahl, G., Rudolph, D., Niemann, B., Guttmann, P., Thieme, J., Schneider, G., David, C., Diehl, M., Wilhein, T., *Optik* **93**, No. 3 (1993) 95-102

12  Henke, B. L., Gullikson, E. M., Davis, J. C., $Z = 1$-92. *Atomic Data and Nuclear Data Tables* **54** (1993) 181-342.

13  Schneider, G., Niemann, B., This volume.

14  Maser, J., Jacobsen, C., Osanna, A., Wang, S., Kalinovsky, A., Kirz, J., Spector, S., Warnking, J., This volume.

15  Kirz, J., Jacobson, C., Howells, M., *Quarterly Review of Biophysics* **28** (1995) 33-130.

16  Jochum, L., Meyer-Ilse, W., *Appl. Optics* **34** (1995) 4944-4950.

17  Heck, J. M., Meyer-Ilse, W., Attwood, D. T., This volume.

18  Stead, A. D., Anastasi, P. A. F., Brown, J. T., Majima, T., Meyer-Ilse, W., Neely, D., Page, A.M., Rondot, S., Shimizu, H., Tomie, T., Wolfrum, E., Ford T. W., This volume.

19  Magowan, C., Brown, J. T., Liang, J., Heck, J., Coppel, R. L., Narla M., Meyer-Ilse, W., submitted

20  Rosenthal, P. J., Wollish, W. S., Palmer, J. T., Rasnick, D., *J. Clin. Invest.* **88** (1991)1467-1472.

21  Dluzewski, A. R., Rangachari, K., Wilson, R. J. M., Gratzer, W. B., *Exp. Parasitol.* **62** (1986) 416-422.

22  Rosenthal, P. J., McKerrow, J. H., Aikawa, M., Nagasawa, H., Leech, J. H., *J. Clin. Invest.* **82**, 1560-1566.

23  Ford, T. W., Page, A. M., Meyer-Ilse, W., Stead, A. D., This volume.

24  Ford, T. W., Cotton, R. A., Page, A. M., Tomie, T., Majima, T., Stead, A. D. , *SPIE* **2523** (1995) 212-220.

25  Balhorn, R., et al., This volume.

26  Brown, J. T., Magowan, C., Balhorn, R., Heck, J., Ford, T., Stead A., Meyer-Ilse, W., This volume.

27  Stead, A. D., Brown, J. T., Judge, J., Meyer-Ilse, W., Neely, D. Page, A. M., Wolfrum, E., Ford, T. W, This volume.

# X-Ray Microscopy in Aarhus

Joanna Abraham[1], Robin Medenwaldt[1], Erik Uggerhøj[1], P. Guttmann[2], T. Hjort[3], J. Jensenius[3], T. Vorup-Jensen[3], F.Vollrath [4], E. Søgaard[5], J. Tyge Møller [6]

[1] ISA, Institute for Storage Ring Facilities, University of Aarhus, DK-8000, Denmark
E-mail: jabraham@dfi.aau.dk or: robin@dfi.aau.dk (R. Medenwaldt)
[2] Forschungseinrichtung Röntgenphysik, Georg-August-Universität Göttingen and Berliner
Elektronenspeicherring-Gesellschaft für Synchrotronstrahlung mbH (BESSY), Germany
[3] Department of Medical Microbiology and Immunology, University of Aarhus,
DK-8000, Denmark
[4] Department of Zoology, University of Aarhus, DK-8000, Denmark
[5] Department of Chemistry, University of Aalborg, Esbjerg, Denmark
[6] Department of Geomorphology, University of Aarhus, DK-8000, Denmark

**Abstract.** We have seen an ever increasing number of collaborative projects since the start of the Aarhus XM in 1992, with such diverse materials as human spermatozoa, freshwater micro-organisms, metal-induced cysts, spider orb silk, iron-precipitating bacteria, and sludge collected from water purification filters.

## 1 The Microscope

The Aarhus X-ray microscope [1] has been in operation since 1992 and has evolved to be part of a user facility at ISA. It operates biannually for ten weeks, which is sufficient (though not optimum) for extensive studies of objects in selected scientific projects, some of which are described below. The majority of those projects involve investigations of wet samples from fields in biology, medicine, and soil sciences.

Objects are illuminated by synchrotron radiation focused by a condenser zone plate (the Göttingen KZP 7 type) through a monochromizing pinhole. Although the usual wavelength is 2.4 nm, the configuration allows a continuous wavelength change throughout the whole water window. A second zone plate images the object on a CCD camera (Photometrics). The CCD chip (Tectronix) is peltier cooled, thinned and back illuminated with 1024 by 1024 pixels of 24 µm size. In combination with a micro zone plate of 30 nm outermost zone width, the achievable resolution is 30 nm at an X-ray magnification of 1600. Some micro zone plates for the Aarhus XM have been fabricated in Göttingen [2]. Since 1996, however, micro zone plates made by Steven Spector from Stony Brook have been in use. All zone plates are germanium structures on silicon backings [3].

Objects are located under atmospheric pressure surrounded by helium gas in order to minimise X-ray absorption. For dry samples, almost any kind of holder can be mounted in the microscope. Wet samples are placed between two silicon foils of

**X-Ray Microscopy and Spectromicroscopy**
Eds.: J. Thieme, G. Schmahl, D. Rudolph, E. Umbach
© Springer-Verlag Berlin Heidelberg 1998

150 nm thickness in a special chamber that is sealed with o-rings. With flexible tubes and syringes, liquids can be pumped in and out, thereby adjusting the layer thickness of the medium and/or exchanging the medium. With typical liquid layer thicknesses of 5-15 μm, samples can be kept in the chamber for many hours without drying out.

For object finding and prefocusing, there are several options. A light microscope with an x-y stage calibrated with an x-y stage in the XM is used to get an overview of the sample, the image of which can be shown on a video screen and can be stored on video or in a computer. After placement of the object in the XM, the holder can be rotated for prefocusing in a video microscope. The prefocused position matches the necessary final position in the X-ray beam within a micron. The X-ray micrograph can be recorded on the CCD or through a two stage micro channel plate in combination with a phosphor screen and video camera. The latter option is used for adjustment and alignment and when dynamic, real-time effects are made visible with X rays. Although the resolution in this case is only 150 nm and the noise is high, this option is often used to survey larger areas around objects or when looking at living, moving samples such as sperms. These have to be immobilised before imaging, which can be done by irradiating them for a second.

Imaging times of wet samples are typically 10-20 s. On this timescale, the mechanical stability of the XM is high. The set-up, where the frame of the microscope is rigidly bolted to a vibration damping table resting on pneumatic vibration isolators, has shown that vibrations are below the detectable limit and thus have negligible influence on the imaging properties of the XM.

## 2 The Projects

### 2.1 Spider Silk

Spider silk is extremely strong, weight for weight it is stronger than nylon or steel. It is extremely elastic more so than any commercially made rubber. It can be extended many times its own length but contracts easily and immediately to its former length. However, it has been little studied unlike the cocoon silk of moths such as *Bombyx mori*, which has a long history (at least 5000 years). The cocoon silk of *Bombyx mori* has played an important role in industry and consequently has been extensively studied. The wealth of information available on cocoon silk made it an obvious starting point in the study of spider silk. As genetical techniques advanced, it became clear that spider silk evolved independently and that there was no likely ancestor with the cocoon silk. Now it has been shown that dragline spider silk has a very different structure from that of the silkworm cocoon silk. This is not surprising when one considers that spider silk is mechanically far superior.

Light microscopy and scanning electron microscopy studies suggest a skin-core microstructure, but others disagree. Some hypothesise that one strand consists of a fibril structure, others say not. These and many other contradictory findings have created a gap in the information that is now beginning to be addressed by taking advantage of new techniques.

Fritz Vollrath from the Department of Zoology hypothesises that the thread of the *Nephila* spider silk has a previously unsuspected structural organisation consisting of a structured fibril wall surrounding a fibrilless core that can explain its extraordinary tensile strength [4]. We have started a project in this direction and thus have some preliminary results. Fig.1 shows threads of *Nephila* spider silk, where some of the fibrils are visible at the ends of a broken thread. The protein polymers are densely packed and absorbency is high, which demonstrates the problems involved in its study. The question about a possible empty core inside this particular silk could be answered in an easy way with X-ray microscopy, simply by comparing the X-ray transmission of the silk with theoretical values. Thereby we showed that the thread was not hollow. Future investigations are planned and will involve different preparation techniques and manipulation of the silk. For example, by treating the silk with urea solutions, thus causing the silk to swell.

A different kind of silk, the hackled spider silk (Fig. 2), is much thinner than *Nephila* silk and consists of a network.

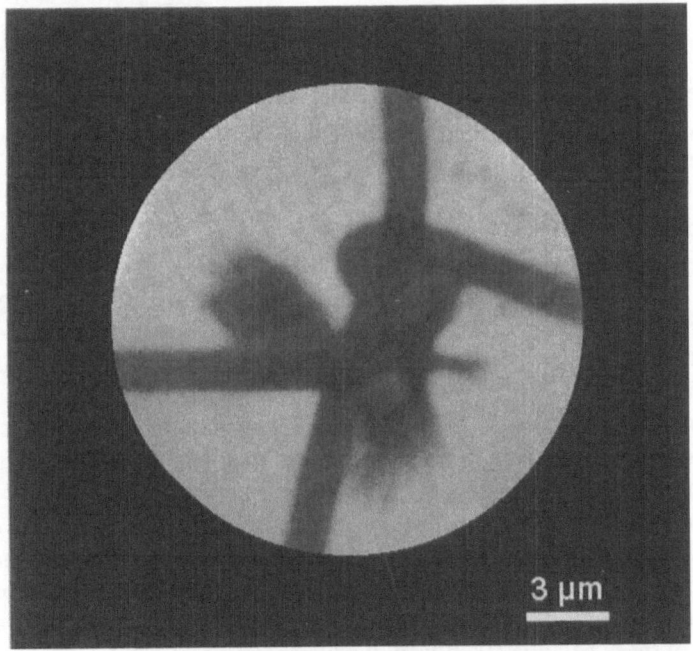

**Fig. 1.** X-ray micrograph of *Nephila* spider silk. $\lambda$=2.4 nm, t=10s

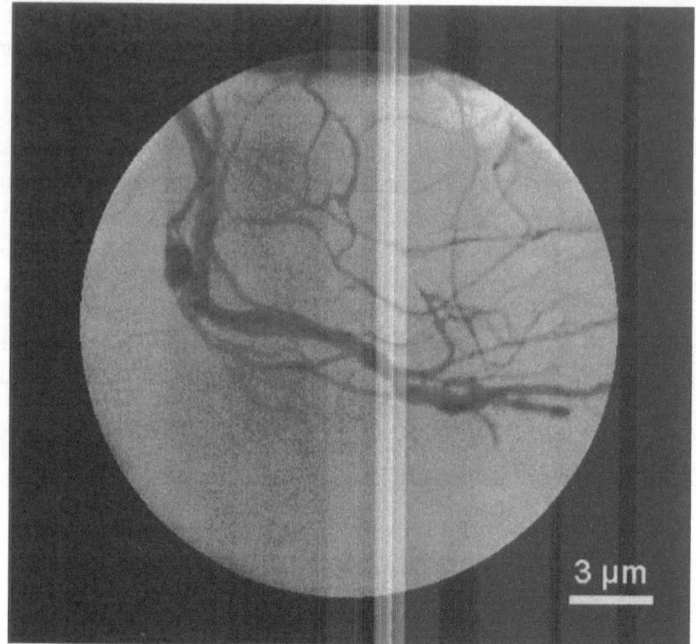

**Fig. 2.** X-ray micrograph of hackled spider silk. λ=2.4 nm, t=2s

## 2.2 Colloidal Chemistry

The water treatment works in and around Esbjerg, Denmark, receive high levels of iron that has made its way into the water system. The conventional method of removing this iron from the water has been chemical, but now it is becoming clear that a biological approach is far more efficient. In Esbjerg, they have four water purification plants, three of which use the chemical method and one uses iron-precipitating bacteria. We are examining samples of sludge from these plants with the aim of gaining a better understanding of how the sludge is structured.

Despite their known efficiency in iron removal, the microbes responsible have been little studied. At Astrup water purification works, Esbjerg, the organism responsible is *Leptothrix,* a bacteria known for its iron precipitation abilities. However, the biology of such organisms is less well understood. Fig. 3a shows an XM image of these bacteria taken from the water treatment plant, Esbjerg.

Also, at the Department of Geomorphology, Aarhus University, the process of iron precipitation in Danish wetlands (probably connected with the occurrence of bacteria) is being studied. Natural spring water with a high content of iron has been imaged with the Aarhus XM (Fig. 3b). These samples show bacteria and the shells in which they inhabit.

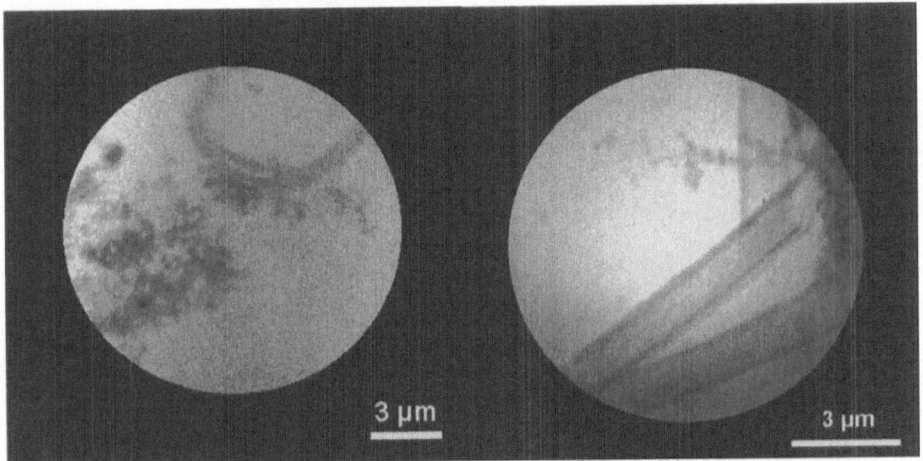

**Fig. 3.** X-ray micrographs of the sheaths of iron-precipitating bacteria *Leptothrix*.
*Left* (a): The microbe responsible for iron removal at Esbjerg water purification works.
*Right* (b): Bacteria seen in a sample taken from Danish wetlands. λ=2.4 nm, t=20s and t=60s

## 2.3 Spermatozoa

Often in the past it was the woman who was persecuted if a couple remained child-less, and even today there is a stigma attached to the situation, be it voluntary or not. However, in recent years, it is male infertility that has been in the spotlight with controversial claims that sperm counts have declined in recent years and that male infertility has increased. It is not surprising then, that there is much work to elucidate the mechanisms and processes involved in the developmental stages of the spermatozoon.

The sperm is a highly specialised cell. The head is packed with genetic information and an acrosomal vesicle containing hydrolytic enzymes that will help the sperm penetrate the egg's outer coat and so fertilise it. Mitochondria are strategically placed at the base of the tail where they can effectively power the flagellum (Fig. 4a).

Sperm maturation is associated with a series of changes in the membranes surrounding the head region. When the sperm are deposited into the vagina they do not have the ability to fertilise an egg. However, by the time they reach the egg in the oviduct they will have acquired the capacity to fertilise. Little is known about the mechanism of capacitation and so far no morphological changes have been observed during this process. It has been reported that the tail motion changes after capacitation which lead us, together with the Department of Medical Microbiology and Immunology at the University of Aarhus, to study mitochondrion morphology through the developmental process.

Sperm are fragile, they have a single plasma membrane which is easily damaged by conventional electron microscopy preparation techniques. With the XM we could look at fully-hydrated sperm that had intact membrane structures.

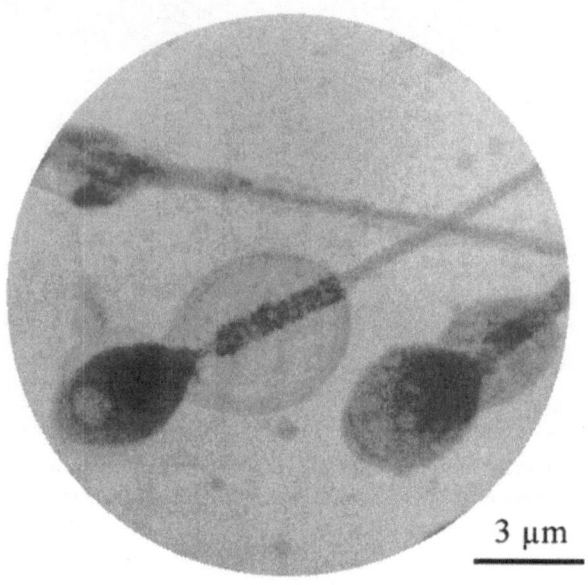

3 μm

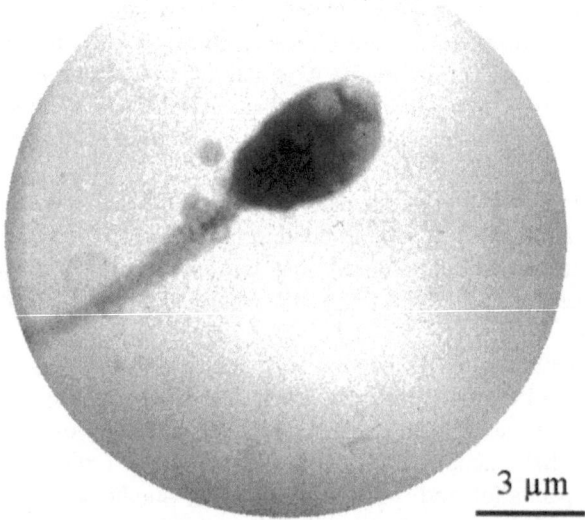

3 μm

**Fig. 4.** X-ray micrographs of spermatozoa taken with the Göttingen XM at BESSY.
*Top* (a): Fresh ejaculated sperm. Note the densely packed mitochondria.
*Bottom* (b): Capacitated sperm. Note the less dense mitochondria. λ=2.4 nm, t=2s

Figure 4a shows sperm from fresh ejaculate. Important to note are the mitochondria, densely packed around the base of the flagellum. In Fig. 4b of a capacitated sperm, note that the mitochondria are now not as densely packed and are vacuolated with an increased volume. This observation has not been reported previously but could be linked to the increased tail movements seen in the capacitated sperm.

## 2.4 Filamentous Blue-Green Algae

Filamentous blue-green algae are primary producers, using sunlight for photosynthesis. They have an important role to play in the lake ecosystem. However, some species can become a problem if nutrient levels rise above a critical level. When they grow in very large numbers, toxic blooms can be formed, choking the lake and ultimately killing the other organisms present. Fig. 5 shows a blue-green algae seen in a sample of lake water collected from Aarhus University lake. Note that sensitive structures such as the mucilaginous sheath are fully hydrated and intact. Fig. 6 shows a pennate diatom.

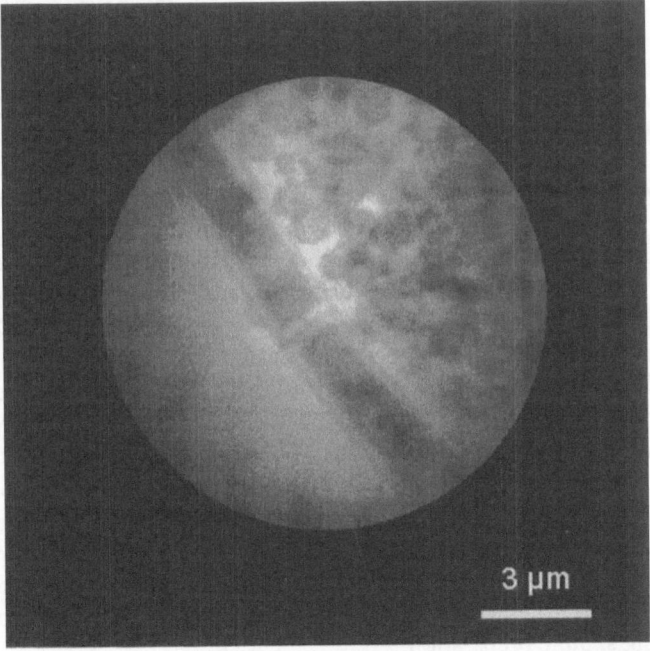

**Fig. 5.** X-ray micrograph of blue-green algae, note the fully-hydrated mucilaginous sheath.
$\lambda$=2.4 nm, t=8s

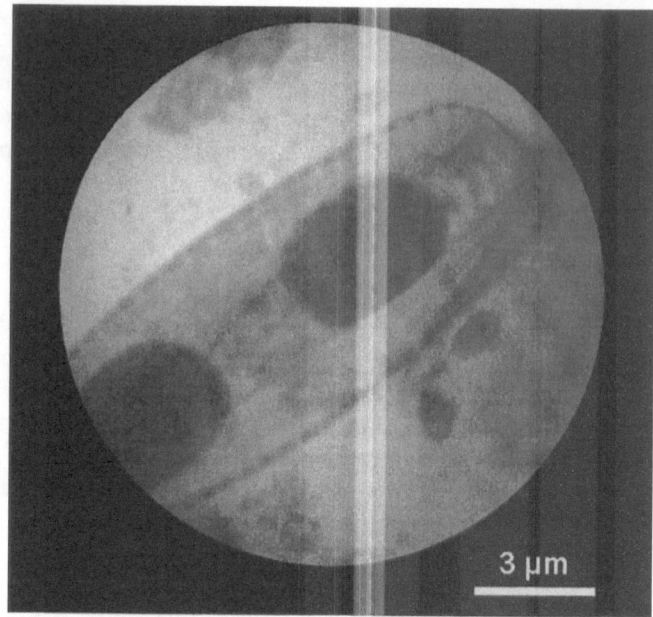

**Fig. 6.** X-ray micrograph of Pennate diatom. λ=2.4 nm, t=12s

## 2.5 Protozoa and Encystment

Protozoa are single-celled or colonial, eukaryotic organisms. The cells vary considerably in size but are usually 1-250 μm in diameter. They are a diverse group, both morphologically and in their environmental adaptations, so have occupied a wide range of ecological niches. In recent years, research using protozoa has flourished not only to forward knowledge of the protozoa themselves, but because biologists have recognised that these organisms provide excellent subjects for studying biological phenomena at the cellular level. This increased research activity has primarily been aimed at elucidating the structure and understanding the functioning of protozoa as cells. There is a great deal of interest in planktonic protozoa and their functional role in both marine and freshwater environments. World-wide, protozoa form a significant part of planktonic biomass and, more important, have a major role in the flow of energy and recycling of nutrients. So far, ultrastructural studies of micro-organisms have been limited to conventional electron microscopy, which despite its high resolution capabilities, also produces many artefacts during sample preparation.

Encystment (cyst formation) is a stage in the life cycle of many invertebrates used to avoid adverse conditions. Generally, encystment is induced by starvation, depletion of oxygen, increased salinity or dehydration. The cyst may survive in the dormant state for many years. Excystment takes place once conditions become favourable again. Fig.7 shows a protozoan cyst taken from a mixed culture of protozoa.

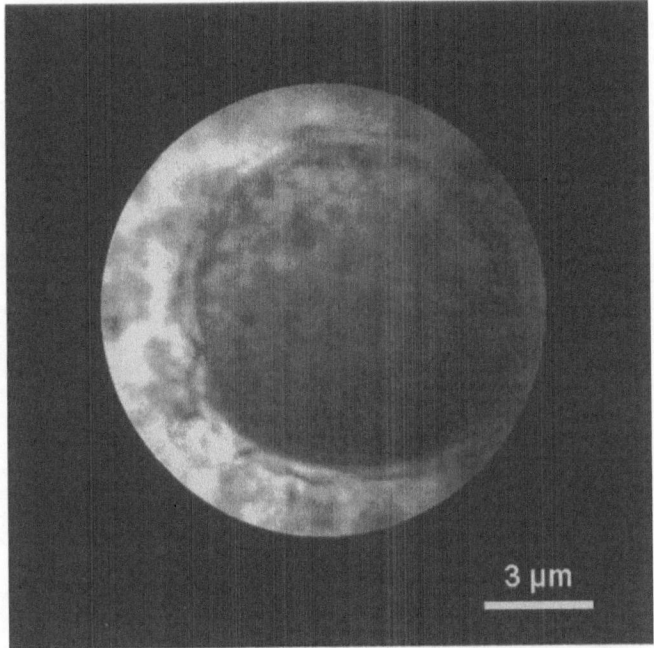

**Fig. 7.** X-ray micrograph of a protozoan cyst which has an undulating outer membrane.
λ=2.4 nm, t=16s

Metals such as copper can be toxic to protozoa even at low concentrations [5]. Once the metal enters the cell, it can be accumulated and disrupt the elemental distributions within the cell [6]. Therefore, some species exploit the avoidance mechanism of encystment in order to survive during elevated external levels of copper. *Chilomonas paramecium* starts to form a cyst within minutes of exposure to elevated external levels of copper. Metal-induced encystment (cryptobiosis) of flagellated protozoa has been studied here at ISA by LM and XM. *Chilomonas paramecium* is approximately 20 μm long and 8 μm wide in favourable conditions. However, the cell becomes more spherical within 10 minutes of copper exposure. When the cell is centrifuged and resuspended in nutrient poor medium the cell also forms a cyst. The rate of encystment can be followed by measuring the cell dimensions. When exposed to copper the process is much faster. Fig. 8a shows an XM image of an untreated cell. The cell appears very dense with many organelles. After only 10 minutes in a solution with 10 ppm copper, the cells lose their cellular integrity and round-up (Fig. 8b). This project aims to study structural changes during the process of encystment in real time, to quantify the rate of metal-induced cyst wall formation and examine the structural organisation within the cell during this process. Further studies will determine if the process is metal dependent.

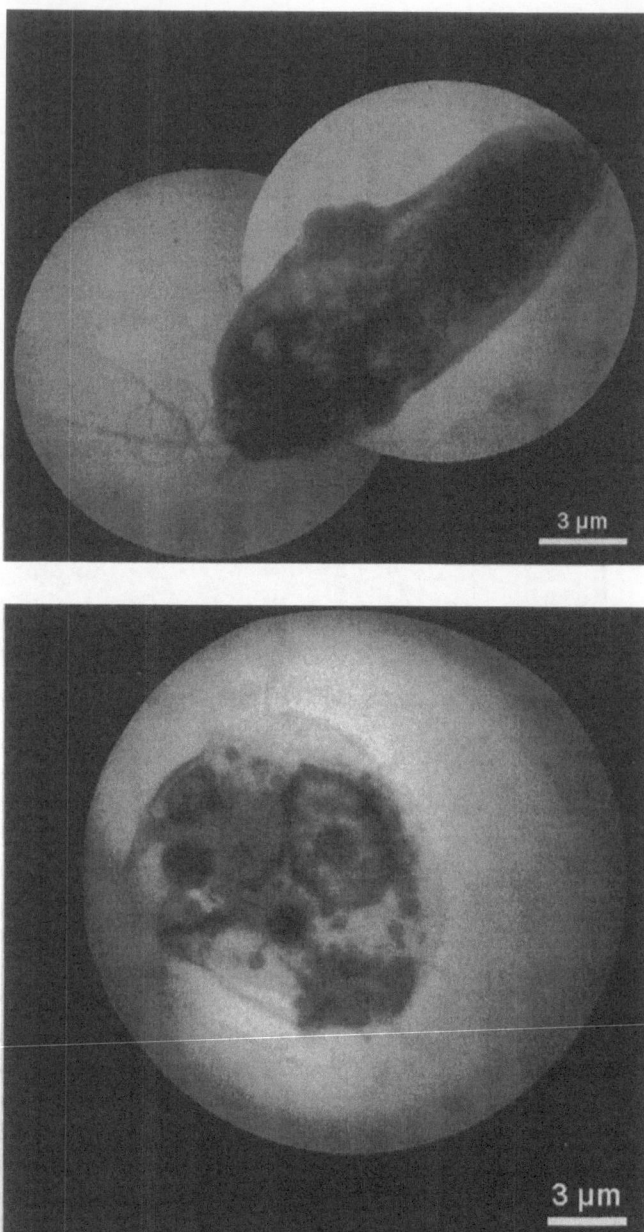

**Fig. 8.** X-ray micrographs of *Chilomonas paramecium*. *Top* (a): untreated cell, *bottom* (b): cell after 35 min. exposure to copper. λ=2.4 nm, t=12s

# 3 Conclusions

As a user facility, the Aarhus XM continues to be developed and improved, whilst the number of collaborative research projects steadily increases each year. In the forthcoming synchrotron radiation period, it is planned to incorporate phase contrast and stereo X-ray microscopy.

# Acknowledgements

We thank the Göttingen XM group led by Prof. G. Schmahl and the St. Brook group by Prof. J. Kirz for their advisary and practical help.

# References

1    R. Medenwaldt, C. David, N. Hertel, and E. Uggerhøj, p.323 in *X-Ray Microscopy IV*, Proceedings of the 4-th International Conference (Chernogolovka, Russia, 1994).
2    J. Thieme, C. David, N. Fay, B. Kaulich, R. Medenwaldt, M. Hettwer, P. Guttmann, U. Kögler, J. Maser, G. Schneider, D. Rudolph, and G. Schmahl, p.487 in *X-Ray Microscopy IV*, Proceedings of the 4-th International Conference (Chernogolovka, Russia, 1994).
3    R. Medenwaldt and M. Hettwer, J.X-ray Sci.Technol. **5**, 202-206 (1995).
4    F. Vollrath, T. Holtet, H.C. Thøgersen and S. Frische, Proc. R. Soc. Lond. B, 263, 147–151 (1996).
5    J.V. Abraham, Ph.D. Thesis, Manchester University, (1994).
6    J.V. Abraham, R.D. Butler and D.C. Sigee, Micron and Microscopica Acta, 23, (3), 343–344 (1992).

# Cryo X-Ray Microscopy Experiments with the X-Ray Microscope at BESSY

G. Schneider and B. Niemann

Forschungseinrichtung Röntgenphysik, Georg-August-Universität Göttingen,
Geiststraße 11, D-37073 Göttingen, Germany
E-Mail: gschnei1@gwdg.de

**Abstract.** The transmission X-ray microscope (TXM) at the electron storage ring BESSY was used to image frozen-hydrated objects at cryogenic temperatures and atmospheric pressure. After shock freezing in liquid ethane different initially living biological objects such as cells, chromosomes and algae were investigated at 2.4 nm wavelength in amplitude contrast mode. The high X-ray absorption of cellular structures and the 10-fold lower absorption of frozen water yield natural element contrast. The X-ray images show details inside the biological objects down to 30 nm size, which is about one order of magnitude better than conventional visible light microscopy.

## 1  X-Ray Imaging of Hydrated Biological Objects

X-ray microscopy allows to image complete hydrated specimens of about 10 $\mu$m thickness in the water window wavelength region between the inner-shell absorption edges of oxygen at 2.34 nm and carbon at 4.37 nm [1]. It was demonstrated that these objects can be imaged both in amplitude and in phase contrast mode [2, 3]. About 100 $\mu$m thick specimens can be investigated at wavelengths around 0.3 nm [4]. Therefore, using a TXM structural and functional informations about intact hydrated cells can be obtained. The small numerical aperture of present X-ray optics leads to longer depth of focus compared to high-resolution visible light objectives. Thus, the TXM can advantageously be used to record series of images at different viewing angles of complete, non-sectioned and unstained cells for tomographic 3D reconstruction. Recently the first tomographic reconstruction was obtained from a data set of dried samples taken with the TXM at 2.4 nm wavelength [5].

By comparison, a transmission electron microscope (TEM) can detect the elastically and inelastically scattered electrons for image recording. If only the elastically scattered electrons are detected and inelastically scattered electrons are filtered, phase contrast images can be obtained by defocussing and making use of the spherical aberrations. However, the contrast transfer function depends critically on the object's thickness [6]. This method is only used successfully for imaging sections – dried or frozen – of about 0.1 $\mu$m thickness. For thicker specimens the interpretation of phase contrast images becomes much more difficult.

X-Ray Microscopy and Spectromicroscopy
Eds.: J. Thieme, G. Schmahl, D. Rudolph, E. Umbach
© Springer-Verlag Berlin Heidelberg 1998

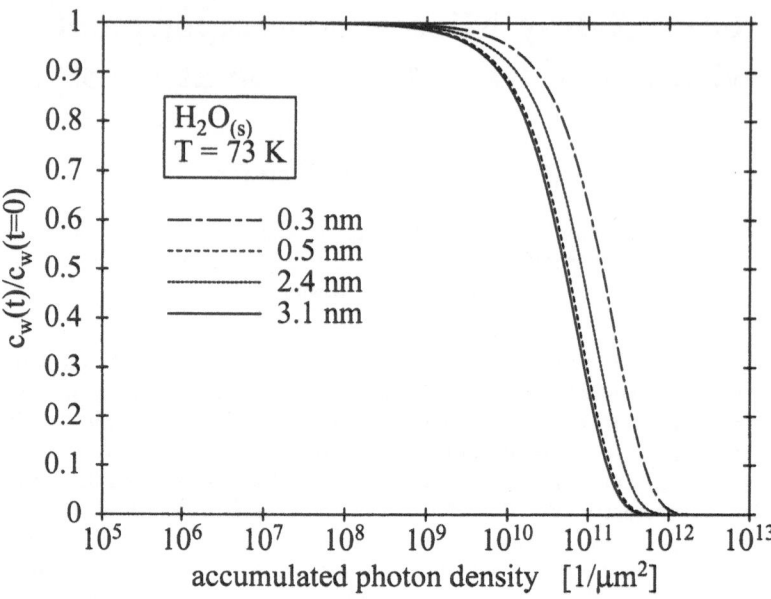

**Fig. 1.** The relative number $c_w(t)/c_w(t = 0)$ of intact water molecules at 73 K under irradiation with photons of 0.3, 0.5, 2.4 and 3.1 nm wavelength. The number of intact water molecules decreases significantly at photon densities of about $10^{10}$ $photons/\mu m^2$ which corresponds to dosages of about $10^8$ Gy in vitrified water at 2.4 nm wavelength.

Therefore, it is difficult to acquire 3D information about a complete cell, because several thin sections of an object have to be prepared and imaged separately. Furthermore, images of unstained biological material embedded in vitrified water show significantly lower contrast than images of dried samples.

Different authors calculated that the X-ray dosage applied during imaging increases inversely with the fourth power of the resolution [7, 8, 9]. As a result hydrated biological objects with structures of 50–10 nm in size become visible with dosages in the range of $10^5$–$10^8$ Gy. However, biological structures are sensitive to X-rays. To avoid structural changes during X-ray imaging the ultrastructure of biological objects can be stabilized by different preparation techniques known from electron microscopy. Stabilization is achieved by embedding dried objects in organic polymers or by cryo-fixing the objects. The cryo technique allows to stabilize the hydrated objects more closely to the living, wet state and is therefore favorable.

The diffraction of X-rays in the object depends on the elemental distribution and this should remain constant during the exposure time to avoid visible structural changes. A model describing the X-ray induced kinetics was developed to evaluate the critical photon density and dosage for significant changes of the elemental distribution in irradiated frozen-hydrated biological objects [9, 10]. In this model diffusion of molecules and Brownian motion of cell particles are negligible at cryogenic temperatures and the vitrified water acts as a stabilizing matrix for the organic structures. The upper dosage limit tolerable during imaging is obtained as soon as the irradiated vitrified water decomposes significantly into hydrogen and oxygen gases. Results derived from this model are presented in Fig. 1 for different wavelengths. Here the quantum yield for water molecule decomposition is about 1 water molecule per absorbed 200 eV energy at a temperature of 73 K [11]. Radiation damage of vitrified water becomes significant in the photon density range of $10^9$–$10^{10}$ $photons/\mu m^2$ (see Fig. 1) which corresponds to dosages of $10^8$–$10^9$ Gy in protein at 2.4 nm wavelength. Therefore, we introduced the cryo technique to X-ray microscopy [12]. Experiments with the cryo-TXM proved that cryogenic samples are structurally stable even at dosages of $10^{10}$ Gy without introducing structural changes which become visible at the resolution limit of the TXM [13]. This dosage limit is about $10^3$–$10^4$ fold higher than the dosage actually applied for resolving 30–40 nm features. The experiments indicate that it is possible to resolve frozen-hydrated structures with even less than 10 nm size without inducing significant structural changes.

## 2  Cryo Object Chamber

The X-ray optical arrangement and the cryogenic object stage adapted to the TXM are shown in Fig 2. A condenser zone plate acts as a dispersive and focussing element which monochromatizes the incoming polychromatic radiation and focusses it directly onto the object [14, 15]. The X-ray objective produces a magnified image on a back-thinned charge coupled device (CCD-camera) [16].

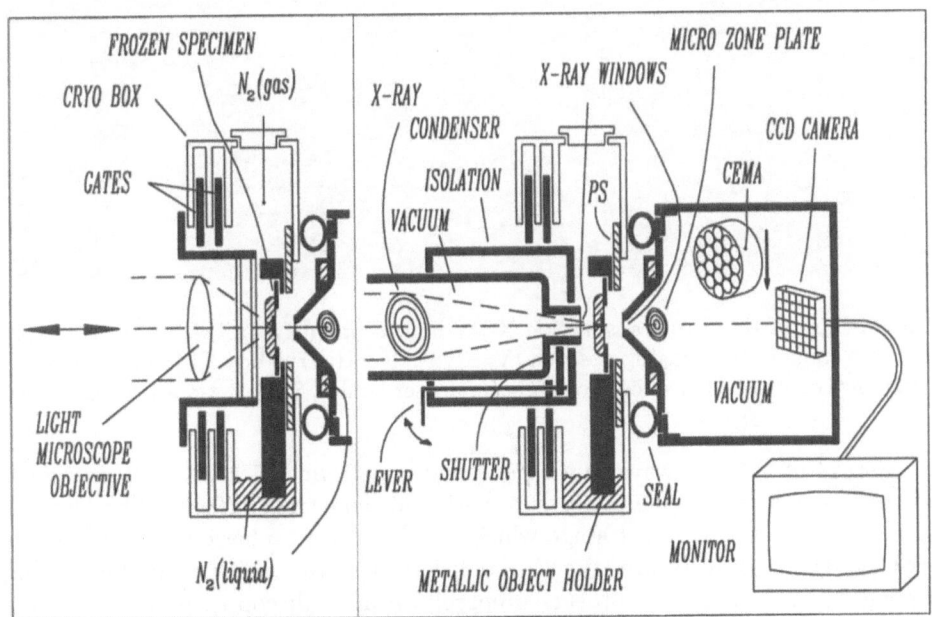

**Fig. 2.** Schematic of the arrangement which is adapted to the TXM for vitrified objects. Left: Prefocusing the object with the LM. Right: Imaging position.

**Fig. 3.** Part of the cryo transmission X-ray microscope (cryo-TXM) at BESSY with the cryo stage in the imaging position (V=cryo valve, B=cryo box, LM=differential interference contrast light microscope, C=condenser holder, S=x, y, z-positioning stage).

X-rays of 2.4 nm wavelength penetrate a 200 $\mu$m thick layer of cryogenic nitrogen gas at atmospheric pressure with sufficient transparency. Therefore, it is possible to use an object stage which is held at atmospheric pressure. In this case cryogenic objects are directly cooled by the gas which simplifies taking tomographic data sets of objects prepared on a small copper strip or objects mounted into a thin glass capillary. For this purpose the TXM optics have to be located in two separated evacuated vessels: At the object stage the X-rays penetrate a window behind the X-ray condenser, pass through the cryogenic nitrogen atmosphere containing the object and return through another window in front of the X-ray objective into the vacuum system of the TXM [17]. The object stage with the cryogenic specimens is enclosed in a thermally insulating cryo-box. It is adjustable along and perpendicular to the optical axis. The bottom of the cryo-box is filled with liquid nitrogen which evaporates and thus keeps the objects at cryogenic temperatures. The drift rate of the cryo-stage was measured to be about 1 nm/s. It does not significantly influence the resolution of the microscope in practice. Positioning and prefocussing of the cryogenic specimens are performed using a conventional light microscope (LM). For this purpose the cryo-box contains a transfer system with two gates. The transfer system is opened and a glass-tube with a thermally insulating front-end window is inserted. Using the LM with a long-distance objective, it is possible to look through the window onto the cryogenic object, to adjust and to prefocus it. For X-ray imaging the light microscope and the glass-tube are replaced by the X-ray condenser arrangement.

## 3   X-Ray Micrographs of Cryogenic Objects

If hydrated biological objects are frozen slowly at atmospheric pressure large ice crystals are formed, which destroy the ultrastructure. However, with cooling rates higher than 10000 K/s vitrified water can be formed [18]. In order to obtain these high cooling rates the water layer thickness surrounding the specimens should not exceed 10 $\mu m$. This thickness also guarantees sufficient transparency for 2.4 nm radiation. In practice the water layer thickness is regulated under light microscopical control. Afterwards shock-freezing is done with liquid ethane at 100 K. Note that no cryo-sectioning is necessary and therefore sectioning artifacts cannot occur.

X-ray images of frozen-hydrated objects were taken with micro zone plates with outermost zone widths of 40 nm and 30 nm. The first order diffraction efficiencies of their nickel zones are 15% and 10% at 2.4 nm wavelength, respectively [19, 20]. The contrast of all images is due exclusively to the objects' elemental composition; no fixatives and staining compounds were added. Smallest features in frozen-hydrated cells down to 30–40 nm size become visible in the cryo-TXM with these objectives in the amplitude contrast mode (see Fig. 4). This image demonstrates the high natural element contrast of cell structures in vitrified water. Various different structures in the cytoplasma, the nuclear membrane and the nucleus are visible; the chromatin arrangement inside the thick intact nucleous can be studied with significantly higher resolution than with conventional optical light microscopes. In addition, proteins of interest can be detected with specific gold labeled antibodies in an X-ray microscopical image through their strong X-ray absorption.

At room temperature some living objects move quite rapidly, e.g. the alga *Euglena spirogyra*. This movement prevents the visualization of internal structures with high resolution. Thus, cryo-processing was used for fixation (see Fig. 5). The organization of the chloroplasts – which are important organelles of photosynthesis – can be analyzed in the intact alga. Figure 6 shows an amplitude contrast image of a part of the polytene chromosome *Chironomus thummi*. The organization of the chromatin fibers can be studied especially in the interband regions. Figure 7 shows an amplitude contrast image of *Caenorhabditis elegans*. The cuticulla can be seen as ring-shaped structures at the edge of the about 15 $\mu m$ thick object. Inside the unstained object single cells as indicated by the arrows are resolved. The structures outside the thick object are due to ice crystallization. Figure 8 shows an X-ray micrograph of an alga *Chlamydomonas rheinhartii*. Different organelles like the chloroplast and the pyrenoid are visible in the image, but it is especially interesting to note that the alga contains several X-ray dense phospholipid vesicles. With other microscopical techniques it is impossible to demonstrate the distribution of the vesicles inside the alga by solely using the natural element contrast.

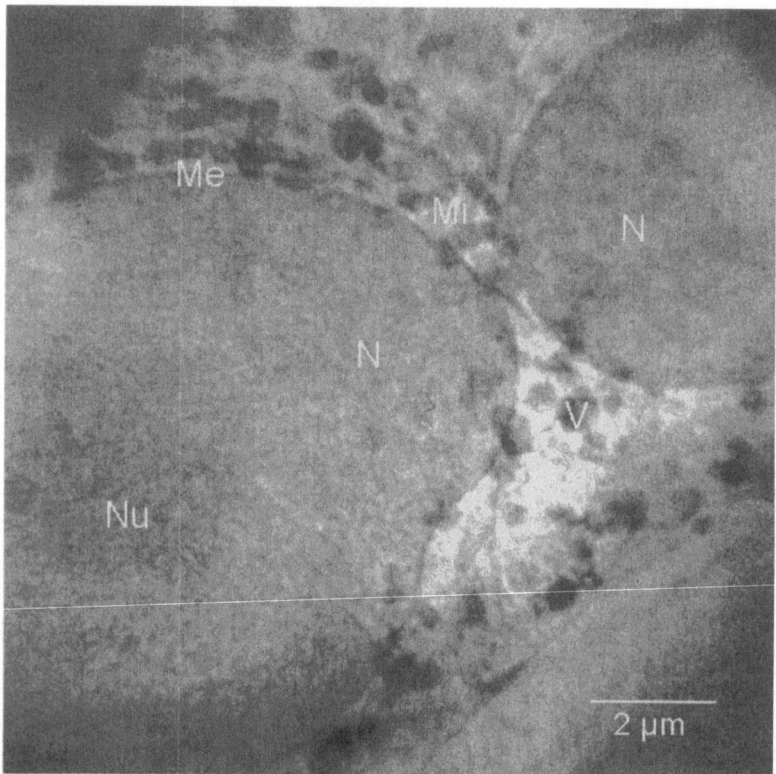

**Fig. 4.** X-ray micrograph taken at 2.4 nm wavelength of an intact Ptk2 cell – a kidney cancer cell of a rat kangaroo. The living cell was shock frozen in liquid ethane. The nuclear membrane (Me), the mitochondria (Mi) and the cell nucleus (N) with the nucleolus (Nu) are resolved inside the cell. Furthermore, several X-ray dense vesicles (V) in the cytoplasm can be identified. Exposure time in amplitude contrast: 5 s, accumulated dosage in vitrified water: about $3 \times 10^6$ Gy.

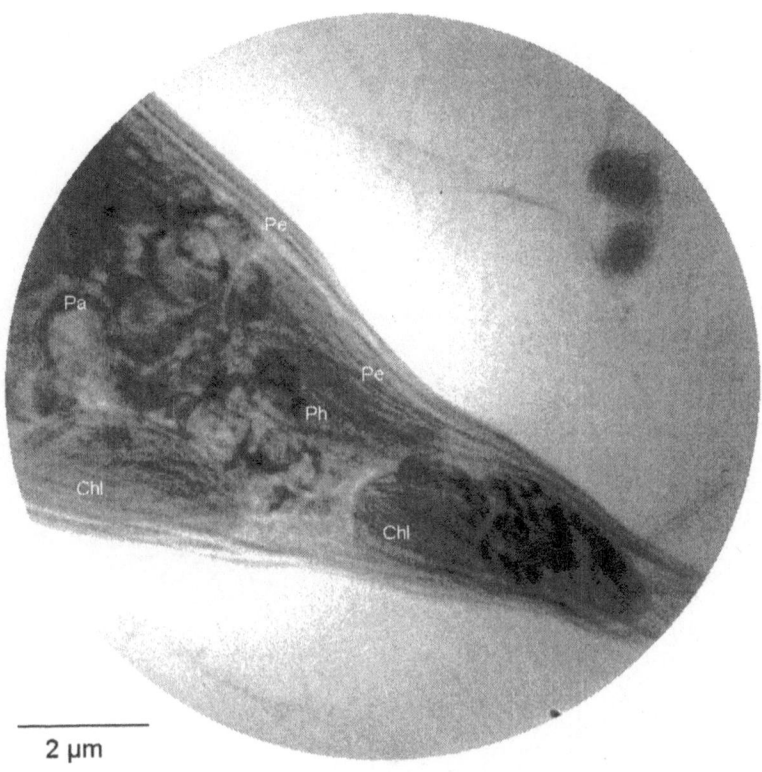

2 µm

**Fig. 5.** X-ray micrograph taken at 2.4 nm wavelength showing a part of the frozen-hydrated alga *Euglena spirogyra*. The chloroplasts (Chl), the paramylon warts (Pa), the pellicular (Pe) and phospholipid vesicles (Ph) are visible. Exposure time: 6 s, accumulated dosage in vitrified water: about $2 \times 10^6$ Gy.

## 4    Summary and Conclusions

The Göttingen TXM has been expanded to allow imaging of specimens at cryogenic temperatures. In contrast to the cryo stages for electron microscopes which operate in vacuum we have developed a system which allows to image specimens in cold nitrogen gas at atmospheric pressure. This simplifies exchanging samples in the cryo-TXM because no air to vacuum transfer gate is necessary.

In these investigations cryogenic biological objects were imaged with an X-ray microscope at 2.4 nm wavelength. The experiments have shown the high structural stability which was estimated from model calculations. In this report amplitude contrast X-ray images were presented. An improved image contrast is obtained at lower doses with the cryo-TXM working in the pure and especially in the optimized phase contrast mode [21]. By adding a tomographic stage to the cryo-TXM it will be possible to study the 3D arrangement of the organelles close to the living state. In addition, cryo X-ray microscopy will also allow investigations of kinetic processes: The cooling time to produce a vitrified sample

**Fig. 6.** X-ray micrograph taken at 2.4 nm wavelength in the amplitude contrast mode of a part of the polytene chromosome *Chironomus thummi*. Bands are highly condensed and fibers are clearly visible in the interband regions.

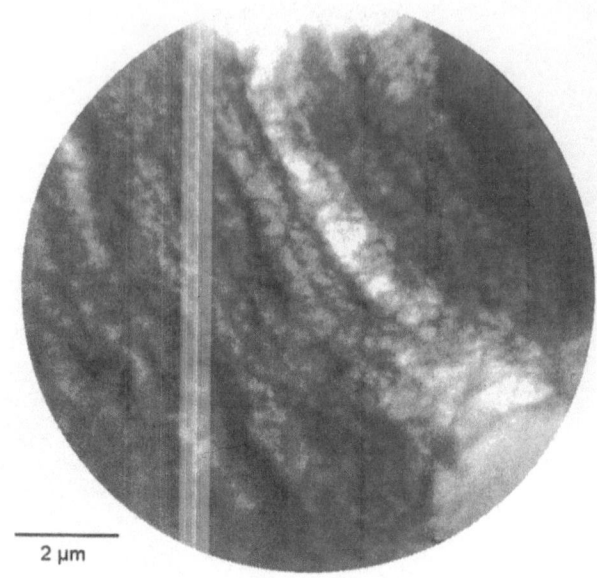

2 µm

**Fig. 7.** X-ray micrograph taken at 2.4 nm wavelength in the amplitude contrast mode of a part of *Caenorhabditis elegans*. Single cells as indicated by the arrows are visible inside the object. In addition, the ring-shaped structures known as Cuticulla can be identified.

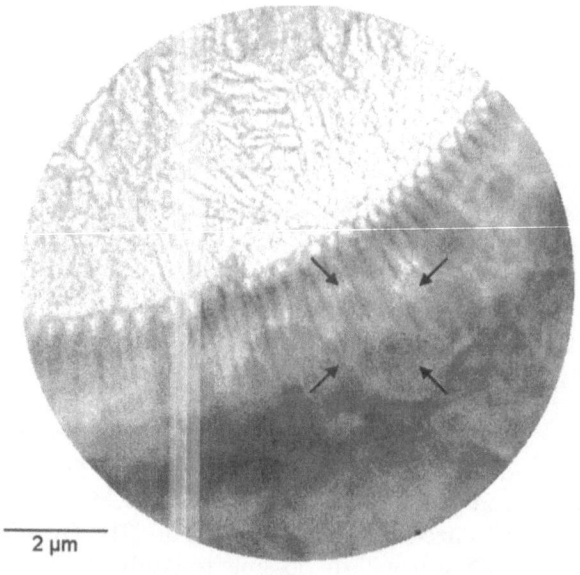

2 µm

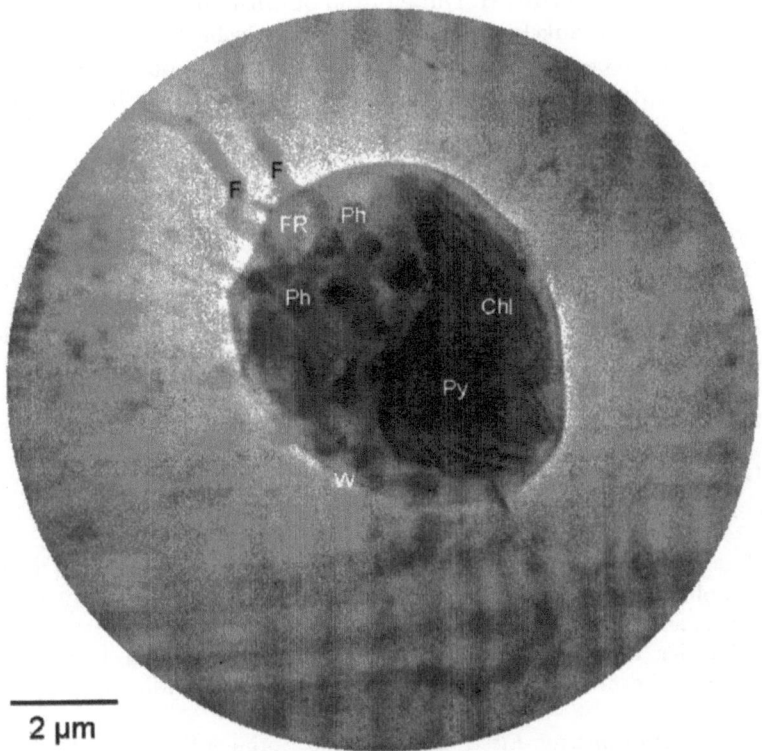

2 µm

**Fig. 8.** X-ray micrograph taken at 2.4 nm wavelength of an alga *Chlamydomonas rheinhardtii*. The wet object was shock frozen in liquid ethane. Two flagellae (F) of 300 nm diameter, the site of the flagellar roots (FR), the chloroplast (Chl) with the pyrenoid (Py) and some X-ray dense phospholipid vesicles (Ph) can be identified inside the cell wall (W). Exposure time in amplitude contrast: 5 s, accumulated dosage in ice: about $4 \times 10^6$ Gy.

is in the order of a millisecond, thus an active process in a cell can be stopped by freezing at a well determined state. We conclude that the cryo-TXM allows numerous different investigations which are impossible or hard to achieve with other methods.

## Acknowledgements

G. Schmahl and D. Rudolph are gratefully acknowledged for their encouragement and support. The authors are indebted to P. Guttmann for his experimental help and the staff of BESSY for excellent working conditions. The image of the polytene chromosome was taken in cooperation with M. Robert–Nicoud and J.B. Sibarita from the University Joseph Fourier, Grenoble. Samples of *Caenorhabditis elegans* were supplied by members of the III. Zoologisches Institut–Entwicklungsbiologie, University of Göttingen. Additionally we wish to thank P. Nieschalk, J. Herbst, H. Düben from our institute and the optical work-shop from the Max-Plank-Institut für biophysikalische Chemie for machining the cryo stage and technical assistance. This work was supported by grants from German Federal Minister for Research and Technology (BMFT) under contract number 05 644 MGA.

# References

1. H. Wolter, Ann. Physik, 6.Folge, Bd.10, (1952) 94–114
2. G. Schmahl, D. Rudolph, G. Schneider, P. Guttmann, B. Niemann, Optik 97, No.4, (1994) 181 - 182
3. G. Schmahl, D. Rudolph, P. Guttmann, G. Schneider, J. Thieme, and B. Niemann, Rev. Sci. Instrum. 66 (2), (February 1995) 1282–1286
4. G. Schmahl, D. Rudolph, in: X-Ray Microscopy Instrumentation and Biological Applications eds. P. C. Cheng, G. J. Jan, Springer-Verlag, (1987) 231–238
5. J. Lehr, Optik 104, No. 4, (1997) 166–170
6. C. Dinges, Dissertation, Technische Hochschule Darmstadt, (1996)
7. D. Sayre, J. Kirz, R. Feder, D. M. Kim, and E. Spiller, Ultramicroscopy 2, (1997) 337–349
8. D. Rudolph, G. Schmahl and B. Niemann, in: Modern Microscopies, Techniques and Applications, A. Michette, and P. Duke eds., Plenum Press, (1990) 59–67
9. G. Schneider, Dissertation, University of Göttingen, (1992)
10. G. Schneider, in X-ray Microscopy IV, V. V. Aristov and A. I. Erko (Eds.), Bo-gorodskii Pechatnik Publishing Company, Chernogolovka, Moscow Region, (1994) 181–195
11. L. Reimer: *Transmission Electron Microscopy*, Springer Series in Optical Sciences Vol.36, (1989) 450–451
12. G. Schneider and B. Niemann, X-Ray Science, 2, Summer 1994 newsletter, Center for X-ray science, King's College, London, (1994) 8–9
13. G. Schneider, B. Niemann, P. Guttmann, D. Rudolph and G. Schmahl, Syn-chrotron Radiation News, Vol. 8, No. 3, (1995) 19–28
14. G. Schmahl and D.Rudolph, Optik 29, (1969) 577–585
15. B. Niemann, D. Rudolph, and G. Schmahl, Optics Communications 12, (1974) 160–163
16. T. Wilhein, D. Rothweiler, A. Tusche, W. Meyer–Ilse, in: X-ray Microscopy IV, Aristov, V. V. & Erko, A. I. (Eds.), Bogorodskii Pechatnik Publishing Company, Chernogolovka, Moscow Region, (1994) 470–474
17. B. Niemann, G. Schneider, P. Guttmann, D. Rudolph and G. Schmahl, in X-ray Microscopy IV, V. V. Aristov and A. I. Erko (Eds.), Bogorodskii Pechatnik Publishing Company, Chernogolovka, Moscow Region, (1994) 66–75
18. P. Echlin: *Low-Temperature Microscopy and Analysis*, Plenum Press, New York, (1992)
19. G. Schneider, T. Schliebe, and H. Aschoff, J. Vac. Sci. Technol. B 13(6), (Nov/Dez 1995) 2809–2812
20. T. Schliebe, G. Schneider and H. Aschoff, Microelectronic Engineering 30, (1996) 523–516
21. G. Schneider, G. Schmahl, T. Schliebe, M. Peuker, and P. Guttmann, (1996) this conference

This article was processed using the LATEX macro package with LLNCS style

# Development of a Cryo Scanning Transmission X-Ray Microscope at the NSLS

Jörg Maser, Chris Jacobsen, Janos Kirz, Angelika Osanna, Steve Spector, Steve Wang, Jan Warnking

Physics Department, State University of New York at Stony Brook, Stony Brook, NY 11794–3800, USA

**Abstract.** We have developed a cryo Scanning Transmission X-ray Microscope at the X1A beamline at the NSLS. The system is designed to image hydrated biological objects of a thickness of several micrometers at temperatures of around 110 K at a resolution of ultimately 30 nm or less. A description of the setup of the cryo-STXM is given. We have started to commission the system and present some results, including first images of biological samples.

## 1 Introduction

The Scanning Transmission X-ray Microscope (STXM) at the National Synchrotron Light Source (NSLS) at Brookhaven National Laboratory (BNL) uses soft X-rays of wavelengths between 2 nm and 5 nm from the X1 undulator [10]. It is capable of imaging specimens of several micrometer thickness, mostly of biological objects, with a spatial resolution of currently around 50 nm. Its main operation mode is bright field absorption contrast. It is also well suited to performing spectroscopic measurements such as elemental and chemical state mapping and spectroscopy of small sample areas ($\approx 0.2\mu$m) [2, 20, 4]. Its flexible setup allows the use of other image acquisition modes such as dark field contrast [5], differential phase contrast [14], microdiffraction imaging [15], or of other contrast mechanisms such as X-ray induced luminescence [9, 3] and dichroism contrast [1].

A major application of X-ray microscopy is the investigation of hydrated, biological samples (see e.g., [11]). Many of these objects suffer structural damage from the irradiation by X-rays (see e.g., [19, 16]). To address the problem of radiation damage, we have developed and are commissioning a cryo-STXM [13]. This system is designed to image hydrated objects at a temperature of around 110 K in expectation of increased radiation hardness of the samples at these temperatures. Cryo techniques have been extensively researched by electron microscopists, and a significant increase of radiation hardness of biological objects has been demonstrated (see e.g., [6, 7]). Theoretical calculations show that a significant reduction of structural damage can also be expected in cryo X-ray microscopes [16], and recent experiments have confirmed this [17].

X-Ray Microscopy and Spectromicroscopy
Eds.: J. Thieme, G. Schmahl, D. Rudolph, E. Umbach
© Springer-Verlag Berlin Heidelberg 1998

There are several advantages in using the cryo method in a STXM. Increased radiation hardness will allow us to image radiation sensitive objects at high spatial resolution without chemical fixation, and hence without artifacts stemming from this preparation method. The ability to record multiple images of the same sample area will allow us to obtain elemental and chemical state contrast at high spatial resolution, and to experiment with tomographic methods on biological objects. Finally, use of fast-cooling methods will allow us to freeze dynamical processes with a time constant in the millisecond range.

## 2    Instrumental Setup of Cryo-STXM

### 2.1    Concept of Cryo-STXM

In most X-ray microscopes in use, the sample is placed in an ambient environment. This allows easy access to the sample area, a rather fast exchange of the sample, and is well suited for imaging samples in a wet state. It also makes changes to the sample area or a change of detectors relatively straightforward and enhances the flexibility of the system. On the other hand, absorption of the X-rays by air or a replacement gas such as helium can reduce the usable spectral range of the system, and even small variations in the ratio of the replacement gas and residual air in the beam path can cause a considerable increase of noise in spectroscopic measurements. Also, when cooling the sample to low temperatures, convectional heat exchange between the cold sample and its environment significantly contributes to thermal drifts. Therefore, we have decided to place the sample in a high vacuum. This reduces the sensitivity of the cryo-STXM to thermal drifts, gives us a good control of contamination of the cold sample, and allows us full use of the water--window spectral range.

### 2.2    Optical Setup

A Fresnel zone plate (FZP) is used to focus the spatially coherent part of the undulator beam into a diffraction-limited spot on the sample (see e.g., [13]). A central stop on the zone plate and an order sorting aperture (OSA) are used to block out unwanted light. For imaging, the sample is scanned through the spot. The transmitted signal is collected by a detector and the measured value stored pixel by pixel in a computer. For focusing, the FZP and the OSA are moved together along the optical axis. For spectral scans, the monochromator is scanned instead of the sample, and FZP and OSA are moved along the optical axis to keep the sample in focus at all times.

### 2.3    Hardware Setup

The cryo-STXM is contained in a vacuum chamber which is pumped by an ion pump. A turbo station is used for evacuation and for prepumping of the sample in an airlock prior to insertion into the chamber. The vacuum chamber sits on a frame which is supported by kinematical mounts. These allow positioning of the chamber in 3 directions and along 3 rotational axes. The stages for the coarse scanning motion of the sample and for detector positioning are outside the vacuum, with bellows for bringing the motion inside the chamber. The vacuum chamber is presently separated from the beamline by a $Si_3N_4$ window and can, therefore, be vented without venting the downstream section of the beamline. Two viewport doors give fast access to the optics and to the detector area. The ultimate pressure of the system is designed to be on the order of $10^{-7}$ torr.

## Setup of the Scanning Motions

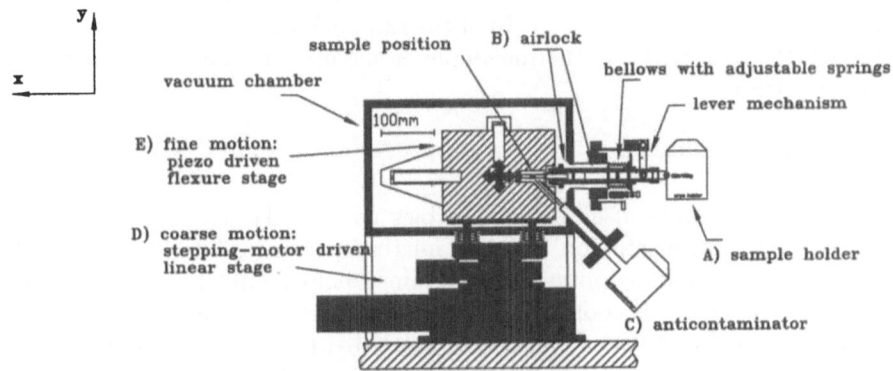

**Fig. 1.** Front view of the cryo-STXM. The X-rays impinge perpendicular to the plane of the figure. The sample is mounted on a $LN_2$-cooled sample holder (A). The holder interfaces with an airlock (B) which can move in $x$ and $y$ against a bellows and is guided by a lever mechanism. The sample holder rests in a socket on the fine scanning stage (E). For coarse scans, stepping motors (D) drive the fine scanning stage and with it sample holder and airlock in $x$ an $y$. For fine scans, only sample holder and airlock are moved. A cold trap (C) reduces contamination of the sample.

Figure 1 shows a front view of the vacuum chamber with the sample holder (A), the airlock (B) and the scanning stages (D and E). For image acquisition, the sample is scanned in a plane perpendicular to the optical axis. A coarse scanning motion (D) allows imaging of large fields, and a fine scanning motion (E) is used to obtain high spatial resolution. The coarse motion is provided by stepping-motor driven linear translation stages, located outside the vacuum, with a minimum step size of 1 $\mu$m and a range of several millimeters. To avoid off-center loads, a true vertical translation stage (Newport UZM160) is used for the vertical motion. The fine scanning motion is performed by an in-vacuum aluminum flexure stage which is mounted atop the coarse stage through bellows. Fine scans are driven by Queensgate piezoelectric actuators with capacitance micrometers which allow minimum steps of a few nm and have a range of 70 $\mu$m.

First tests give a mechanical resolution of the fine stage of better than 20 nm. The flexure stage has been designed using a concept developed by S. Lindaas, M. Howells and C. Jacobsen [12].

The sample is held in a cryo holder (A) which is inserted into the vacuum chamber through an airlock (B). The airlock is suspended at two points from a lever mechanism with pivot points formed by Lucas hinges. The airlock is connected to the vacuum chamber with bellows. This allows it to move freely in a plane orthogonal to the optical axis. After inserting the cryo holder into the vacuum chamber, it is mechanically connected to the airlock with a lock nut, such that airlock and cryo holder move as one unit without slip. The tip of the cryo holder ends in a sapphire ball which is placed in a cone-shaped hole on a socket on the scanning stage. It is held in place by vacuum force which is partly balanced by adjustable springs to yield a small preload. The socket on the scanning stage, in which the cold tip of the sample holder rests, is made of invar to reduce thermal drifts. A thermocouple is installed on the socket to monitor its temperature.

## Cryo Sample Holder

Placing the sample in a vacuum enables us to make use of equipment developed for cryo transmission electron microscopes (cryo-TEM). To hold the sample, we use a modified cryo-TEM specimen holder built by E. A. Fischione Instr. [13]. The holder is cooled by $LN_2$ to a temperature of around 110 K. It is designed to interface with the airlock of a JEOL-100 CX TEM; we use such an airlock to allow a quick transfer of the sample into the vacuum chamber. The temperature of the cryo holder is continuously monitored, and can be raised by activating a heating circuit. This allows us to maintain temperatures between 110 K and 370 K. Controlled heating is particularly useful to investigate manifestation of radiation damage or ice crystal formation by comparing images of samples taken at different temperatures.

The specimen is placed on an electron microscope grid, which in turn is mounted on a seat in the tip of the holder. Additional seats are provided in the tip to allow mounting of a test pattern and of a pinhole for alignment purposes and resolution tests. A shutter, operated by an out-of-vacuum actuator, can be placed around the sample to protect it from contamination during all phases of sample transfer and between taking images. To further reduce contamination, we have also installed a cold trap around the sample. This so-called "anticontaminator" (Fig. 1C) consists of two copper blades, cooled to a temperature of below 110 K, which are placed around the sample.

## Zone Plate and OSA

The FZP and OSA each sit on a support which is mounted to 3-axis translation stages. This allows rather convenient positioning of both components with respect to each other. We use remotely controlled, vacuum-compatible PMAs (Piezo Micrometer Actuators, New Focus) as actuators. The PMAs allow positioning at a speed of around 1 mm/min which has proven sufficient for all practical purposes. Since the OSA is closest to the cold sample, its support is made of invar, and a thermocouple for temperature monitoring attached. We have also kept open the option of heating the OSA to prevent thermal drifts, but the drifts encountered so far are comfortably small.

Both the FZP and the OSA translation stages are mounted on a vacuum-compatible linear $z$-stage which is driven by a DC motor along the optical axis for focusing and spectral scans. The stage is bolted to the bottom of the vacuum chamber. Alignment of the $z$-motion parallel to the optical axis is accomplished by tilting the vacuum chamber with its kinematical mounts.

## Detector Arrangement

Figure 2 shows a side view of the vacuum chamber with the scanning stages for the sample (A, B) and the positioning stage for the detector (E). The detector assembly for the transmitted X-ray flux (F) is mounted on a platform in the vacuum chamber, downstream of the sample area. The platform can be positioned in $x$, $y$ and $z$ with manual micrometer stages outside the chamber. The X-ray detector is mounted inside a small steel can. Electrical connections and water cooling are provided through a long, flexible tube to the detector. X-rays are admitted to the evacuated front area of the detector through a $Si_3N_4$ window. The whole detector can is mounted on a slide mechanism on top of the platform and can be moved in and out of the beam (in $x$) with a wobble stick.

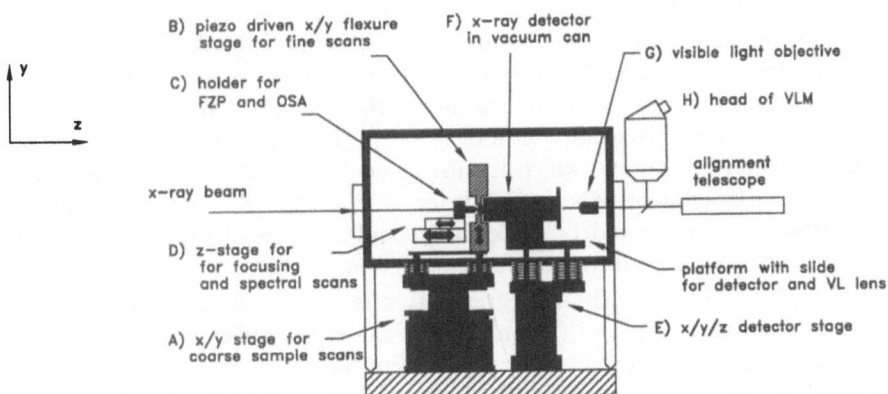

**Fig. 2.** Side view of the cryo-STXM. Coarse (A) and fine (B) scanning stages move the sample through the micro probe. FZP and OSA (C) are moved parallel to the optical axis by a separate z-stage (D) for focusing and spectral scans. The X-ray detector (F) is contained in a steel can inside the vacuum chamber. It can be moved out of the beam and replaced by infinity-corrected visible light objectives (G) using slides on the platform. The platform is positioned with a micrometer stage (E). The head of the visible light microscope (H) is placed outside vacuum.

To preview the sample optically, we have installed a Nikon OptiPhot microscope head (H) with infinity-corrected objectives inside the vacuum (G). The lenses are mounted on the same in-vacuum slide as the detector box, and are moved in position when the detector is moved out of the beam. The position of the detector is defined by a kinematical mount, while the position of the lenses is defined by stops. Once the detector is aligned on the optical axis, all components can be moved in and out of position with the wobble stick using the kinematical mount and the stops, and no readjustment of the detector position with the micrometer stage is needed.

In our first experiements, we have used a photomultiplier tube (PMT) with a phosphor-covered entrance window as an X-ray detector. The phosphor converts the X-rays to visible light, and the PMT converts the visible light pulses to a TTL signal. We plan to use an avalanche photo diode system (APD) with active quenching circuit from EG&G in the future [13].

# 3   Operation and Initial Results

## 3.1   Sample Preparation and Transfer

The objective during sample preparation and transfer is to cool a hydrated sample without formation of ice crystals, and keep it at a temperature where no crystallization can take place during sample transfer and imaging. Ice crystallization during cooling can be prevented by either fast cooling methods or by high pressure cooling methods. In order to prevent formation of ice after cooling, the sample has to be kept at temperatures of around 130 K or below during all steps of sample preparation, transfer and observation. At low temperatures on this order, the sublimation rate of water in vacuum is also very slow which helps prevent freeze drying of the sample during imaging. We chose to cool the sample with a plunge freezing method (see e.g., [7]), where the sample is rapidly plunged into liquid ethane, which in turn is cooled by $LN_2$ to a temperature of around 90 K. Cooling speeds on the order of $10^4 - 10^6$ K/s can be obtained by this method, and it should be possible to cool samples of several micrometers thickness with minimal formation of ice crystals.

After plunge cooling, the EM grid with the sample is mounted to the cryo holder in a workstation, and the shutter on the cryo holder is closed to prevent contamination of the sample. The workstation is cooled with $LN_2$, and all operations take place under cold boil-off nitrogen gas. For transfer of the sample holder into the vacuum chamber, an airlock is used. The cryo holder is removed from the workstation and inserted into the airlock, where the area around the tip is immediately pumped down. During pumpdown, the sample temperature is monitored. After few minutes of prepumping, a gate valve to the main chamber is opened, the cryo holder inserted onto the socket on the scanning stage and secured with the locknut to the airlock. If the FZP, OSA and detector are aligned, the shutter can be opened and experiments with the sample can begin.

## 3.2 Measurements and First Images of Biological Samples

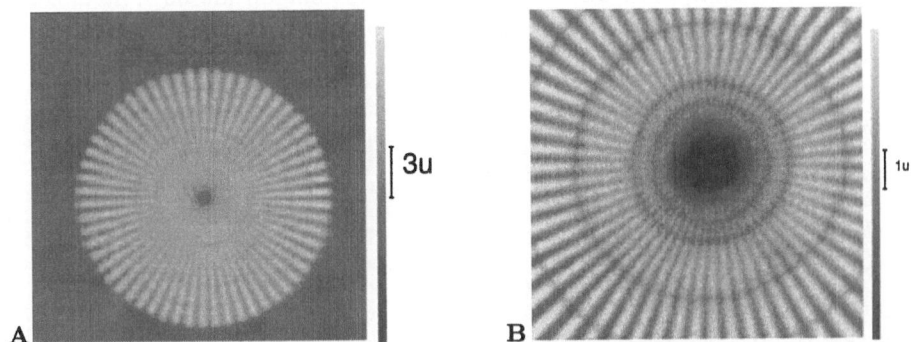

**Fig. 3.** Ge test pattern, imaged with a zone plate of $dr_N = 60$ nm, $D = 160$ $\mu$m. Figure 3A shows good orthogonality and linearity of the scanning stage. Figure 3B shows the central part of the test pattern. Periodic lines and spaces as small as 40–50 nm can be detected, corresponding well to the frequency cutoff of the zone plate used.

The cryo-STXM was commissioned in May and June 1996, and experiments resumed after completion of the upgrade of the X1A beamline. Most of our beam time was dedicated to testing general functions such as vacuum issues, alignment and sample transfer, and to characterize the cryo-STXM mechanically using a test pattern. However, we have been able to take some first images of plunge-cooled biological specimens as well.

All experiments were performed at a wavelength of $\lambda = 2.4$ nm. We used a zone plate of germanium with a diameter of 160 $\mu$m and an outermost zone width of 60 nm, corresponding to a focal length of 4 mm at $\lambda = 2.4$ nm. This reduced the risk of mechanically damaging sensitive components during commissioning, and helped to ease the alignment of the optical components. The zone plate was made by S. Spector using electron beam lithography and a tri-layer etching process [18]. The exit slits of the new X1A beamline were set to 20 $\mu$m in the vertical direction and 25 $\mu$m in the horizontal, corresponding to a product of zone plate diameter $D$ and full angular acceptance $\theta$ of $D \cdot \theta \approx 0.5 \ \lambda \cdot$ rad, providing spatially coherent illumination of the zone plate.

Figure 3 shows images of a test pattern which was also made by S. Spector. A full view of the test pattern, Fig. 3A, demonstrates that the scanning stage is orthogonal and linear. Figure 3B is taken from the central region of the test pattern at a higher resolution (step size: 25 nm). We can detect periodic lines and spaces of around 40–50 nm, which corresponds well to the frequency cutoff at of $1/0.06\mu m$ for the zone plate used. The test pattern is made of 180 nm thick Germanium, the low contrast of this object at $\lambda = 2.4$ nm being the reason for a relatively high noise level in the image.

To measure thermal drifts, we have cooled the test pattern to 130 K, imaged it repeatedly over the course of one hour and recorded the change of position over time. We have found an average drift of around 1.3 nm/sec, with a maximum of 2.5 nm/sec. This corresponds to a shift of few pixels per total image. Since the image is recorded in a serial fashion, drifts of this magnitude have no effect on the resolution, and an acceptably small effect on the linearity of the image.

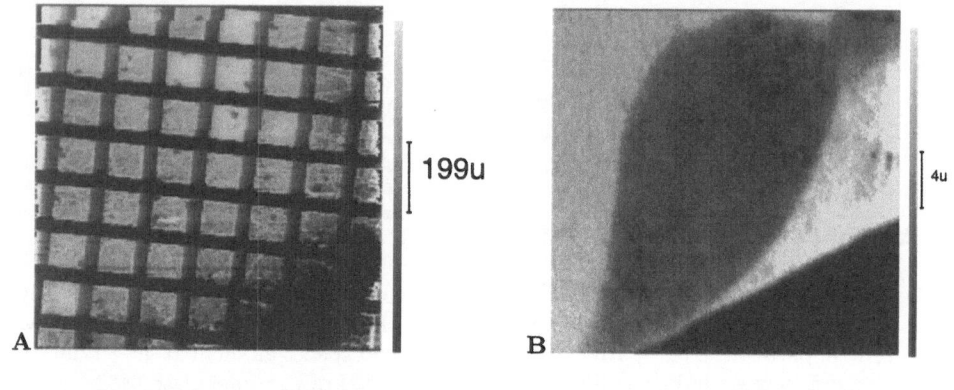

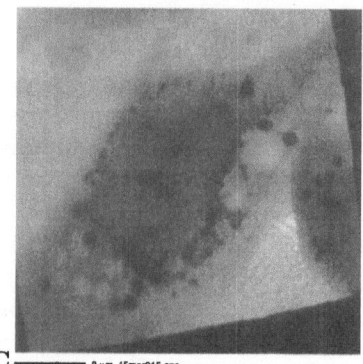

**Fig. 4.** First images taken with the cryo-STXM of biological samples. Figure 4A is taken with the coarse scanning stage from a 1 mm × 1 mm area of the sample. Figure 4B and Fig. 4C are taken taken with the fine scanning stage with a step size of 100 nm. Figure 4B shows an image of a frozen hydrated V79 fibroblast (Chinese hamster lung), Fig. 4C an image of a frozen hydrated 3T3 fibroblast (Fig. 4C). Both samples were imaged at a temperature of 110 K.

In Fig. 4, some images of plunge-cooled biological specimens are shown. All images were taken at a wavelength of $\lambda = 2.4$ nm, with the sample at a temperature of around 110 K. Figure 4A shows an image taken with the coarse scanning stage of a large area of the grid with the samples. This allows us to get an overview of the whole sample grid, and to select specific areas for further study. Figure 4B and Fig. 4C show images of frozen hydrated specimens. The samples by plunging into liquid ethane, without prior fixation or other treatment. Figure 4B is taken of a frozen hydrated fibroblast (transformed V79 fibroblast cell from Chinese hamster lung). Figure 4C is taken of a frozen hydrated 3T3 fibroblast. Both samples show good preservation of structural features.

As can be seen from these first experiments with sample preparation, blotting of the specimen prior to plunging allows us to easily obtain an ice thickness of a few micrometer, as required for imaging with soft X-rays. Also, use of both the coarse and the fine scanning stage allows previewing of the whole sample at low dose (Fig. 4A) for selection of different areas of interest, and subsequent imaging of the selected sample areas at higher resolution (Fig. 4B, C).

## 4   Conclusions and Outlook

We have developed and are commissioning a cryo Scanning Transmission X-Ray Microscope. This system is capable of imaging specimens of a thickness of several micrometers at a temperature of around 110 K. The temperature is stable to $\pm 0.5\ °C$. The sample is kept in a vacuum of $10^{-7}$ torr or better, allowing good control of thermal drifts and of contamination, and full use of the water window spectral range for imaging and spectroscopic measurements. An airlock on the vacuum chamber allows a sample exchange within few minutes. The sample temperature is monitored and can be raised in a controlled way with a built-in heater to well above room temperature. The computer control of the system records data from up to 8 analog and 8 digital channels which allows the use of configured detectors, e.g., and recording of the temperature of the sample and other parts of the microscope. An optical microscope with infinity-corrected lenses for previewing of the sample and prefocusing of zone plate and OSA is installed.

The first tests of the cryo-STXM show good orthogonality of the fine scanning stage. In coarse scanning mode, a field of up to several millimeters can be imaged, whereas the fine scanning mode allows a mechanical resolution of 20 nm or better. Alignment and focusing procedures have been successfully tested, and first test of the spatial resolution were done using a zone plate with an outermost zone width of 60 nm. We were able to detect periodic lines and spaces of 40 nm–50 nm in a test pattern, corresponding well to the exptected frequency cut-off of the zone plate. Thermal drifts were measured to be on the order of around 1.3 nm/s which is small enough not to distort the image or reduce the resolution. We have obtained first images of frozen hydrated specimens.

We plan to pursue two lines of work in the near future. The first is dedicated to fully characterizing the cryo-STXM mechanically and optically, and to improving some of the components. This involves installing an avalanche photodiode with active quenching circuit as an X-ray detector, and implementing a rotary motion on the airlock to allow recording of tomographic data sets. We also want to reduce the rate of thermal drifts further. The experimental program will focus at first on taking 2D images of frozen hydrated objects, on improving the sample preparation protocol, and on determining the radiation hardness of biological cryo-objects. We also plan to perform spectromicroscopy on those samples. One application is to take spectra of samples with vitrified and samples with crystalline ice at the oxygen absorption edge with the intention of using differences in the spectra to help determine the degree of crystallinity of frozen hydrated samples. After installing the rotation mechanism, we will also start experimenting with tomography.

# Acknowledgments

We want to thank Sue Wirick for her continuous help at X1A. This work was supported by the Office of Health and Environmental Research, Department of Energy, under contract FG02-89ER60858, by the National Science Foundation with Presidential Faculty Fellow award RCD-9253618 (CJ), and by the Alexander von Humboldt Foundation through a Feodor Lynen Fellowship (JM). The experiments were performed at the NSLS which is supported by the Department of Energy.

# References

1. H. Ade and B. Hsiao. *Science*, **262**, 927–975, (1993).
2. H. Ade, X. Zhang, S. Cameron, C. Costello, J. Kirz, S. Williams. *Science*, **258**, 927–975, (1992).
3. A. I. von Brenndorff, M. M. Moronne, C. Larabell, P. Selvin, and W. Meyer–Ilse. In: [8], pp. 338–344.
4. C. Buckley, N. Khaleque, S. J. Bellamy, M. Robbins, and X. Zhang: Proceedings of this conference.
5. H. Chapman, J. Fu, C. Jacobsen, S. Williams. *J. Microsc. Soc. Am.*, **2**, 53–62, (1996).
6. J. Dubochet. M. Adrian, H. Chang, J. Homo, J. Lepault, A. W. McDowall, P. Schulz. *Quart. Rev. Biophys.*, **21**, pp. 129–228, (1988).
7. P. Echlin: Low-temperature Microscopy and Analysis, Plenum Press, New York, (1992).
8. A. I. Erko and V. V. Aristov, editors. X-ray Microscopy IV, Chernogolovka, Moscow Region, (1994), Bogorodski Pechatnik.
9. C. Jacobsen, S. Lindaas, S. Williams, and X. Zhang. *J. Microsc*, **172**, 121–129, (1993).
10. C. Jacobsen, E. Anderson, H. Chapman, J. Kirz, S. Lindaas, M. Rivers, S. Wang, S. Williams, S. Wirick, and X. Zhang. In [8], pp. 304–322.
11. J. Kirz, C. Jacobsen, M. Howells. *Quart. Rev. Biophys.*, **28**, pp. 33–130, (1995).
12. S. Lindaas, M. Howells, C. Jacobsen and A. Kalinovsky: *J. Opt. Soc. Am.* **A 13**, 1780–800, (1996).
13. J. Maser, H. Chapman, C. Jacobsen, A. Kalinovsky, J. Kirz, A. Osanna, S. Spector, S. Wang, B. Winn, S. Wirick, X. Zhang. X-ray Microbeam Technology and Applications, **SPIE 2516**, 78–89, (1995).
14. G. R. Morrison. In: [8], pp. 479–484.
15. D. Sayre, H. N. Chapman. *Acta Crystallographica*, **A51**, 237–252, (1995).
16. G. Schneider. In: [8], pp. 181–195.
17. G. Schneider, B. Niemann, P. Guttmann, D. Rudolph, G. Schmahl: *Synch. Rad. News*, **8**, pp. 19–29, (1995).
18. S. Spector, C. Jacobsen, D. Tennant: Proceedings of this conference.
19. S. Williams, X. Zhang, C. Jacobsen, J. Kirz, S. Lindaas, J. Van't Hoff, S. S. Lamm. *J. Microscopy*, **170**, 155–165, (1992).
20. X. Zhang, Rod Balhorn, Joe Mazrimas, J. Kirz. *J. Struct. Biology* **116**, 335–344, (1996).

This article was processed using the LaTeX macro package with LLNCS style

# The X-Ray Microscopy Facility Project at the ESRF

Jean Susini and Ray Barrett

European Synchrotron Radiation Facility,
BP220, F-38043 Grenoble cedex, France

**Abstract.** A beamline dedicated to X-ray imaging and spectro-microscopy in the 0.2–6.0 keV energy range is under construction at the ESRF. This beamline is installed on a low beta straight section equipped with three phaseable undulators and will consist of two branch lines: the first, a scanning microscope including various contrast modes; the second, an imaging full-field microscope using Zernike phase contrast. The scanning microscope will be equipped with 2 fixed-exit high resolution monochromators (crystal/multilayer monochromator and plane grating monochromator). Both microscopes will use high resolution zone plates as focusing elements. The design of this beamline is discussed with emphasis on the optical layout and on heat load management.

## 1 Introduction

Third generation synchrotron sources produce a beam of unprecedented quality: the extremely low emittance coupled with high brilliance together with the versatility of new insertion devices, offer the capability to control brightness, spectrum, polarisation, coherence and size of the beam. This means that X-ray microscopy techniques which have been extensively used in the soft X-ray region [1] can now be extended, with the anticipation of very high performance to higher photon energies. This will enable new investigations: study of thicker specimens, access to K absorption edges of elements of major interest in the biological and materials sciences, in particular from Potassium to Chromium, access to M and L edges of heavy metals (i.e. Au or Ag) for specimen labelling, and the use of X-ray fluorescence for trace element mapping.

To exploit these characteristics an X-ray microscopy beamline is being built at the European Synchrotron Radiation Facility, whose source properties offer the required coherence and brilliance [2]. Approved in 1993, the project started at the end of 1994, the first beam will be delivered in December 1996 and the beamline will be open to external users by the end of 1997. Scanning and imaging microscopes will be built on two independent branch lines. In parallel, collaboration contracts have been placed between the ESRF and two European laboratories for the fabrication of high energy zone plates [3–6].

After a brief overview of the ESRF source properties, the conceptual design of the X-ray microscopy beamline will be presented. The problem of building a soft X-ray beamline which should cover an energy range (0.3–6 keV) on a high energy machine will be addressed. Finally, the optical layout will be described in more detail.

**X-Ray Microscopy and Spectromicroscopy**
Eds.: J. Thieme, G. Schmahl, D. Rudolph, E. Umbach
© Springer-Verlag Berlin Heidelberg 1998

## 2 The Source

The ESRF is a high energy machine with an electron energy of 6.02 GeV and emittances (1% coupling) of $\varepsilon_h = 4 \times 10^{-9}$ mrad (horizontal) and $\varepsilon_v = 4 \times 10^{-11}$ mrad (vertical). Currently, two types of straight sections, low or high $\beta$, are available. The considerations taken into account in the choice of the source type are summarised in Table 1.

**Table 1.** Considerations involved in choice of the straight section.

| Straight section | High $\beta$ | Low $\beta$ |
|---|---|---|
| Optical functions | $\beta_h = 27.0$m, $\beta_v = 13.3$m | $\beta_h = 0.58$m, $\beta_v = 3.14$m |
| Source parameters | $\sigma_h = 330\mu$m, $\sigma_v = 23\mu$m<br>$\sigma'_h = 12.1\mu$rad, $\sigma'_v = 7.5\mu$rad | $\sigma_h = 59\mu$m, $\sigma_v = 11\mu$m<br>$\sigma'_h = 83\mu$rad, $\sigma'_v = 8.2\mu$rad |
| Pros | - Total power can be limited by the use of an aperture.<br>- Symmetric beam<br>- Requires small optics. | - Better phase-space matching<br>- Wider beam @ 30m<br>Ølower power density<br>Øpossibility for simultaneous operation of the two branch lines |
| Cons | - Phase-space mismatch<br>- High power density | - Asymmetric beam<br>- Requires longer horiz. deflecting optics |

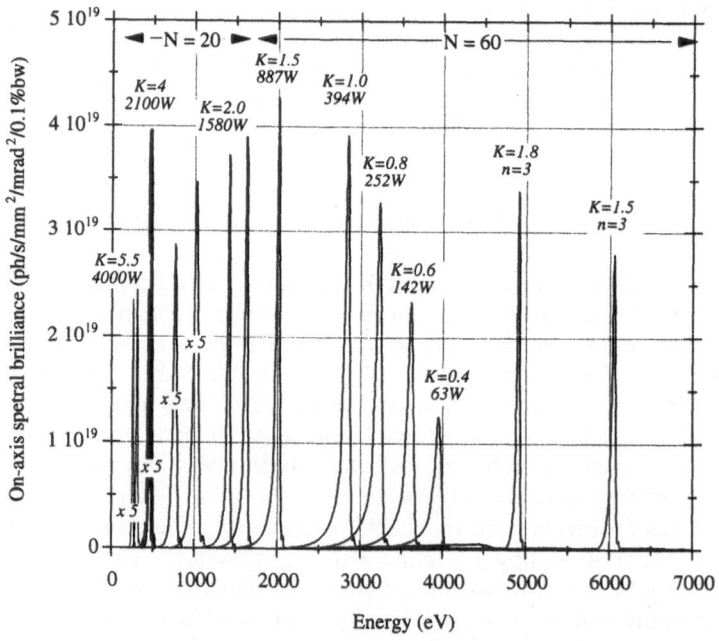

**Fig. 1.** On-axis spectral brightness vs. energy for various K values.
The integrated total power emitted by the IDs is given.

A number of imaging experiments require at least partially coherent illumination. Therefore, the main criterion for the choice of the source is the flux per coherent phase-space volume which can also be expressed as the number of wave modes contained in the photon beam. The low beta straight section, ID21, of the ESRF storage ring has been chosen as the source best suited for our applications due to the higher phase-space density (proportional to $\lambda/\sigma.\sigma'$). The principle disadvantage of this source is the large horizontal beam divergence which gives rise to a very flat elliptical beam shape which does not match the spherical acceptance of the zone-plate.

The ID21 straight section is 4.8 m long and is equipped with three identical 1.6 m long, 80 mm period, phaseable insertion devices (ID) allowing the number of poles to be changed between 20, 40 and 60. The first harmonic is tuneable from 0.2 to greater than 4.5 keV corresponding to a gap range from 15 to 80mm. Due to the flexibility of the ESRF strategy of constructing undulators in three independent sections, the total emitted power can be maintained at a level which can be easily handled by the first mirror without affecting significantly the coherent brilliance: a 4.8 m device would be used for the highest energy experiments, while a single 1.6 m ID would be used for soft X-ray experiments. In diffraction limited imaging conditions, the factor 9 in the peak brilliance lost by the fact than only 1 ID is used is largely compensated by the effect of the lateral coherence width increase with wavelength at a given distance. In all cases, the total power through an aperture of 5x5 mm$^2$ placed at 28 m from the source is always maintained below 700 W.

## 3 Optical Layout

The main  design features of the optical layout of the beamline are:
- Preservation of the high coherent brightness of the source.
- Scanning (STXM) and imaging (TXM) microscopes are to be built on two independent branch lines which could be used simultaneously: the STXM requires only a fraction of the beam (corresponding to a few wave modes) with no loss of performance, the remaining being used by the TXM.  Wide band-pass, horizontally deflecting, X-ray multilayers are used to steer the beam on the TXM branch line. The Bragg angle of 4 degrees allows a large physical separation between the two endstations.
- On the STXM branch, a small pinhole aperture (5 μm to 50 μm) will be used as a secondary source, set at about 1m from the zone-plate. This aperture and the microscope are mechanically linked to the same support in order to minimise relative movements (mechanical vibrations) between the two components. Even in an experiment requiring spatial coherence, the optimum amount of phase-space to accept is rather more than a single mode and involves a trade-off of high coherence for diffraction limited focusing – against flux – which is required for observation of weak contrast phenomena and fluorescence experiments. Moreover, an overfilling of the aperture used as secondary source and the zone-plates allows the effects of beam instability to be minimised. The aperture size and the distance between the pinhole and the microscope are variable to accommodate a wide variety of microscopy, spectro-microscopy and related coherence experiments over a large energy range.

The beamline will be operated in windowless mode in order to maximise the photon flux at the lower energies and to minimise the degradation of the inherent coherent brightness of the source.

## 3.1 Total Power Management by Use of two X-Ray Mirrors

It is important to limit the heat load on those optical components which strongly influence the energy resolution, namely the focusing mirrors and the two monochromators. Consequently, a 2-bounce horizontally deflecting mirror device consisting of two parallel silicon mirrors is used for :

i)   Power filtering: 90% of the unwanted part of the spectrum is absorbed by the first mirror. The maximum integrated power is damped from 700 W to less than 50 W after two reflections.

ii)  Suppression of the higher-order harmonics of the insertion device: the cut-off angle can be tuned from 5 to 20 mrad allowing harmonic rejections better than $10^{-3}$ for any energy ranging from 1 to 6 keV. At these small grazing angles the position of the exit beam is effectively fixed.

iii) Separation of the bremsstrahlung radiation from the synchrotron radiation: the combination of the mirror device with a bremsstrahlung stop and a collimator allows a tremendous reduction of the shielding required for the downstream beam transport. If the "pink" beam is efficiently "cleaned" and collimated there is no shielding required downstream of the collimator. This strategy has not only an obvious financial impact upon the beamline design cost but also allows relatively free access to the microscope during the experiments.

Furthermore, horizontally deflecting grazing incidence mirrors have several advantages :

i)   preservation of the vertical emittance of the source which determines both the degree of coherence of the photon beam and the energy resolution of the downstream monochromators.

ii)  The photon beam in the horizontal plane is very wide. Therefore, the heat load along the mirror length is much more homogeneous than it would be in the case of a vertically deflecting mirror and, the thermal gradient along the mirror length is minimised.

The two first mirrors are side water cooled. A gallium-indium layer between the mirror and the copper block provides both a good thermal contact and an efficient mechanical decoupling. The first mirror which removes much of the unwanted power, is 130 mm thick while the mirror is cooled along a 10 mm strip on the top. Finite-element thermo-mechanical simulations show that this cooling geometry allows the thermal deformation to be kept below 10 μrad rms for a maximum absorbed power of 700 W.

## 3.2 Layout of the STXM Branchline

The optical design was influenced by several constraints:

For spectroscopic applications the very wide energy range cannot be covered with a single device, therefore 2 high resolution monochromators are planned. It is intended that the operations necessary to change between these two basic operating modes (see Fig. 2) should be automated as much as possible and allow imaging over the full spectral range to be achieved for the same sample with a minimum of manual realignment.

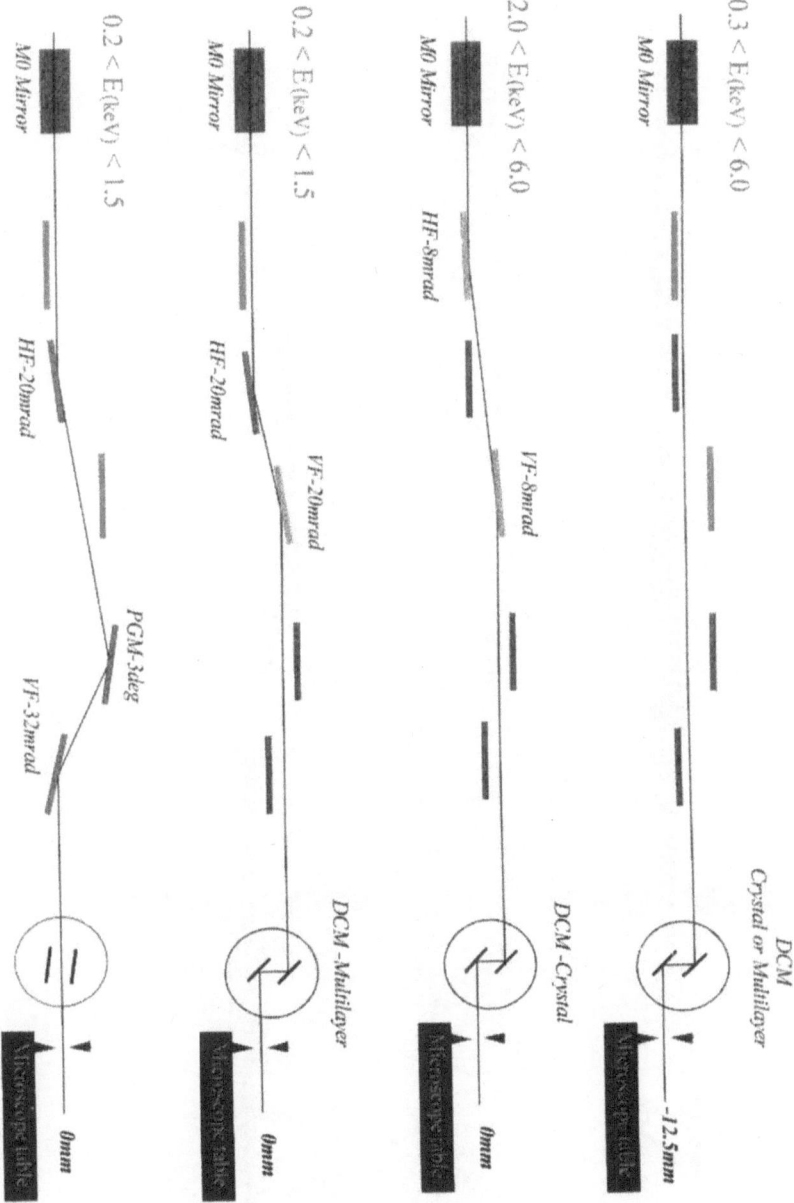

**Fig. 2.** Various optical configurations of the beamline

## Crystal Monochromator (2.0–6.0 keV)

A fixed exit 2-crystal monochromator which can be equipped with Silicon and multilayers will be used in the 2–6 keV range (Fig. 3). The Bragg angle is tuneable from 3 to 70 degrees. The energy resolution is about $+5.10^{-4}$ for Silicon and $10^{-2}$–$10^{-3}$ for multilayers. The optics will be water cooled. In order to ensure a good

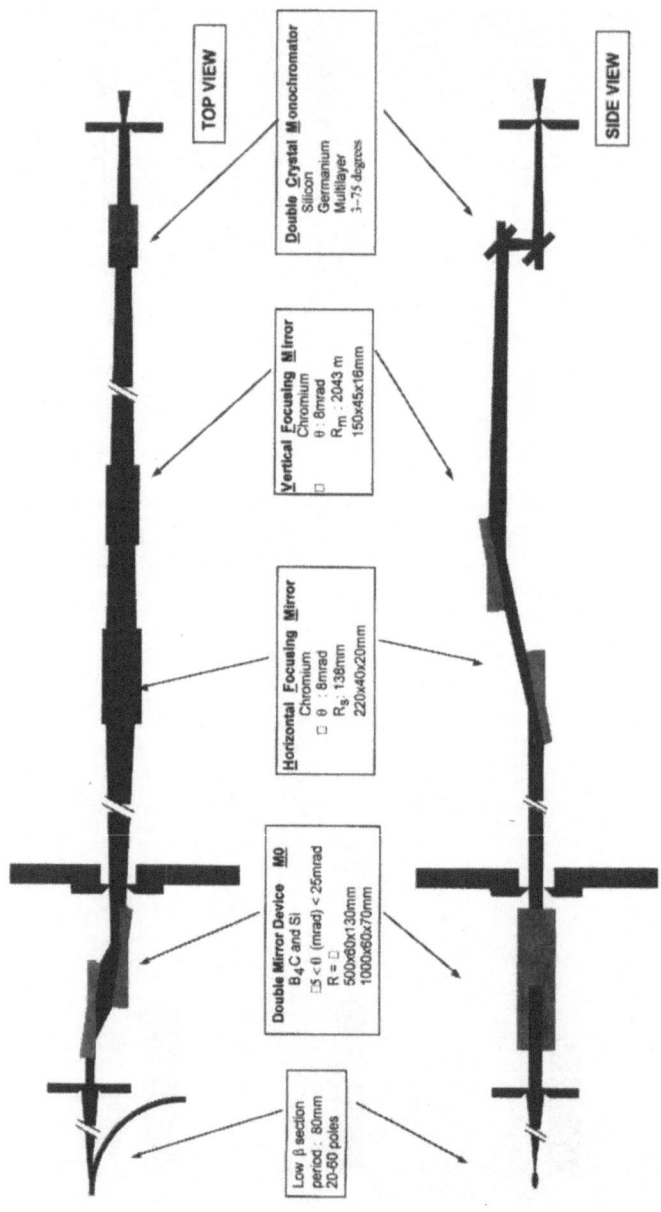

**Fig. 3.** Parameters of the double crystal monochromator and associated focusing mirrors

stability of the outgoing beam during an energy scan, the monochromator is positioned as close as possible to the entrance aperture of the microscope (about 1 m). In this operation mode the source is imaged onto the pin-hole plane via a sagitally focusing cylindrical mirror in the horizontal plane and a vertically focusing spherical mirror.

## Plane Grating Monochromator (0.2–1.5 keV)

The grating monochromator design (see Fig. 4) associates plane gratings, which are tuned by a simple rotation, with a vertically focusing spherical mirror (VFM). In this design it is the source (conjugated by the VFM) which acts as the entrance slit of the monochromator. The pinhole apertures for the STXM act as the exit "slits" of the grating monochromator and are fixed at 8.23m from the VFM. In the horizontal plane, a sagitally focusing mirror (HFM) is used to image the source onto the same pinhole apertures. Three holographic gratings have been optimised to cover the full energy range [7]. In order to achieve a good overall efficiency, in particular, at high energies, the deviation angle has been set to 6 degrees. Assuming slope errors of the order of 1μrad on each of the vertically deflecting components the average calculated resolving power of the PGM should be around 6000 over the operating energy range.

## Stigmatic Design

The exit pinhole of the grating monochromator, used as secondary source for the scanning microscope, being fixed over the full energy range, the different focusing mirrors must have the correct radius of curvature. This is particularly critical because the short distance between the zone-plate and the source (pinholes) makes the microscope performance very sensitive to astigmatism [8]. The maximum tolerable longitudinal separation between the vertical and horizontal sources (corresponding to the foci of the focusing mirrors) is only 10cm. Therefore, the vertical focusing mirror is bendable (bimorph mirror) in order to match its image plane exactly with those of the horizontally focusing mirrors. A beam position monitor will be set at the secondary source plane [9].

## The Scanning Microscope

The STXM is conceived to work with zone plate type optics which currently offer the best proven performance for this type of application. The development of high resolution zone plates capable of efficiently focusing X-rays up to 6keV has been addressed in two different collaborative research programs established with laboratories at King's College London and Göttingen [3-6]. Clearly a single zone plate will be incapable of providing efficient focusing over the entire spectral range available and one of the challenging aspects of the microscope design has been to facilitate the remote exchange of the zone plate allowing close to optimum focusing conditions regardless of the operating energy. The X-ray microscope will be placed in an environmental chamber allowing operation in air, helium or vacuum ($10^{-4}$–$10^{-6}$ mbar). The entire microscope can be moved along the beam axis relative to the fixed exit pinhole which acts as the secondary source. This movement allows the pinhole–zone plate distance to be varied from 0.3 to 1.5 m and allows the illumination conditions of the zone plate to be adapted to the experiment. Taking into account the mirror reflectivities, monochromator band pass and undulator characteristics the estimated photon flux in a 50nm probe is of the order of $10^9$–$10^7$ ph/s.

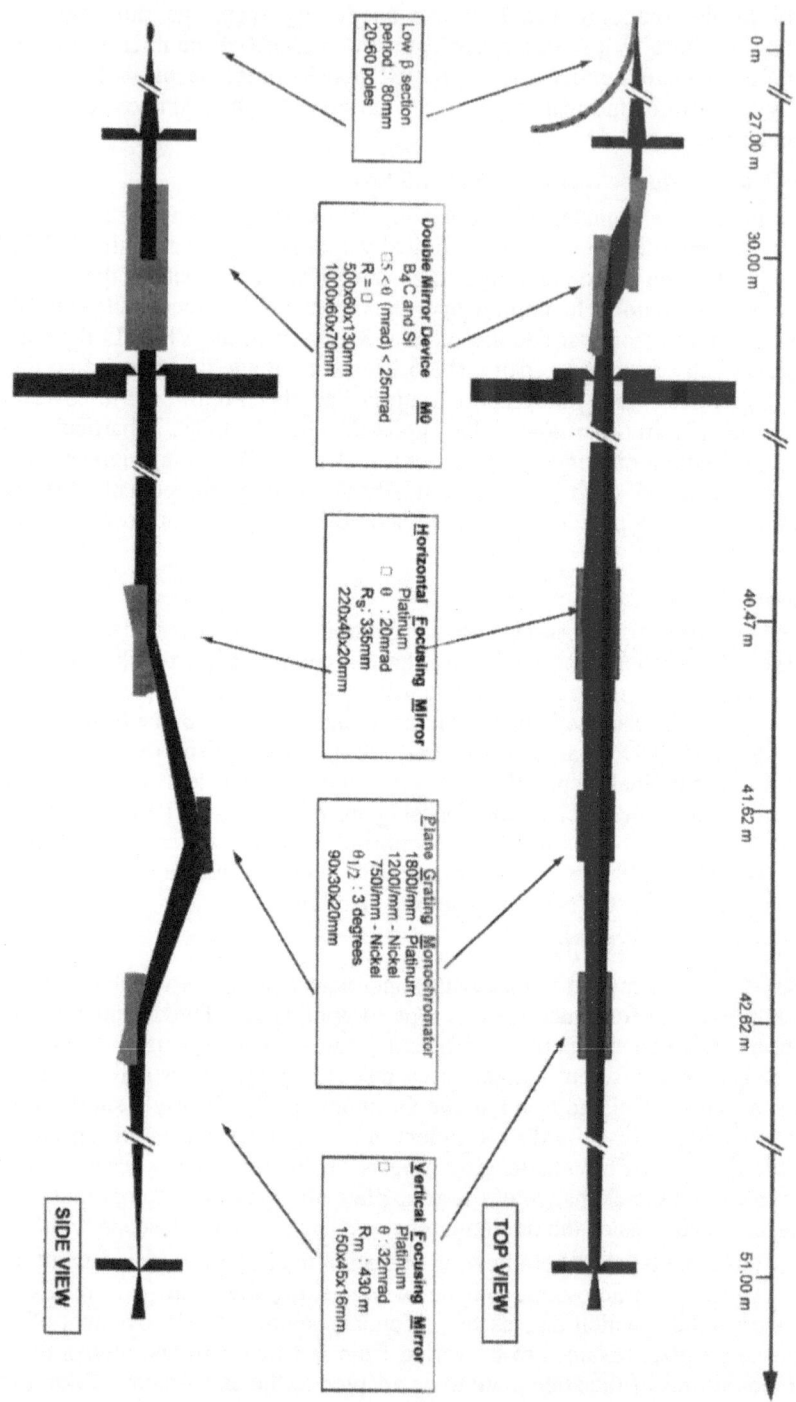

**Fig. 4.** Parameters of the plane grating monochromator and associated focusing mirrors

The wide spectral operating range of the microscope is attractive for spectro-microscopy. Whilst in its simplest form this might consist of taking multiple images of a single sample region at different incident energies, an interesting extension is to perform highly spatially resolved XAS scans on small regions of the sample. The spatial resolution of this mode is potentially limited by the probe size, convergence and the sample thickness, but requires careful mechanical design due to the energy dependence of the zone plate focal length.

The sample will be scanned using a combination of piezo driven flexure and mechanical stages giving a total scan area of $10 \times 10$ mm$^2$. The current aim is to approach pixel rates of 1 kHz. A manual sample rotation will be available, primarily for fluorescence mode imaging, but with the possibility of future upgrade to motorised movement for micro-tomography measurements

The microscope design is intended to offer maximum flexibility for the use of various different detector types. Currently it is planned for absorption measurements to use alternatively, proportional gas detectors, PIN photodiodes and avalanche photodiodes. A high energy resolution Germanium solid state detector will also be available for fluorescence measurements. A standard sample holder has been designed upon similar principles to that currently used on the full field imaging microscope at the ALS [11]. This kinematically mounted system should allow regions of interest to be identified and recorded on a standard light microscope prior to transfer into the X-ray microscope and rapidly aligned to the probe scan. It is intended for the same holder to be used on the TXM allowing rapid transfer between the two microscopes. The microscope will be controlled using essentially standard ESRF VME based electronics running OS9 with a user interface running on Unix workstations.

## 3.4 Layout of the TXM Branchline

The main difficulty of the implementation of a TXM branchline using the ID21 source is associated with the relatively small emittance and the relatively large energy range. The consequent problems of creating a suitable condenser system for the microscope are discussed elsewhere [10]. Zone plate optics will be used for the objective lens with similar characteristics to those planned for use in the scanning microscope. It is probable however, for reasons of imaging dose, that a stronger emphasis will need to be placed upon the efficiency of these zone plates, potentially at the expense of resolution. It is planned initially to use a channel cut Si 111 monochromator for this branch thus allowing narrow bandpass imaging to be performed. The microscope will enable absorption contrast (bright field), dark field and Zernicke phase contrast imaging to be performed. The estimated flux at the detector plane for bright field imaging with a Si 111 monochromator is of the order of $10^{10}$ ph/s. Cryo-cooling is foreseen to reduce the effects of beam damage and thus permit multiple imaging of dose sensitive samples such as required in tomographic measurements. It is anticipated that the primary detectors to be used will be CCD cameras. It is probable that in the longer term two such detectors will be available depending upon the imaging energy to be used. In the case of the lower energies this will probably be a back-thinned CCD using direct conversion whilst for higher energies (to preserve dynamic range) a phosphor coupled system is envisaged.

# 4 Conclusions

The conceptual design of the X-ray microscopy facility at the ESRF has been described. Priority has been given to the spatial resolution (100–50nm), resolving power of the monochromators (E/$\Delta$E • 4000–8000), flux at the sample position ($10^9$–$10^7$ ph/s in a probe of 50nm diameter for the STXM) and flexibility (trade-off of energy resolution for flux, multiple contrast modes, two microscopes). The beamline will exploit the potential of the imaging applications in the 2–6 keV energy region which are, so far, relatively unexplored. In particular, the access to the K and L edges and emission lines of the medium elements will offer new capabilities: higher penetration depth, fluorescence contrast. For many samples, cryomicroscopy will be necessary to cope with the very high flux and the related radiation damage.

## Acknowledgements

The authors are very grateful to R. Baker, G. Marot and R. Guetta for their for their very active involvement in the design of the beamline components, E. Delcamp and F. Polack for the optimisation of the parameters of the grating monochromator. We thank C. Buckley and B. Niemann for a number of very constructive discussions.

## References

1    J. Kirz, C. Jacobsen, and M. Howells, *Soft X-Ray Microscopes and their Biological Applications*, Quarterly Reviews of Biophysics, 28 (1), (1995).

2    M.R. Howells, R. Burge, C.J. Buckley, A. Miller, G. Morrison, D. Rudolph, G. Schmahl, J. Thieme and M. Vollbrecht, *Conceptual design for an X-ray microscopy facility at the ESRF*, ESRF Technical Report, (1992).

3    B. Kaulich, *Nanostructures with high aspect ratios for phase zone plates for a phase contrast X-ray microscope at ESRF for wavelengths around 0.3nm*, Proceedings of X-ray Microscopy IV, editors A.I. Erko and V.V. Aristov, Chernogolovka (1994).

4    B. Kaulich, *Phase zone plates for X-ray microscopy at the ESRF*, this conference.

5    P. S. Charalambous and A. Firsov, *Optimization of the process parameters for the fabrication of high resolution diffraction optical elements*, Proceedings of X-ray Microscopy IV, editors A.I. Erko and V.V. Aristov, Chernogolovka (1994).

6    P. Charalambous and R.E. Burge; *Zone plate fabrication at King's College, London*, this conference.

7    E. Delcamp, B. Lagarde, F. Lagarde and J. Susini, *Optimization software for X-UV monochromators*, Proceedings SPIE conference, 2856, Denver (1996).

8    McNulty, A. Khounsary, Y.P. Feng, Y. Qian, J. Barraza, C. Benseon, and S. Shu, *A beamline for 1-4 keV microscopy and coherence experiments at the Advanced Photon Source*, Rev. Sci. Instrum., 66 (9), September 1995.

9    G.S. Dermody, C.J. Buckley, J. Susini, and R. Barrett, *Design considerations for a prototype beam position monitor for the X-ray microscopy beamline ID21 at the ESRF*; this conference.

10   W. Meyer-Ilse, *X-ray microscopy in Berkeley*, this conference.

11   S. Oestreich and B. Niemann, *Design of a condenser for an X-ray microscope on a low-β section undulator at the ESRF*, this conference.

# The X-Ray Microscopy Project at BESSY II

P. Guttmann, G. Schmahl, B. Niemann, D. Rudolph, G. Schneider, J. Bahrdt[1]

Forschungseinrichtung Röntgenphysik, Georg-August-Universität Göttingen,
Geiststraße 11, D-37073 Göttingen, Germany
[1]Berliner Elektronenspeicherringgesellschaft mbH (BESSY), Lentzeallee 100,
D-14195 Berlin, Germany

**Abstract.** It is planned, to use an undulator at BESSY II as source for a transmission X-ray microscope (TXM) and a scanning transmission X-ray microscope (STXM). The undulator with a magnetic period of 41 mm (U41) will be installed in a low $\beta$ - section and will allow to cover the water window wavelength region in the first harmonic. The characteristics of this undulator will be presented.

In addition, improved X-ray microscopes will be discussed. The different properties of the undulator radiation compared to the dipol radiation used at BESSY I will lead to a new design of the condenser optics. Possible arrangements of these optics will be given. An overview of the planned X-ray microscopy area at BESSY II and of the planned experiments will be shown.

## 1 Introduction

In Berlin-Adlershof a new synchrotron radiation source (BESSY II) is under construction and will be used for the X-ray microscopy project. BESSY II [1] is an electron storage ring of the third generation i.e. it is optimized for the operation with insertion devices. The nominal electron energy will be 1.7 GeV and the nominal stored electron current will be 100 mA. The geometry of the storage ring is given by 16 straight sections. In table 1 the main storage ring parameters are given. Figure 1 shows the schematic layout of the BESSY II laboratory [2].

The need for high brilliance of the source is one of the requirements for X-ray microscopy. Therefore, it is planned to use an undulator beamline for the X-ray microscopy project at BESSY II.

**Table 1.** Electron storage ring parameters of the BESSY II facility [1].

| | | | |
|---|---|---|---|
| nominal energy: | 1.7 GeV | ring circumference: | 240 m |
| nominal electron current: | 100 mA | number of dipols: | 32 |
| natural emittance $\varepsilon_n$: | $6.1 \cdot 10^{-9}$ $\pi$·rad·m | number of straights: | 16 |
| revolution time: | 800 ns | free space for insertion devices: | |
| puls length: | 18 ps | short: 3.430 m | long: 4.250 m |

**X-Ray Microscopy and Spectromicroscopy**
Eds.: J. Thieme, G. Schmahl, D. Rudolph, E. Umbach
© Springer-Verlag Berlin Heidelberg 1998

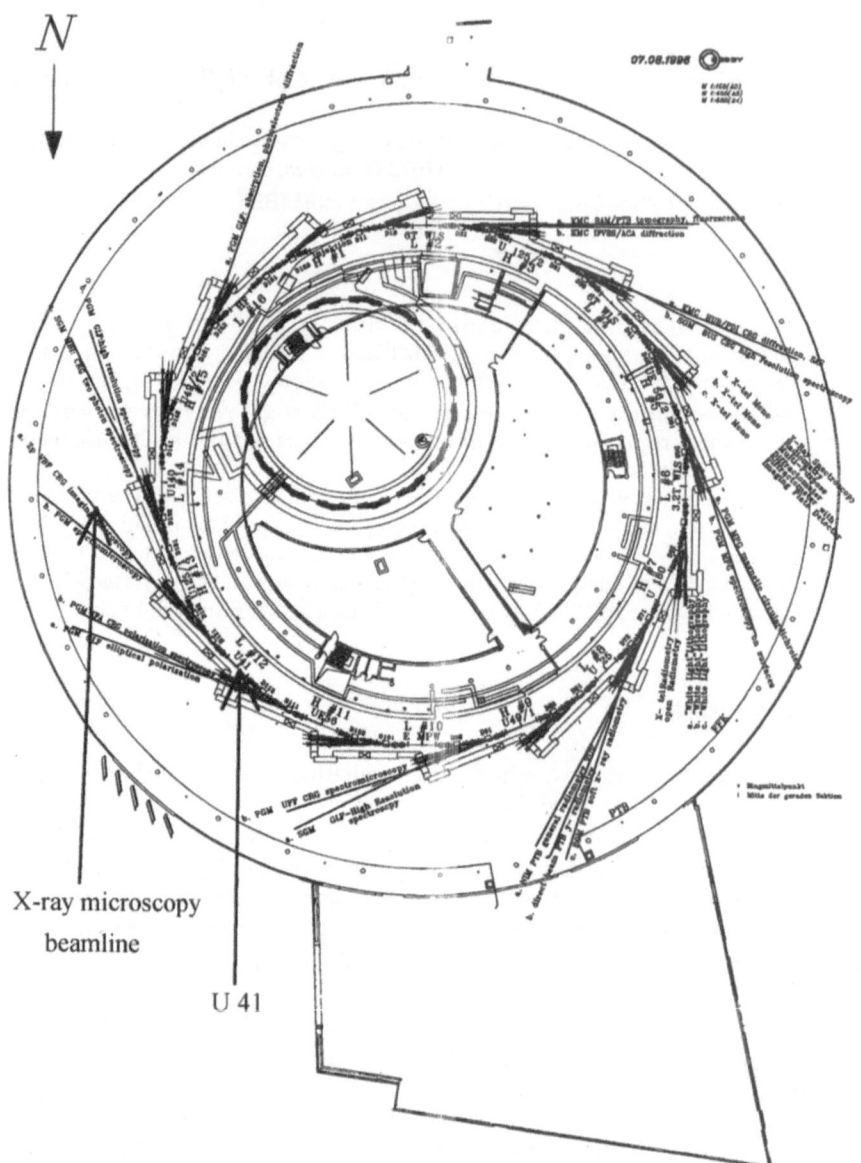

N

X-ray microscopy
beamline

U 41

**Fig. 1.** Schematic layout of the BESSY II facility [2].

## 2 Undulator U 41

For the X-ray microscopy experiments at BESSY II X-ray wavelengths in the region of the so called water window between 2.283 nm and 4.365 nm (O-K- resp. C-K-absorption-edge) will mainly be used. The undulator should allow to cover this wavelength region in the first harmonic. This will lead to a minimum thermal load to the first optical element. An undulator U 41 with a magnetic periodic length of 41 mm fulfills the requirements. The properties of this undulator are given in table 2.

**Table 2.** Properties of undulator U 41 placed in a low $\beta$- (short straigth) section [3].

| | | | |
|---|---|---|---|
| period length $\lambda_0$: | 41 mm | electron energy: | 1.7 GeV |
| number of periodes N: | 80 | ring current: | 100 mA |
| length L: | 3.321 m | | |
| $\beta_x$: | 1.12 m/rad | horiz. emittance: | $6.1 \cdot 10^{-9}$ m·rad |
| $\beta_y$: | 1.46 m/rad | vert. emittance | |
| magnetic gap $g_{mag}$: | $\geq 15$ mm | (3% coupling): | $1.8 \cdot 10^{-10}$ m·rad |
| vacuum gap $g_{vac}$: | 11 mm | | |
| $K_{max}$: | 2.556 | $B_0$: | 0.668 Tesla |
| $P_{T,max}$: | 270 Watt | $dP/d\Omega_{max}$: | 484 Watt/mrad$^2$ |
| first harmonic: | 7.19 nm - 2.08 nm | 172 eV - 596 eV | |
| K range: | 2.4 - 0.5 | | |
| third harmonic: | 2.40 nm - 0.93 nm | 518 eV - 1340 eV | |
| K-range: | 2.4 - 1.0 | | |
| fifth harmonic: | 1.44 nm - 0.68 nm | 863 eV - 1810 eV | |
| K range: | 2.4 - 1.3 | | |

Detailled calculations have been performed to evaluate the behaviour of the undulator U 41 in a high and low beta section [4,5]. By using a phase sensitive raytracing method and incorporating the depth of field effects it was found that the brilliance will be comparable in both cases but in the low beta section the source will be rounder (see Fig. 2). The second result was that the thermal load  - which is proportional to the angular flux density - to the first optical element will be lower in the case of the low beta section (see Fig. 3). Therefore, the U 41 will be installed in the low beta section L #12 (see Fig. 1).

a)

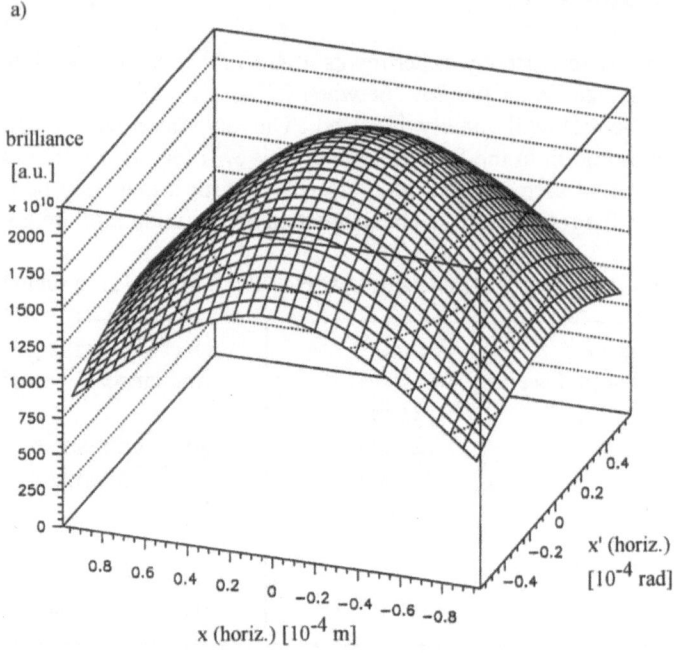

b)

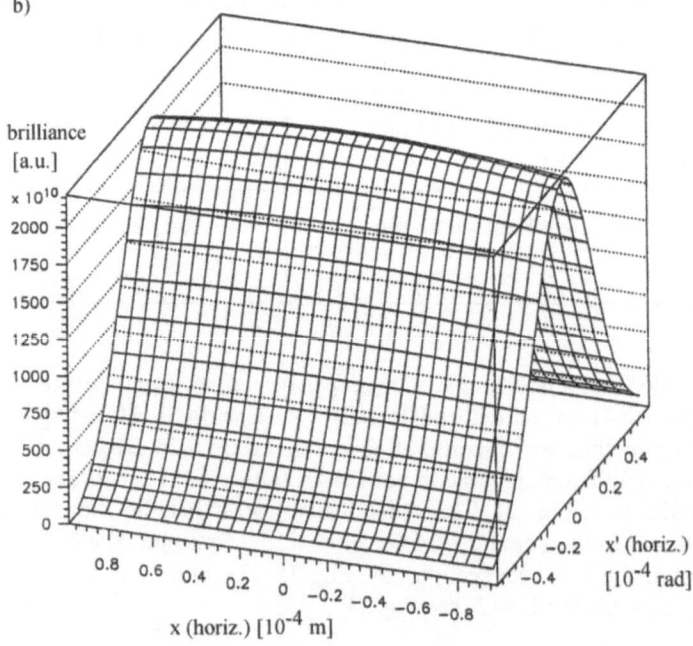

**Fig. 2.** Brilliance of U 41 at hv = 500 eV for y = 0 m and y' = 0 rad in
a) low β- (short straight) and  b) high β- (long straight) section.

a)

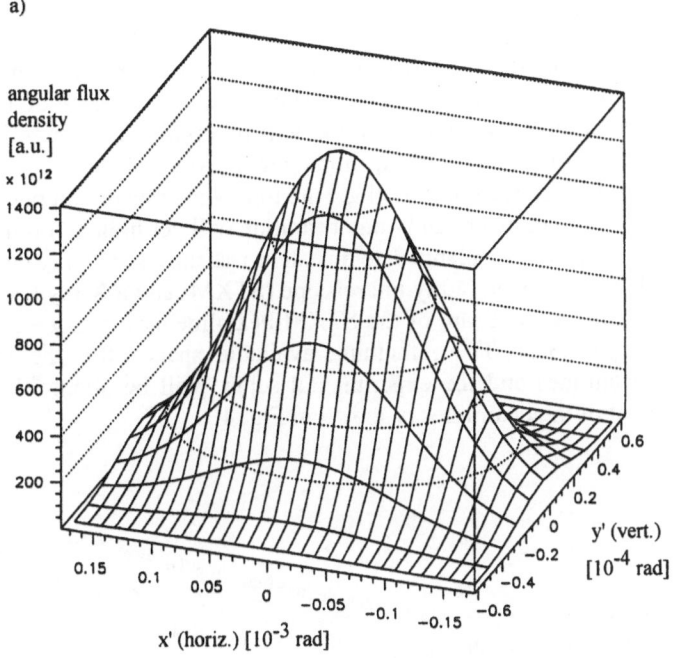

b)

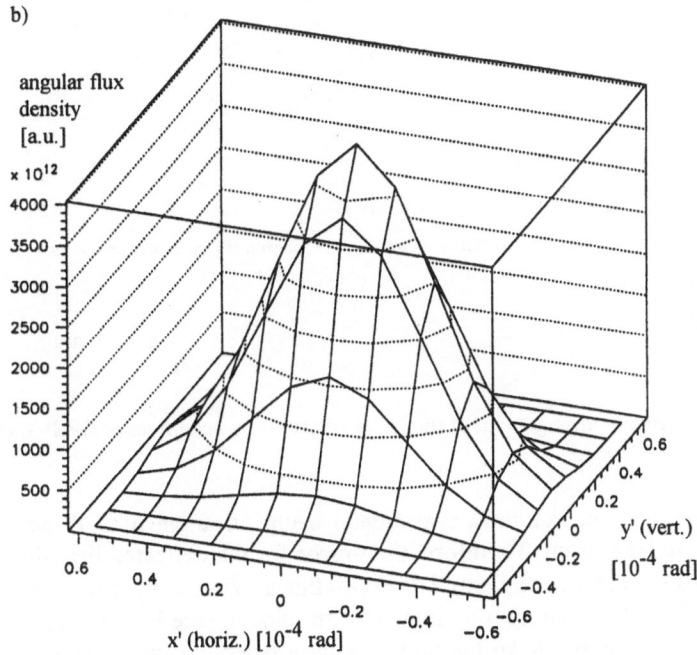

**Fig. 3.** Angular flux density of U 41 at hv = 500 eV in a) low β- (short straight) and
b) high β- (long straight) section.

# 3 X-Ray Microscopy Area at BESSY II

The beamline of the undulator U 41 will be used for X-ray microscopy experiments as well as for spectromicroscopy experiments. Therefore, in a distance of 17000 mm from the middle of the undulator the beamline will be divided into three beamlines (see Fig. 4). A premirror chamber with two movable mirrors designed by the BESSY company will be used. Two beamlines will get the light via a mirror and one beamline will supplied with the direct beam. Only one beamline will get light at a time. For the X-ray microscopy project it is planned to install a transmission X-ray microscope (TXM), a scanning transmission X-ray microscope (STXM) and a X-ray test chamber (XTC). The XTC will use the beam of the STXM-beamline by moving a mirror into the beam (see chapter 3.3). The remaining available space after the pre-mirror chamber for the beamlines and the experimental setup will be about 22.5 m. The planned arrangement of the beamlines is shown in Fig. 4.

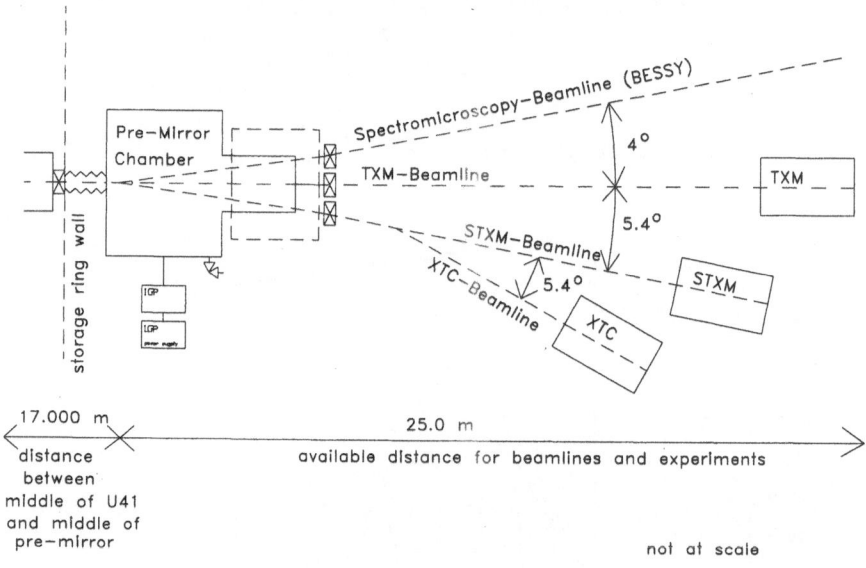

**Fig. 4.** Schematic arrangement of the X-ray microscopy beamlines at BESSY II.

To get enough space for the experimental setup an angle of 5.4° between the beamlines should be chosen. Therefore, the mirror for the deflected beamline will be used at an grazing incidence angle of 2.7°. With this grazing angle a short wavelength cut-off is realized by coating the mirror with nickel (see Fig. 5). At $\lambda = 2.4$ nm the the total power coming from the undulator at a beam current of 100 mA will be about 24 W [6]. The pre-mirror will absorb the main part of the thermal load.

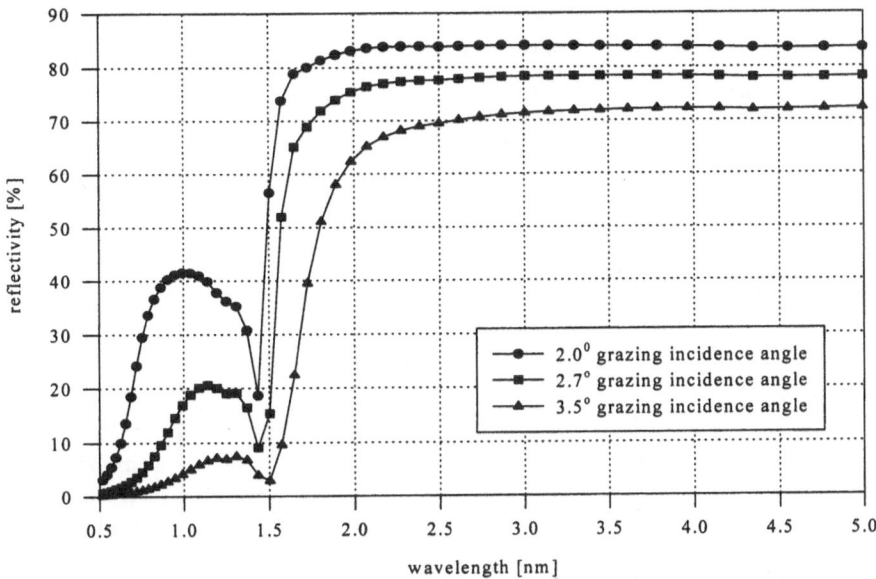

**Fig. 5.** Reflectivity of the first mirror coated with nickel at different grazing incidence angles.

## 3.1 The Transmission X-Ray Microscope (TXM)

For the TXM at the U 41 a gain in flux of about a factor 60 at $\lambda = 2.4$ nm will be reached at BESSY II compared to BESSY I. This gain can be used to shorten the exposure time to values in the 0.01 sec region, to increase the monochromaticity for getting a higher contrast and a better resolution or to do a combination of both. The TXM will be built up at the beamline with the direct beam. To reduce the thermal load to the first optical element of the X-ray microscope a mirror will be used. Using a mirror coated with nickel under $2.7°$ grazing incidence angle a short wavelength cut-off will be realized, too. With a total power of 24 W at $\lambda = 2.4$ nm, 100 mA beam current [6] the remaining thermal load to the next optical element (condenser zone plate) will be about 90 mW/mm$^2$ using the first harmonic. Experiments have shown that the condenser zone plate can withstand a thermal load of 150 mW/mm$^2$ [7].

Due to the undulator beam geometry a new condenser concept will be necessary. One possibility is to use a condenser system with two zone plates as discussed in [7] and shown in figure 6. The first zone plate collects the radiation of the concentrated undulator beam. The rather large diameter of the second zone plate (condenser zone plate) is necessary to get the monochromaticity necessary for diffraction limited imaging.

In a microscope the best resolution can be get with incoherent illumination of the object. Consequently, the aperture of the condenser has to be adapted to that of the micro zone plate [8]. Several possibilities are under discussion and are partly described in [9].

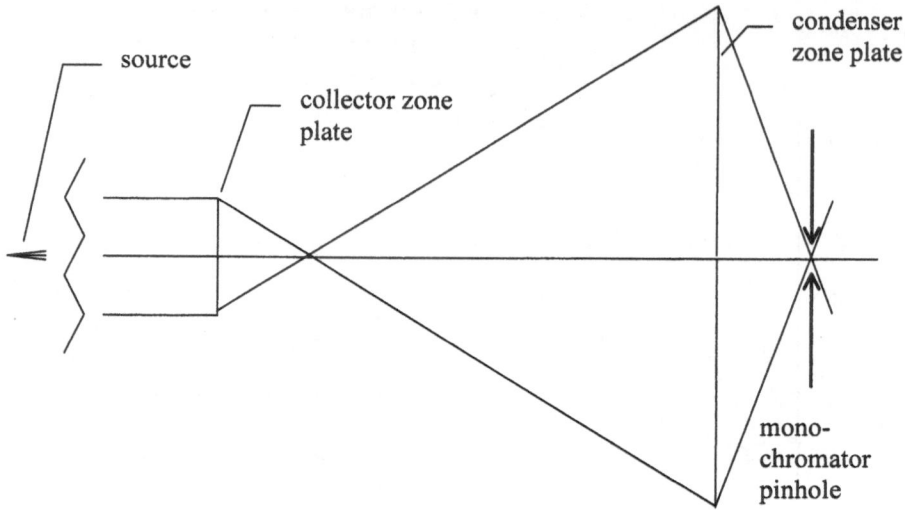

**Fig. 6.** One of the possible arrangements of a condenser/monochromator system with two zone plates for the TXM [7].

The X-ray microscope itself will be modified in some details but will remain in principle as described in [10].

### 3.2 The Scanning Transmission X-Ray Microscope (STXM)

The STXM uses only spatially coherent photons. The gain in spatially coherent photons/sec in the scan-spot will be about $7\cdot10^4$ at $\lambda = 2.4$ nm given by the gain in brilliance compared to BESSY I. This means that integration times per pixel shorter by the same factor will become possible. Due to the expected high photon rate confocal scanning X-ray microscopy will be possible. A 500 x 500 pixel image will take e.g. 25 sec [11].

The schematic arrangement of the STXM beamline, especially of the mono-chromator, is shown in figure 7 and is described in detail in [11,12]. The pre-mirror at 2.7° grazing incidence angle is used for short wavelength cut-off below $\lambda = 1.5$ nm. The monochromator consists of a plane grating (blaze type, 600 lines/mm, total angle of deflection of 5°) combined with a knife edge mounted above its center. The gap between the edge and the grating determines the bandwidth. A monochromaticity as high as $\lambda/\Delta\lambda = 1600$ at $\lambda = 2.4$ nm can be achieved. A micro zone plate forms the scan-spot. About $10^9$ photons/sec are estimated to be in a scan-spot of 30 nm diameter [11]. The object under investigation is moved mechanically in the focal plane of the micro zone plate.

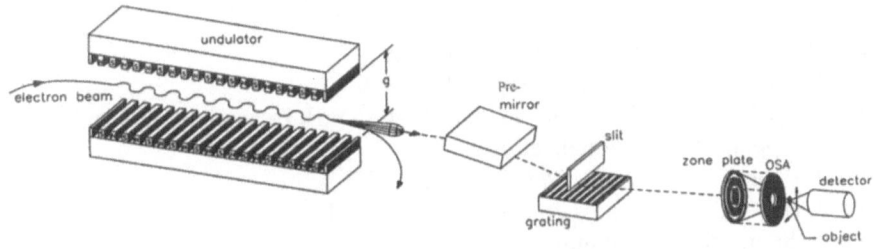

**Fig. 7.** Schematic arrangement of the scanning transmission X-ray microscope at
BESSY II [11].

### 3.3 The X-Ray Test Chamber (XTC)

The XTC is necessary for the calibration of optical elements and detectors used for
X-ray microscopy. It is planned to use the same monochromator as for the STXM.
Therefore, a movable mirror should be in the STXM-beamline between the
monochromator and the scanning stage to deflect the beam into the XTC-beamline.
The XTC itself will be a versatile instrument like that one at BESSY I.

## 4 Outlook

With the improved transmission X-ray microscope and scanning transmission X-ray
microscope at BESSY II many different studies will be performed: Imaging and
morphological studies of specimens in their natural wet state will be possible in
amplitude and phase contrast in the fields of biophysics, biology, medicine, colloid
chemistry, environmental sciences, soil sciences and solid state physics. Besides
elemental mapping, quantitative analysis, XANES mapping, luminescence analysis
and dynamical studies will be done. Microtomography for high resolution 3-D
imaging, also in combination with cryo X-ray microscopy, will be carried out. Higher
resolution with up to 10 nm is planned.

## Acknowledgements

This work has been funded by the German Federal Minister of Education and
Research (BMBF) under contract number 05 644 MGA. Special thanks are given to
the following BESSY members for their help: A. Gaupp, K. Blümer, M. Scheer,
F. Senf and W. Braun.

# References

1    BESSY homepage: http://www.bessy.de.
2    BESSY Newsletter No. 6, September 1996.
3    BESSY GmbH, Beamline Handbook BESSY II, Dezember 1995 and Gaupp, private communication.
4    J. Bahrdt, Appl. Optics **34** (1995), 114–127.
5    J. Bahrdt, Appl. Optics **36** (1997), 4367–4381.
6    F. Senf, private communication.
7    M. Dirksmöller, "Der Undulator U 3 des Berliner Elektronenspeicherrings für Synchrotronstrahlung BESSY II als Lichtquelle für ein Röntgenmikroskop", Diplomarbeit, Universität Göttingen (1992).
8    S. Oestreich, B. Niemann: Design of a condenser for an X-ray microscope on a low-$\beta$ section undulator source at the ESRF, this volume.
9    B. Niemann: High Numerical-Aperture X-Ray Condensers for Transmission X-Ray Microscopes, this volume.
10   G. Schmahl, D. Rudolph, P. Guttmann, G. Schneider, J. Thieme, B. Niemann, Rev. Sci. Instrum. **66** (1995), 1282–1286.
11   Irtel von Brenndorff, B. Niemann, D. Rudolph, G. Schmahl, J. Synchrotron Rad. **3** (1996), 197–198.
12   Irtel von Brenndorff, B. Niemann, D. Rudolph, G. Schmahl: A monochromator for a scanning transmission X-ray microscope at the U 41 undulator beamline at BESSY II, this volume.

# Imaging Soft X-Ray Microscopy with Zone Plates in Parallel Use of Optical Microscope for Wet Bio-Specimens in Air at UVSOR

Norio Watanabe[1], Atsuhiko Hirai[2], Kuniko Takemoto[3], Yoshio Shimanuki[4], Mieko Taniguchi[5], Eric Anderson[6], David Attwood[6], D. Kern[7], Sumito Shimizu[8], Hiroshi Nagata[8], Kenzo Kawasaki[4], Sadao Aoki[1], Yasuyuki Nakayama[2], Hiroshi Kihara[3]

[1] Institute of Applied Physics, University of Tsukuba, Tennoudai 1-1-1, Tsukuba, Ibaraki, 305, Japan, E-mail: watanabe@kirz.bk.tsukuba.ac.jp
[2] Department of Physics, Ritsumeikan University, Kusatsu, Shiga 525-77, Japan
[3] Physics Laboratory, Kansai Medical University, Hirakata, Osaka 573, Japan
[4] Department of Oral Anatomy, Tsurumi University, Tsurumi, Yokohama 230, Japan
[5] Department of Physics, Nagoya University, Chikusa-ku, Nagoya 464, Japan
[6] Center of X-ray Optics, Lawrence Berkeley Laboratory, Berkeley, CA 94720, USA
[7] IBM T. J. Watson Research Center, Yorktown Heights, NY 10598, USA
[8] Nikon Corp. Nishi-ooi, Shinagawa-ku, Tokyo 140, Japan

**Abstract.** Soft X-ray microscope, a specimen holder of which was placed in air, was constructed at UVSOR, synchrotron radiation facility at Institute for Molecular Science, Japan. This made possible to investigate a specimen without impairing the vacuum of the microscope and to prefocus a specimen with an optical microscope incorporated in the microscope. Dry and wet specimens could be observed at a wavelength of 0.94 nm.

## 1 Introduction

We have been developing a soft X-ray microscope with zone plates to observe hydrated biological specimens with higher resolution than that of an optical microscope. Our previous report showed that a 63 nm line and space pattern could be resolved at a wavelength of 3.2 nm, and dry and wet biological specimen could be observed[1].

In 1995, the soft X-ray microscope was improved in the following points. (a) A specimens stage was put in air gap, which was separated from condenser and objective vacuum chambers by SiN windows. This made possible to investigate a specimens without breaking the vacuum of the beam line. (b) An optical microscope was set to adjust and prefocus a specimens. (c) A cooled CCD camera system was used as a detector. This type of a soft X-ray microscope was first developed by Göttingen X-ray microscope groups [2]. This report describes our new microscope chamber and the results of imaging experiments.

## 2 Optical System

The optical system was the same type as the Göttingen X-ray microscope [2]. Figure 1 shows the optical system of the microscope. Synchrotron radiation from the bending magnet source BL8A at UVSOR (750 MeV, 200 mA, Institute for Molecular Science, Okazaki, Japan) was used. Soft X-rays from the source were monochroma-

X-Ray Microscopy and Spectromicroscopy
Eds.: J. Thieme, G. Schmahl, D. Rudolph, E. Umbach
© Springer-Verlag Berlin Heidelberg 1998

tized by a condenser zone plate (CZP) and incident on a specimen. The image of a specimen was enlarged by an objective zone plate (OZP), and focused on a cooled backside illuminated CCD camera (Astromed Corp. CCD: SITe Corp. SI502A). To prevent zero-th order radiation of the OZP from reaching the imaging area on the detector, the image of a specimen was focused on the off-axis area outside of the zero-th order radiation.

Focusing tests were performed at wavelengths of 1.3 nm, 2.4 nm; and 3.2 nm. Images were observed at only 1.3 nm in wavelength[1]. Then, images were observed at a wavelength of about 1 nm. The synchrotron source has the electron beam size of $\sigma_x$=0.39 mm (horizontal) and $\sigma_y$=0.26 mm (vertical) [3]. Assuming the diameter of the source to be 0.80 mm, the source image size at a pinhole plane was calculated to be 120 μm (first order) or to be 32 μm (third order) at 1.3 nm in wavelength. The third order radiation of the CZP was mainly used to reduce the source image size at the pinhole plane, and to improve the monochromaticity. The monochromaticity $\lambda/\Delta\lambda$ of the linear monochromator was calculated to be 67 from the relationship $\lambda/\Delta\lambda$ =D/2d, where D is a diameter of a CZP and d is the diameter of the source image at the pinhole plane because the diameter of the source image size was larger than that of the pinhole[4]. The fourth order radiation of 0.94 nm in wavelength was strongly contaminated into the third order radiation. It could not be removed, and soft X-rays of both wavelengths were used for imaging.

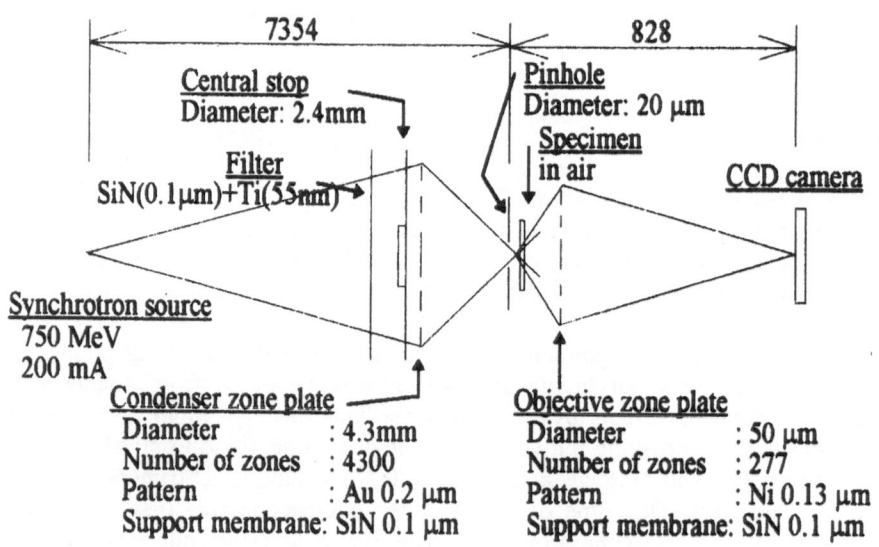

**Fig. 1.** Optical system of soft X-ray microscope at UVSOR

---

[1] After the experiments, the quantum efficiency of the CCD was measured. The efficiency was several percent at a wavelength longer than 2 nm. It is too low for a backside illuminated CCD with no coat. This seems to be main reason that images could not be observed at a wavelength longer than 2 nm.

# 3 Microscope Chamber

Figure 2 shows a schematic diagram of the microscope chamber. The microscope was separated to two vacuum parts, a CZP chamber and an OZP chamber. A specimen stage was placed in air. The CZP chamber could be moved back and forward along the optical axis with a pneumatic cylinder, the air gap between the two chambers could be changed from several hundred microns to several millimeters and a specimen could be easily changed. These stage was placed on a rotating stage and the soft X-ray microscope and an optical microscope could be switched by rotating the stage with a pneumatic cylinder. The optical microscope could be used for searching and prefocusing a specimen.

Figure 3 shows a schematic of the specimen holder. SiN windows of 0.1 µm thickness and 200 by 200 µm area were used for vacuum seals. These windows were stuck on frames by adhesive (Torr Seal, Varian Corp.). The pinhole was also stuck on the frame of the SiN window outside of the vacuum chamber. The pinhole was changed to new one with a period of about one month because it was closed with contamination.

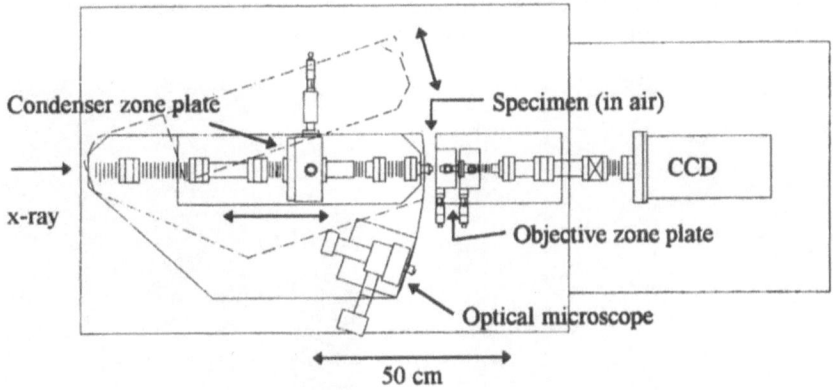

**Fig. 2.** Top view of the microscope chamber

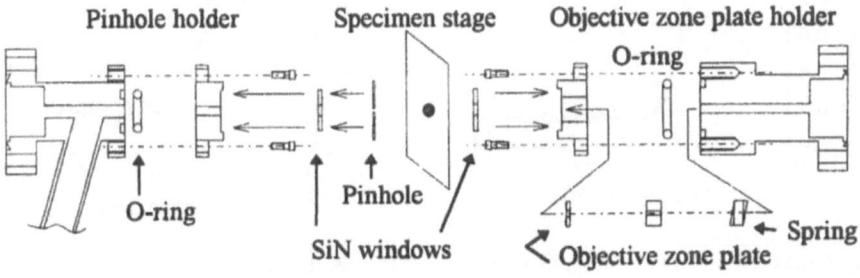

**Fig. 3.** Schematic of the specimen stage in air

## 4 Results and Discussion

Figs. 5 and 6 show images of Cu #2000 mesh at wavelengths of 1.3 nm (the third order radiation of the CZP) and 0.94 nm (the fourth order). These images were taken under the same CZP position and the OZP position was only changed along the optical axis to focus soft X-ray images of 1.3 nm and 0.94 nm in wavelength, respectively. The resolutions were estimated from the edge profiles of these images. The width of the intensity rise from 10 to 90 % was 0.26 μm at a wavelength of 1.3 nm, and 0.50 μm at 0.94 nm. Fig. 6 shows an image of a Ni zone plate as a specimen at a wavelength of 0.94 nm. The specimen had the same specification as the OZP. The diameter of the first zone was 3 μm. The finest resolvable zone width was 0.14 μm. These resolutions were worse compared with a theoretical one of the OZP (1.22 × the outermost zone width = 55 nm). It was considered to be due to the low monochromaticity and the higher order radiation of the CZP. To obtain images without chromatic aberration, it is necessary to use quasimonochromatic soft X-rays of monochromaticity $\lambda/\Delta\lambda$ =277, which is the zone number of the OZP. However, the theoretical monochromaticity was 67 in this experiment. Superposition of several order radiation of the CZP caused degradation of the image quality. The superimposed mesh image (central bright rectangle area) in Fig. 6 is an image focused with the fourth order radiation of 0.94 nm in wavelength.

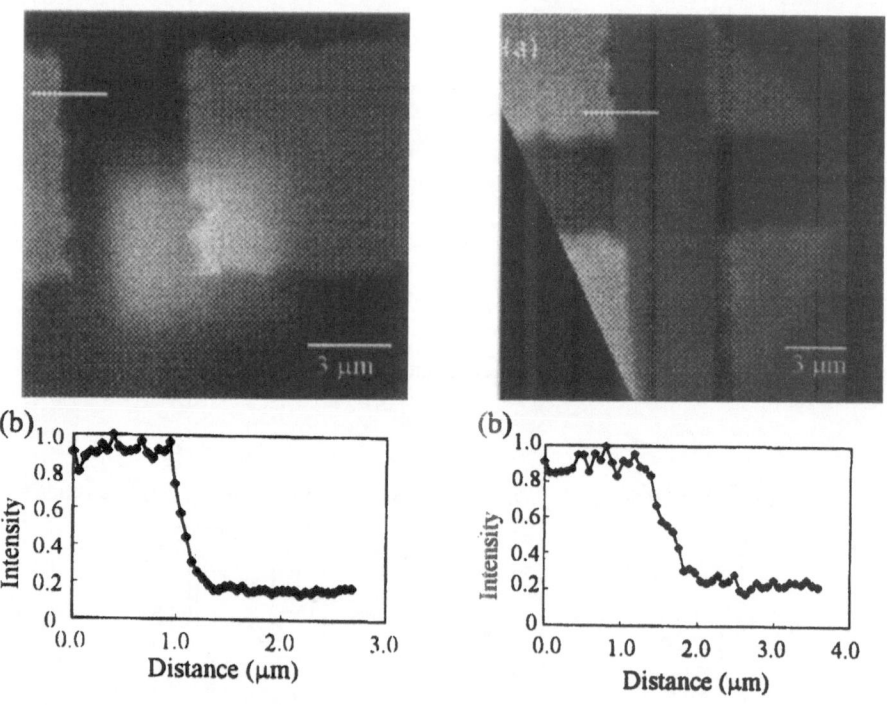

Fig. 4. (a) Cu #2000 mesh image at 1.3 nm in wavelength. (b) The intensity profile of the image at the white bar.

Fig. 5. (a) Cu #2000 mesh image at 0.94 nm in wavelength. (b) The intensity Profile of the image at the white bar.

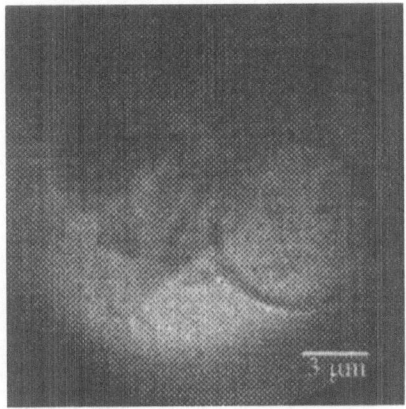

**Fig. 6.** Image of a zone plate as a Specimen. Wavelength: 0.94 nm.

**Fig. 7.** Image of diatoms at 0.94 nm.

Using a specimen holder with polyimide foils, several wet biological specimens could be observed at 0.94 nm in wavelength[5]. Fig. 7 shows an image of a diatom, which was collected from the sea, and killed by sulfuric acid, and observed.

## Acknowledgments

The authors are grateful to the help and encouragement by Dr. T. Kinoshita, E. Nakamura and other staffs of the Institute for Molecular Science.

## References

1    N. Watanabe, S. Aoki, Y. Shimanuki, K. Kawasaki, M. Taniguchi, E. Anderson, D. Attwood, D. Kern, S. Shimizu, H. Nagata, and H. Kihara, in *X-ray Microscopy IV*, eds. V. V. Aristov, and A. I. Eriko, (Bogorodskii Pechatnik Publishing Company, Chernogolovka, Moscow region, Russia 1994), p.333.

2    B. Niemann, G. Schneider, P. Guttmann, D. Rudolph, and G. Schmahl, in X-ray Microscopy IV, eds. V. V. Aristov, and A. I. Eriko, (Bogorodskii Pechatnik Publishing Company, Chernogolovka, Moscow region, Russia 1994), p.66.

3    UVSOR Activity Report 1995 (1996)

4    B. Niemann, D. Rudolph, and G. Schmahl, Opt. Commun. 12, 160 (1994).

5    K. Takemoto, N. Watanabe, A. Hirai, Y. Nakayama, and H. Kihara, this conference.

# High-Resolution Three-Dimensional Imaging with an X-Ray Microscope

J. Lehr

Forschungseinrichtung Röntgenphysik, Georg-August-Universität Göttingen, Geiststraße 11, D-37073 Göttingen, Germany

**Abstract.** The Göttingen transmission X-ray microscope (TXM) at BESSY has been used for threedimensional imaging. As a first test-object sheaths of the bacteria *Leptothrix Ochracea* have been visualized. 33 tilted-view images have been taken at tilt angles between 0 and 160 degrees in steps of 5 degrees. Before reconstruction the images were flat-field corrected and aligned. The rotation axis was determined and the images were rotated accordingly. For tomographic reconstruction a fast MART (Multiplicative Algebraic Reconstruction Technique) algorithm was used. In the recon-structed volume structures of sizes well below 100 nm are made visible.

## 1 Introduction

The circumstances in X-ray microscopy [1, 2, 4] are know to be well suited for three-dimensional investigations for several reasons:

- The refractive index for soft X-rays is close to 1 for all substances. Therefore, scattering reflection at interfaces is negligible and clear images can even be obtained from thick objects.
- The image-field of the Göttingen X-ray microscope is of the same size as the thickness of typical objects, i.e. of the order of 10 μm. For typical objects like cells the three-dimensional structure has to be considered because their lateral dimensions are of the same size as their thickness.
- The attenuation length for soft X-rays in water is also around 10 μm. Hydrated objects of thicknesses around 10 μm have therefore a reasonable transmission.

Whereas the lateral resolution of the X-ray microscope at BESSY is one order of magnitude smaller than the resolution of visible light microscopes, the depth resolution is of the same size or even larger than for visible light microscopes. The reason herefore is that todays X-ray objectives have a relatively small aperture in the range from 0.02 to 0.04 while apertures can be as large as 1.4 in light microscopy. The focal depth $\delta_t$ of an X-ray microscope is $2.44 \, dr_n^2/\lambda$. The objective-zoneplate, which was used for the experiments described below has an outermost structure-width of 39 nm. For this zoneplate $\delta_t$ is 1.5 μm and the ratio $\delta_t{:}\delta_l$ is 33:1.

Instead of being a handicap for three-dimensional investigation like in visible light microscopy, the spatially anisotropic resolution of the X-ray microscope can be made a virtue when combined with the appropriate reconstruction method. Within the depth of focus, the image-formation process can be described by a projection along the optical axis:

**X-Ray Microscopy and Spectromicroscopy**
Eds.: J. Thieme, G. Schmahl, D. Rudolph, E. Umbach
© Springer-Verlag Berlin Heidelberg 1998

$$I(x,y) = I_0(x,y)\exp(-\int_Z \mu(x,y,z)dz) \qquad (1)$$

Here $I(x,y)$ denotes the image-intensity at the point $(x,y)$, $I_0(x,y)$ is the illumination intensity and $\int_Z \mu(x,y,z)dz$ is the line integral of the absorption coefficient at $(x,y)$ along the optical axis.

In X-ray microscopy, depth information can best be obtained from tilted-view images [5]. This leads to the following approaches: The easiest way, which already gives a good impression of the threedimensional organization of specimens is, to take stereo images, which has successfully been demonstrated for the Göttingen X-ray microscope at BESSY [6]. In order to get real threedimensional data sets that can be used for quantitative measurement of surface and volume it is necessary to implement the method of tomographic reconstruction [7] on the X-ray microscope.

## 2 Tomographic Reconstruction from TXM-Images

3D-reconstruction from X-ray microscopy images imposes a limited data problem. This means that 3D data has to be reconstructed from few projections from a limited tilt-angle. The tilt-angle is limited, because the effective water layer increases with the tilt angle, if objects are prepared on a flat substrate. For tilt-angles larger than 80 degrees, exposure-times would become too long. Since complete automization of the image acquisition process at the TXM at BESSY is not yet achieved, the number of tilted-view images that can be taken in practice is limited. Limited-data reconstruction techniques [9] that have been tested so far were the ART (algebraic reconstruction technique) and the MART (multiplicative algrebraic reconstruction technique) algorithm. Both have proven to suit our needs, where the latter (MART) shows faster convergence and less artifacts when dealing with few projections. Tomographic reconstruction is run on an SGI Indigo$^2$ workstation equipped with a 200 MHz MIPS R4400 processor. Reconstruction of a $256^3$ voxel cube from 33 projections with $256^2$ pixels takes about 20-40 minutes depending on how many iterations are calculated.

## 3 Experimental Results

In a first experiment [10], mineral sheaths of the bacteria *Leptothrix Ochracea* have been used as a dry test-object. This bacterium is known as an iron oxygenating sheath bacterium. It forms long chains of cells held together by a thin tubular sheath which is enriched by iron [11]. These structures have diameters of several microns down to a fractional part of 1 μm. As iron is highly absorbing for soft X-rays in the water window, the contrast is very high and the structures can be visualized particularly well. The objects were prepared on a fine glass-fiber, which was then fixed at one end to a tilting stage. 33 TXM-images were taken at angles from -80 degrees to 80 degrees in steps of 5 degrees. The tilt-angle was adjusted manually to a precision of about ±0.5 degrees. The objective-zoneplate we used has a theoretical lateral resolution below 50 nm and a focal depth of 1.5 μm at 2.4 nm wavelength. Figure 1 (*bottom*) shows a selection of the intensity Data $(I(x,y))_\Theta$ of the tilted-view images.

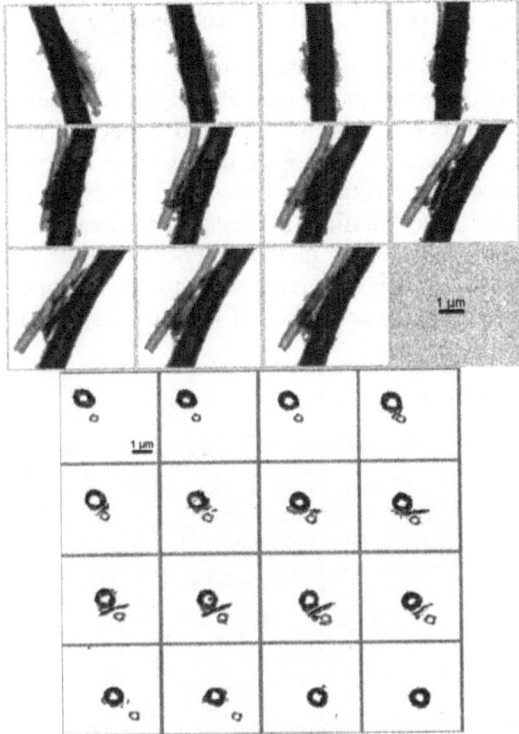

**Fig. 1.** Tilted-view images of mineral sheaths of the bacteria *Leptothrix Ochracea* (*top*).
Slices of the reconstructed volume perpendicular to the rotation axis (*bottom*).
Dark regions in the images correspond to dense object-regions.

The images are background-corrected and normalized to $I_0(x,y)=1$. Before reconstruction the images were aligned and the rotation axis was determined by tracing characteristic points in the object. Already by animating the sequence of tilted-view images a good qualitative impression of the three-dimensional morphology of the object can be obtained. For quantitative measurement however three-dimensional reconstruction of the objects density-values has to be done. The values of the reconstructed volume correspond to $\mu \cdot \Delta s$, where $\Delta s$ is the edge-length of a voxel.

Figure 1 (*top*) shows slices of the reconstructed volume perpendicular to the rotation axis. In the reconstructed volume structures of sizes well below 100 nm have been made visible. A more illustrative representation of the object can be obtained by volume-rendering or surface-rendering of an isosurface (Fig. 2). In the case of the objective-zoneplate we used for the experiments presented here, the condition for using (1) to describe the image formation process is not strictly fulfilled, since the diameter of the reconstructed volume is larger than the depth of focus. But as the results show, our reconstruction-algorithm seems to be quite tolerant to this small error.

## 4 New Possibilities with 3D X-Ray Microscopy

This technique gives access to a wide range of new types of X-ray microscopy investigations. It is now possible to get quantitative data for surface and volume of
objects and distances in objects can be measured. Since we reconstruct the linear
absorption coefficient $\mu(x,y,z)$ of the object, the densities of structures can be determined if their chemical composition is known. This can for example be used to
measure the chemical concentration in pecial cell features, if their chemical
composition is known from other methods. For our test-object, the calculated
absorption coefficients correspond quite well to the values we expect when
assuming that the structures are composed of hydrated $Fe_2O_3$ with densities around
3 gcm$^{-3}$ [11, 12].

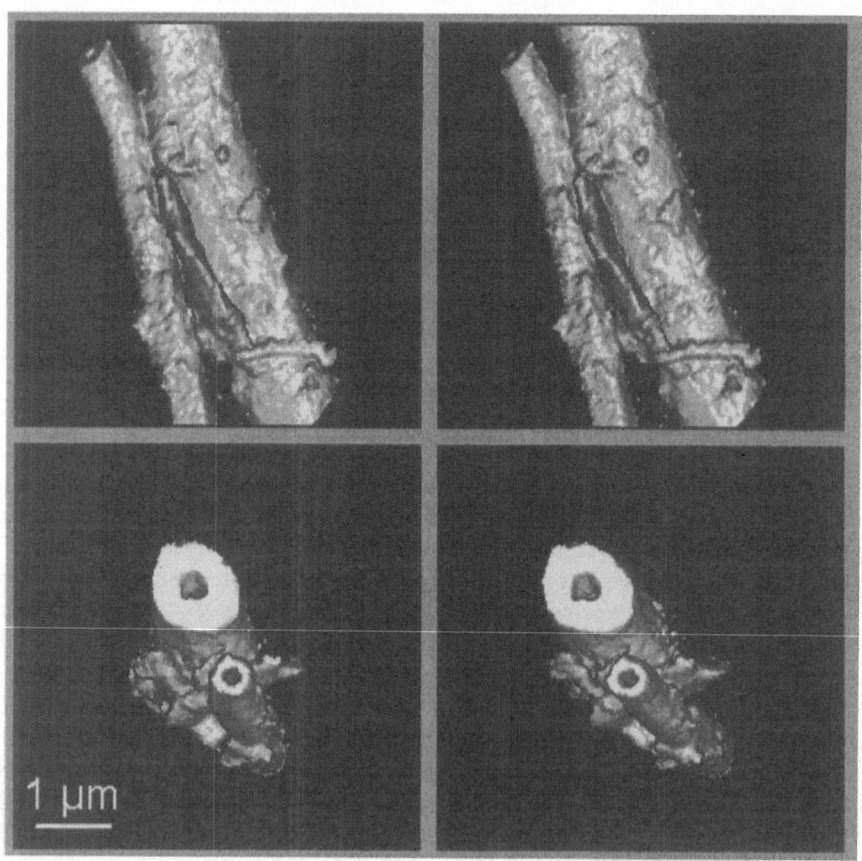

**Fig. 2.** Stereo-pairs of an isosurface-rendering of the reconstructed volume.

# 5 Outlook

To make this method more easily accessible for X-ray microscopy users, several improvements have to be made, mainly concerning the automization of the image acquisition process and the object preparation. For alignment of the image data and determination of the rotation axis up to now we use an interactive method which is based on visual estimation of characteristic points and the sinograms of the projection data. The automization of this process will also be subject of future investigations.

Any method for three-dimensional reconstruction requires that many images of the same object have to be taken. For wet biological specimens at room temperature this means that the dose applied to an object will introduce structural changes in the object [13]. It will therefore be vital to combine the method described above with the cryo fixation method. For two-dimensional imaging of biological specimen this method has already been implemented on the X-ray microscope at BESSY [8].

## Acknowledgements

The author wishes to thank Prof. Dr. G. Schmahl for his continuous support and encouragement and Dr. P. Guttmann for his help at BESSY. For the surface-rendering, a graphics program from Lab. DyOGen, Université Joseph Fourier, Grenoble was used. This work has been funded by the German Bundesministerium für Wissenschaft, Bildung, Forschung und Technologie (BMBF) under contract number 05644 MGA and by the PROCOPE exchange program.

## References

1   G. Schmahl, D. Rudolph, B. Niemann, P. Guttmann, J. Thieme, G. Schneider, C. David, M. Diehl and T. Wilhein: X-Ray Microscopy Studies. Optik 93, pp. 95–102, (1993).

2   G. Schmahl, D. Rudolph, B. Niemann, P. Guttmann, J. Thieme and G. Schneider: Röntgenmikroskopie. Naturwissenschaften 83, (1996), pp. 61–70.

3   V .V. Aristov and A.I.Erko (eds.): X-Ray Microscopy IV. Proceedings of the 4-th International Conference, September 20-24, 1993. Chernogolovka, Russia, 1994.

4   Niemann, G. Schneider, P. Guttmann, D. Rudolph and G. Schmahl: The New Göttingen X-Ray Microscope with Object Holder in Air for Wet Specimens. in: [3] pp. 66–75.

5   W.S. Haddad, I. McNulty, J.E.Trebes, E.H. Anderson, R.A. Levesque, L. Yang. Ultrahigh-Resolution X-ray Tomography. Sciene, vol. 266, 18 Nov. (1994), pp. 1213–1215.

6   J. Lehr: 3D X-Ray Microsopy: High-Resolution Stereo-Imaging with the Göttingen X-Ray Microscope at BESSY. Zoological Studies 34, Supplement (1995), pp. 137–138.

7    Rosenfeld and A.C. Kak: Digital Picture Processing. Academic press (1982). second edition, volume 1, pp. 353–430.

8    Schneider, B. Niemann, P. Guttmann, D. Rudolph and G. Schmahl: Cryo X-Ray Microscopy. Synchrotron Radiation News, vol. 8, No. 3 (1995).

9    D. Verhoeven: Limited-data-computed tomography algorithms for the physical sciences. Applied Optics, vol. 32, No. 20, 10 July (1993), pp. 3736–3754.

10   Lehr : 3D X-Ray Microscopy : High-Resolution Tomographic Imaging of Mineral Sheaths of Bacteria Leptothrix Ochracea. Optik. (in press)

11   Thieme, T. Wilhein, P. Guttmann, J. Niemeyer, K.-H. Jacob, S. Dietrich: Direct Visualization of Iron and Manganese Accumulating Microorganisms by X-Ray Microscopy. in: [3], pp.152–156.

12   T. Wilhein: Gedünnte CCDs: Charakterisierung und Anwendungen im Bereich weicher Röntgenstrahlung. Verlag Shaker, Dissertation Universität Göttingen, (1994).

13   R.C. Weast (ed.): CRC Handbook of Chemistry and Physics. CRC Press, Boca Raton, Florida (1988).

14   G. Schneider: Investigations of Soft X-Radiation Induced Structural Changes in Wet Biological Objects. in: [Aristov V.V, A.I.Erko (eds.), 1994], pp. 181–195.

# X-Ray Microtomography
# Using Interferometric Phase-Contrast

Ulrich Bonse, Felix Beckmann, Frank Busch, Olaf Günnewig

Department of Physics, University of Dortmund, D-44221 Dortmund, Germany,
E-mail: bonse@physik.uni-dortmund.de

**Abstract.** The principle and experimental realisation of X-ray *phase*-contrast in computer assisted microtomography (μCT) is described. Phase detection is accomplished by using the X-ray interferometer [1] in a set-up previously developed for absorption-contrast μCT [2]. (The term *micro*tomography is used for computer assisted X-ray tomography (CT) at the micrometer resolution level.) We show that by employing X-ray phase contrast it is possible to image structural details in biological tissues that consist of only light elements much better than with absorption contrast. Examples of specimens of rat cerebrum and rat trigeminal nerve are given.

## 1 Introduction

Computer assisted tomography (CT) is widely used in the medical field. With the increased availability of synchrotron radiation (SR) and the development of area detectors for x rays which are based on the use of Charge Coupled Devices (CCD), the spatial resolution of CT could be improved down to the micrometer level. At this resolution level CT is commonly called microtomography (μCT). μCT is now applied well beyond the medical field in areas like biology [3], materials science and testing [2], and geology [4]. Several overviews of μCT are available in the literature [5–7].

Besides high brightness and/or high intensity of SR its convenient energy tunability is important in μCT. When using absorption-contrast μCT, by modulating the energy it becomes possible to identify directly the elemental composition or even the elemental binding state of certain regions in the sample [8]. However, with specimens consisting mainly of light elements X-ray phase contrast is superior to absorption contrast in the sense that a thinner layer of material is sufficient to cause 1% change of detected signal. The reason is that phase shift scales with Z and absorption with $Z^d$ where Z is atomic number and $3 \leq d \leq 4$ [9]. We illustrate this point by Fig.1 where we plotted the so-called 'advantage factor' $A_M$ of phase contrast over absorption contrast for 10, 40, and 70 keV x rays as function of atomic number. $A_M$ is defined as

$$A_M = M_a / M_p . \tag{1}$$

**X-Ray Microscopy and Spectromicroscopy**
Eds.: J. Thieme, G. Schmahl, D. Rudolph, E. Umbach

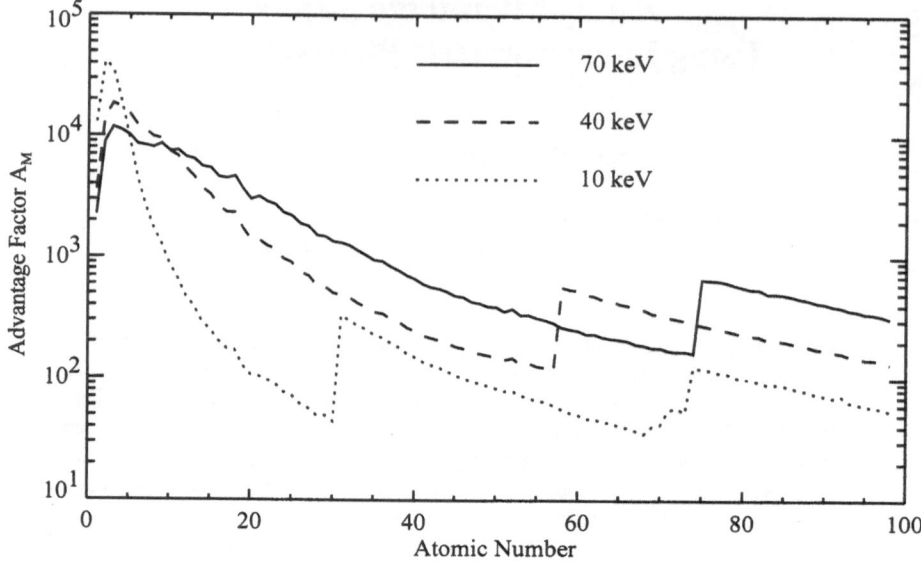

**Fig. 1.** Advantage of X-ray phase-contrast over X-ray absorption-contrast as function of atomic number Z and photon energy. $A_M$ is defined by Eq. 1. Note that especially for light elements $A_M$ is much larger for all energies from 10 to 70 keV. Jumps in curves are due to K- and L-edge absorption. $A_M > 1$ implies larger sensitivity for phase-contrast and hence a smaller radiation dose applied to the specimen at the same detection level.

$M_a$ ($M_p$) is the mass per area in the beam necessary for causing 1% signal change in the case of absorption contrast (phase contrast), respectively. A large value of $A_M$ indicates that phase contrast exceeds absorption contrast. From Fig. 1 we conclude that for the tomographic investigation of organic tissues like nerves, brain or kidney specimens, phase-contrast μCT is very suitable [10]. Samples containing bones (and therefore the medium-Z element Ca) feature enough absorption to be investigated with absorption-contrast μCT very effectively [11].

## 2 X-Ray Phase-Contrast Measurement

The beam geometry in the phase-sensing X-ray interferometer is shown in Fig. 2. A monochromatic SR beam typically 5 mm wide and 3 mm high is incident from the left on the first crystal and there split by Laue-case dynamical diffraction into two coherent beams which after separating in space are Laue-diffracted again and brought to overlap on the entrance surface of the last crystal [1], [12]. Here both beams are diffracted a third time and contribute to either of the beams leaving the interferometer at the right. The specimen of normally cylindrical shape is positioned in the upper of

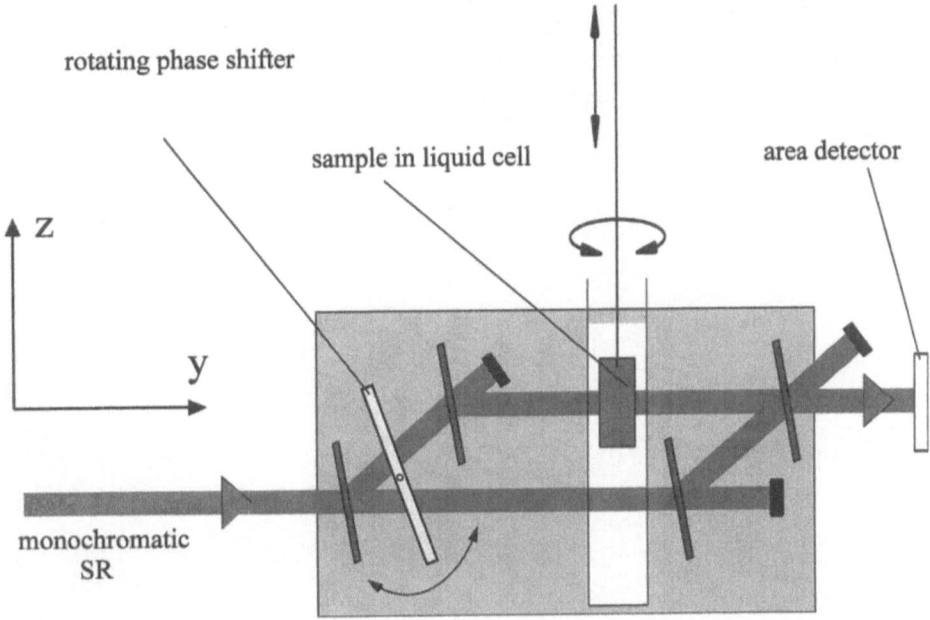

**Fig. 2.** Beam geometry of the phase-sensing X-ray interferometer. Shown is the vertical plane. The four diffracting crystal wafers are all connected with the base (shaded) and are part of a monolithic block of a highly perfect single crystal of silicon. The area detector consists of a scintillator combined with a CCD [2].

the separated beams thereby causing a position dependent relative phase shift of $\Phi(x,z)$ between both beams. We have

$$\Phi(x,z) = \int 2\pi \{ t_\lambda(x,y,z) \}^{-1} \, dy \qquad (2)$$

the phase projection along the beam through the sample. $t_\lambda(x,y,z) \equiv \lambda/[1-n(x,y,z)]$ is '$\lambda$-thickness', i.e. the (in this case locally varying) thickness of material causing a phase shift of $2\pi$. The resulting intensity change at the detector can be expressed as

$$V(x,z) = I(x,z) + K(x,z) \cos( f(x,z) + 2\pi j/P + \Phi(x,z)) \, . \qquad (3)$$

$V(x,z)$ is the X-ray phase radiograph of the specimen for a particular projection direction [13], [14]. $K(x,z)$ ($I(x,z)$) is the coherent (incoherent) contribution to the intensity, respectively. $f(x,z)$ allows for all static phase patterns like the built-in pattern of the empty interferometer etc. The expression $2\pi j/P$ is representing phase shifts of multiples of $2\pi/P$ ($P = 0,1,2,3$) which are intentionally introduced between projections by rotating the phase shifter plate seen at the left in Fig. 2.

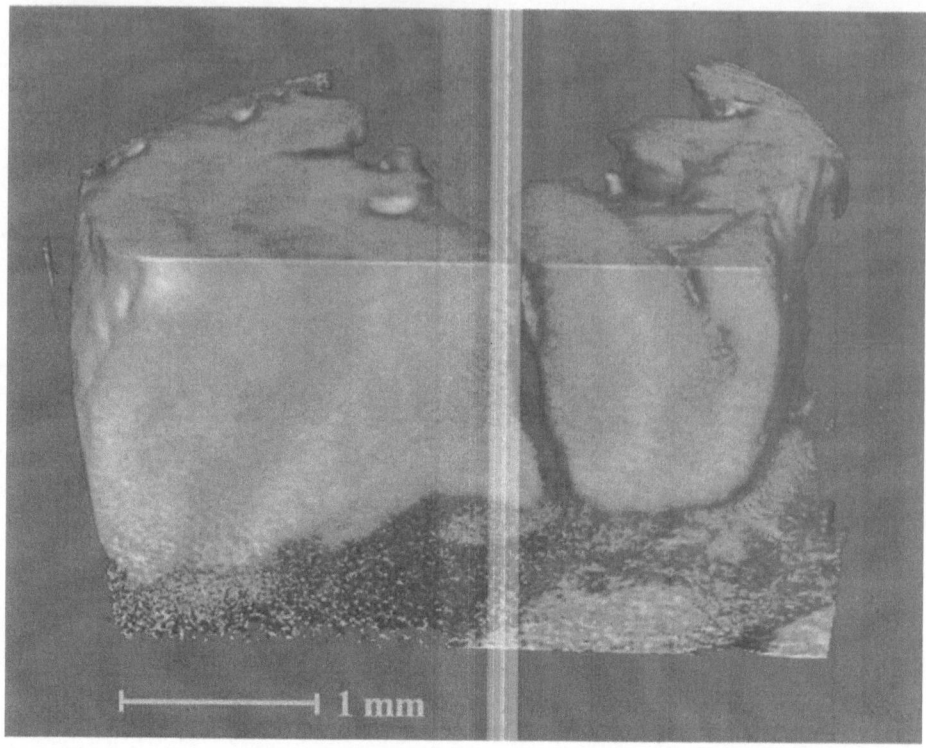

**Fig. 3.** X-ray phase-contrast tomogram of rat cerebrum. Grey and white brain matter can be distinguished. Specimen embedding is transparent.

For a tomographic reconstruction projections in many different directions are needed which are realised by rotating the sample incrementally about the vertical (z-) axis. A typical number are 360 projections at 0.5 degree intervals. In order to eliminate all static patterns also phase radiographs without the specimen are taken:

$$W(x,z) = I(x,z) + K(x,z)\cos(f(x,z) + 2\pi j/P) \tag{4}$$

We calculate sums of phase-weighted projections (given by Eq. 3) of the form:

$$F(x,z) \equiv \sum_{j=0}^{P-1} [I(x,z)+K(x,z)\cos(f(x,z)+2\pi j/P+\Phi(x,z))] \exp(-i2\pi j/P) \tag{5}$$

$$F(x,z)=1/2 \ P \ K(x,z) \exp[if(x,z)+i\Phi(x,z)] \tag{6}$$

when the specimen is in the beam. In a similar way, by using the projections of Eq. 4, we find when the specimen is not in the beam:

$$G(x,z)=1/2 \ P \ K(x,z) \exp[if(x,z)]. \tag{7}$$

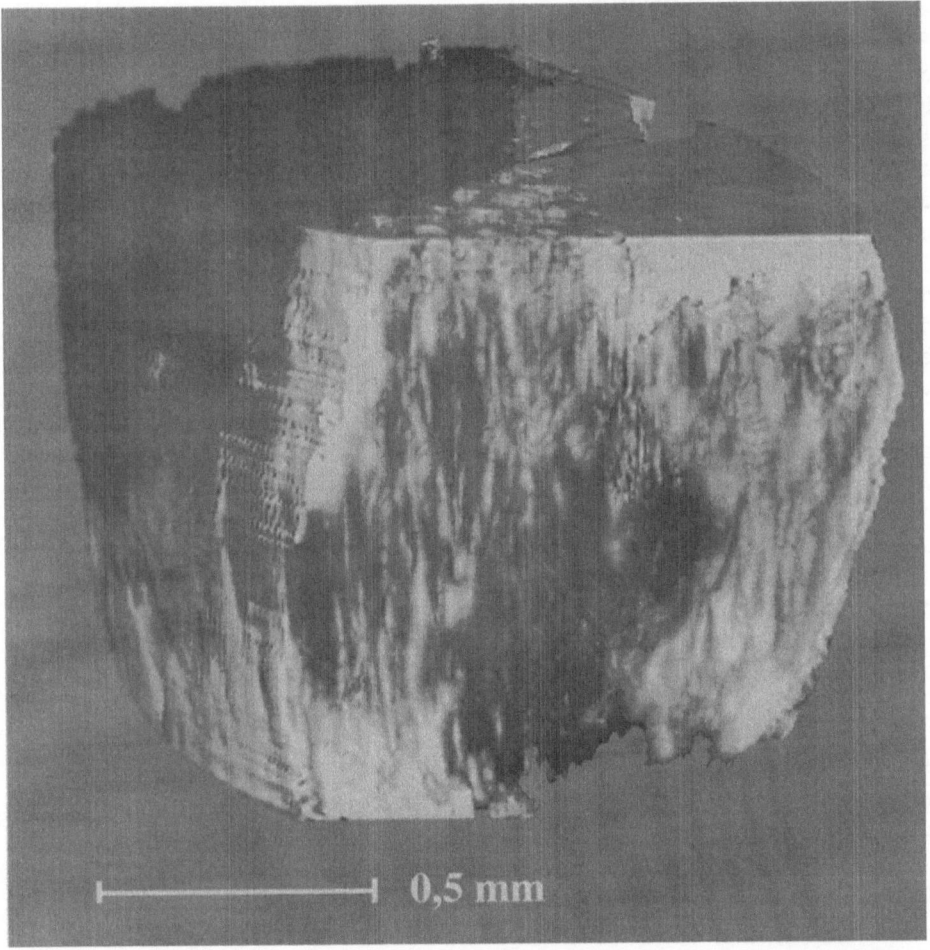

**Fig. 4.** X-ray phase-contrast tomogram of rat trigeminal nerve. Vertical lines cor-respond
to single neurons.

Combining Eqs. 6 and 7 we obtain the desired local phase shift distribution $\Phi$ $(x,z)$ of
the specimen under the given projection angle:

$$\ln\{F(x,z)/G(x,z)\} = i\Phi\ (x,z) - i2\pi k(x,z) \quad \text{where } k(x,z) = \pm\ (0,1,2,...) \quad (8)$$

As is seen from Eq. 8 the phase $\Phi$ $(x,z)$ introduced by the sample is determined only
by multiples of $2\pi$. This ambiguity is lifted by a special algorithm [15] based on the
assumption that there are no discontinuous phase variations inside a realistic
specimen.
Experimental realisations of this technique were achieved recently [5], [16–19].

## 3 Results and Discussion

Figure 3 shows the three-dimensional (3D) rendering of an X-ray phase-contrast microtomogram of rat cerebrum taken with 12 keV x rays. The specimen was embedded in PMMA which is transparent in Fig. 3. Brain tissue is seen in two different grey shades representing white and grey brain matter. In addition, a faint substructure just about the noise level is visible corresponding to nerve-cell aggregations. At the lower edge of Fig. 3, due to a rapid decrease of intensity in the incident beam profile, there is a region where the noise is too large for any structures to be recognised.

Figure 4 shows the 3D rendering of the phase-contrast microtomogram of an intracranal segment of rat trigeminal nerve taken with 12 keV x rays. The embedding material in this case was wax which again is transparent in the picture. White lines correspond to single neurons which run mostly parallel to the nerve axis which is vertical in Fig. 4.

Spatial resolution was determined from measured modulation transfer functions to be about 7.8 µm in the $x,y$- and about 15 µm in the $z$-direction (Fig. 2) which means 7.8 µm resolution in the horizontal and 15 µm in the vertical direction of Fig. 4. The reason for less resolution in $z$-direction is that in the last crystal wafer of the interferometer crystal (Fig. 2) there is increased beam smear due to the Borrmann-fan of dynamical X-ray diffraction in perfect crystals [12]. Reducing the thickness of the last crystal is a possibility to increase vertical spatial resolution.

The two examples show that X-ray phase-contrast imaging is becoming a valuable tool for the investigation of organic tissues containing exclusively light elements.

## Acknowledgements

We thank Prof. M. F. Rajewsky, director of the Institute of Cell Biology (Cancer Research), University of Essen Medical School, and K. Heise and B. Kölsch for kindly providing samples of rat cerebrum and trigeminal nerve. We also like to thank W. Drube, H. Schulte-Schrepping and R. Treusch for assistance during measurements at the BW2 beamline of HASYLAB at DESY, Hamburg. Thanks are also due to the staff of the Laboratory of Eidomatics, Department of Information Science, University of Milano, for the visualization tool XEVA. Financial support by the Bundesminister for Bildung and Forschung, BMB+F, Bonn, under contract No. 05 SPEAAB1 is also gratefully acknowledged.

## References

1    U. Bonse and M. Hart, *An X-ray interferometer*, Appl. Phys. Lett. **6**, 155-156 (1965).

2    U. Bonse, R. Nußhardt, F. Busch, R. Pahl, J.H. Kinney, Q.C. Johnson, R.A. Saroyan, and M.C. Nichols, *X-ray tomographic microscopy (XTM) of fibre-reinforced materials*, J. Materials Science **26**, 4076–4085 (1991).

3    T. Takeda, Y. Itai, K. Hayashi, Y. Nagata, H. Yamaji, and K. Hyodo, *High spatial resolution CT with a synchrotron radiation system*, J. Comput. Assist. Tomogr. **18**, 98-101 (1994).

4    T. Hirano, M. Funaki, T. Nagata, I. Taguchi, H. Hamada, K. Usami, and K. Hayakawa, *Observation of allende and antarctic meteroites by monochromatic X-ray CT based on synchrotron radiation*, in: *Proceedings of the NIPR Symposium on Arctic Meteorites*, No. 3 (ed. National Institute of Polar Research, Tokyo), 270-281 (1990).

5    U. Bonse and F. Busch, *X-ray computed microtomography (μCT) using synchrotron radiation (SR)*, Prog. Biophys. molec. Biol. **65**, 133-169 (1996).

6    W. Graeff and K. Engelke, *Microradiography and microtomography*. In: S. Ebashi , M. Koch, and E. Rubenstein, ed. *Handbook on Synchrotron Radiation*, Vol 4, Elsevier Science Publishers, 361-405 (1991).

7    J.H. Kinney and M.C. Nichols, *X-ray tomographic microscopy (XTM) using synchrotron radiation*, Annu. Rev. Mater. Sci. **22**, 121-52 (1992).

8    J.H. Kinney, Q. Johnson, M.C. Nichols, U. Bonse, and R. Nußhardt, *Elemental and chemical-state imaging using synchrotron radiation*, Appl. Optics, **25**, 4583-4585 (1986).

9    M. Hart and U. Bonse, *Interferometry with x rays*, Physics Today **28**, 26-31 (1970).

10   Beckmann, U. Bonse, F. Busch, and O. Günnewig, *X-ray microtomography (μCT) using phase contrast for the investigation of organic matter*, J. Comp. Assist. Tomogr. **21**, 539–553 (1997).

11   G. Delling, M. Hahn, U. Bonse, F. Busch, O. Günnewig, F. Beckmann, H. Uebbing, and W. Graeff, *Neue Möglichkeiten der Strukturanalyse von Knochenbiopsien bei Anwendung der Mikrocomputertomographie (μCT )*, Der Pathologe **16**, 342-347 (1995).

12   R.W. James, *The optical principles of the diffraction of x rays*, G. Bell and sons LTD, (1967).

13   U. Bonse and M. Hart, *An X-ray interferometer with long separated interfering beam paths*, Appl. Phys. Lett. **7**, 99-100 (1965).

14   M. Ando ,and S. Hosoya, *An attempt at X-ray phase contrast microscopy*, in: G. Shinoda, K. Kohra and T. Ichinokawa, ed. *Proc. 6th International Conference of X-ray Optics and Microanalysis*, Univ. of Tokyo Press, 63-68 (1972).

15   F. Beckmann, *Entwicklung, Aufbau und Test der Phasenkontrast-Mikrotomographie mit Röntgen-Synchrotronstrahlung*, Dissertation, Universität Dortmund (1997).

16   F. Beckmann, U. Bonse, F. Busch, and O. Günnewig, *A novel system for X-ray phase-contrast microtomography*, Hasylab Jahresbericht 1995 **II**, 691-692 (1995).

17   A. Momose, *Demonstration of phase-contrast X-ray computed tomography using an X-ray interferometer*, Nucl. Inst. Meth. Phys. Res. A **352**, 622-628 (1995).

18   A. Momose, T. Takeda, and Y. Itai, *Phase-contrast X-ray computed tomography for observing biological specimens and organic materials*, Rev. Sci. Instrum., **66**, 1434-1436 (1995).

19   A. Momose, T. Takeda, Y. Itai, and K. Hirano, *Phase-contrast X-ray computed tomography for observing biological soft tissues*, Nature Medicine **2**, 473-475 (1996).

20   A.Yu. Nikitin, L.A.P. Ballering, J. Lyons, and M.F. Rajewsky, *Early mutation of the neu (erbB-2) gene during ethylnitrosourea-induced oncogenesis in the rat Schwann cell lineage*, Proc. Natl. Acad. Sci. USA **88**, 9939-9943 (1991).

21   A.Yu. Nikitin, J.-J. Jin, J. Papewalis, S.N. Prokopenko, K.M. Pozharisski, E. Winterhager, A. Flesken-Nikitin, and M.F. Rajewsky, *Wild type neu transgene counteract mutant homologue in malignant transformation of rat Schwann cells*, Oncogene **12**, 1309–1317 (1996).

# Differential Phase Contrast X-Ray Microscopy

G. R. Morrison[1] and B. Niemann[2]

[1]Department of Physics, King's College, Strand, London WC2R 2LS, UK
[2]Georg-August-Universität Göttingen, Forschungseinrichtung Röntgenphysik,
Geiststrasse 11, D-37073 Göttingen, Germany

**Abstract.** The advantages of using a configured detector to obtain differential phase contrast image in a scanning transmission X-ray microscope are described. A prototype system using a CCD detector has been installed on the microscopy beamline at BESSY, and the first images have been acquired.

## 1 Introduction

There is an increasing interest in the development of phase contrast imaging techniques at X-ray wavelengths. These techniques were often first applied in the context of visible light microscopy, and offer the possibility of improved image contrast, particularly when using relatively hard X-rays, or when operating on the low-energy side of an elemental absorption edge, since the phase information can provide complementary specimen information to that furnished by the usual X-ray absorption signal. It is the aim of this paper to describe a method of producing phase contrast X-ray images that can make very efficient use of the available illumination, and will also provide a very flexible set of imaging configurations that extend and enhance the performance of the microscope.

## 2 Phase Contrast Methods in X-Ray Microscopy

The advantages of using phase contrast techniques at X-ray wavelengths were first outlined by Schmahl et al. [1], and since then a number of different schemes have been developed. The first system to be realised was on the transmission X-ray microscope (TXM) at the BESSY synchrotron in Berlin [2], and involved the use of a phase-shifting plate in the back focal plane of the objective zone plate. This method is analogous to the Zernike method of phase contrast that has been in use in visible light microscopy for many years, with the exception that at X-ray wavelengths the phase plate will also attenuate the beams that pass through it. A form of phase contrast imaging is also possible using holographic techniques, since the hologram can, in principle, allow the reconstruction of both the amplitude and phase of the wave scattered by the object [3]. The importance of the phase term in producing image contrast when using coherent beams of 10–50 keV X-rays was recently demonstrated at the ESRF [4], where a simple in-line geometry was used to record the holographic images.

**X-Ray Microscopy and Spectromicroscopy**
Eds.: J. Thieme, G. Schmahl, D. Rudolph, E. Umbach
© Springer-Verlag Berlin Heidelberg 1998

## 2.1   Configured Detectors in STXM

The reciprocity theorem, first outlined for the electron-optical equivalents of the TXM and the STXM [5], can be used to show that, under certain general conditions, the TXM and the STXM produce equivalent imaging conditions. However, the roles of the detector and the source are interchanged between the two instruments, and this has important consequences for the ease with which various imaging modes can be realised in the two instruments. Two examples illustrate very clearly the consequences of the different microscope geometries, when both the TXM and the STXM have the same objective focusing elements with the same diffraction-limited spatial resolution.

First, the spatial resolution actually achieved in an image produced by the TXM is the convolution of the diffraction-limited resolution with a function describing the resolution of the medium used to record the image, whereas in the STXM the spatial resolution is determined by the size of the focused X-ray probe incident on the sample, and this is given by the convolution of the diffraction-limited resolution with a function describing the finite size of the X-ray source. The TXM will achieve its highest possible resolution when the recording medium has a very fine grain, whereas the STXM requires a very small effective source to achieve the same result. Conversely, the size of the effective source in the TXM determines the degree of coherence of the illumination; a small source can lead to nearly coherent imaging conditions, while an extended source usually leads to incoherent imaging conditions. In the STXM, the use of an extended detector that collects all of the transmitted flux leads to incoherent imaging conditions, whereas coherent imaging conditions require the use of a detector that collects only a small fraction of the solid angle of illumination diverging from the specimen plane.

Although the STXM and TXM can produce equivalent imaging conditions, these examples show that one geometry can often result in a more efficient or effective way of realising a particular imaging condition. For example, the use of a detector with a small acceptance angle to produce coherent imaging conditions in a STXM discards all the information available in the rest of the transmitted flux and makes inefficient use of the available flux. The important point is that the two geometries are complementary, and should be used to realise different imaging modes in the most efficient manner possible. However, one particularly appealing feature of the STXM geometry is that the detector response function is (at least in principle) an instrumental parameter that is under the control of the user. The equivalent source intensity distribution in the TXM is, by definition, always a positive quantity. Since the detector response may be either positive or negative, it can be made to vary its functional form in ways that are completely impracticable for an X-ray source. This means that a number of novel imaging modes can be realised in the STXM that have no practical equivalent in the TXM. It is this that leads to the idea of 'configured detector' imaging in the STXM.

## 2.2 Theoretical Development of Configured Detector Imaging

A more detailed discussion of the theoretical development has already been published [6], so only a brief summary will be given here. The STXM detector plane is considered to be in the far field, where the complex amplitude in the detector plane is related by a Fourier transform to that on the specimen exit plane. Positions in the sample plane are given by vector $\mathbf{r}$, while positions in the plane of the detector are referred to by the spatial frequency vector $\mathbf{k}$. For an isoplanatic imaging system, and a specimen that can be described by a multiplicative amplitude transmittance $h(\mathbf{r})$, the complex amplitude of the transmitted wave at the exit surface of the specimen is

$$\psi(\mathbf{r}, \mathbf{r}_o) = \psi(\mathbf{r} - \mathbf{r}_o) h(\mathbf{r}) \tag{1}$$

when the focused probe is centred at $\mathbf{r}_o$. If $R(\mathbf{k})$ denotes the response function of the detector, then the detected signal is given by

$$s(\mathbf{r}_o) = \int |\Psi(\mathbf{k}, \mathbf{r}_o)|^2 R(\mathbf{k}) d\mathbf{k} \tag{2}$$

ignoring all irrelevant phase factors and constants outside the integral. $\Psi(\mathbf{k}, \mathbf{r}_o)$ is the Fourier transform of the wavefunction $\psi(\mathbf{r}, \mathbf{r}_o)$.

In the case of a weak-phase, weak-amplitude specimen the linear amplitude and phase contrast transfer functions for the imaging system can be written in the form

$$T_a(\mathbf{k}) = \mathcal{C}(1, 0, \mathbf{k}) + \mathcal{C}(0, -1, \mathbf{k}) \tag{3}$$

$$T_\phi(\mathbf{k}) = i[\mathcal{C}(1, 0, \mathbf{k}) - \mathcal{C}(0, -1, \mathbf{k})] \tag{4}$$

where

$$\mathcal{C}(m, n, \mathbf{k}) = \int \Psi(\mathbf{k}_1 - m\mathbf{k}) \Psi^*(\mathbf{k}_1 - n\mathbf{k}) R(\mathbf{k}_1) d\mathbf{k}_1 \tag{5}$$

Simple symmetry arguments show that, if the probe wave-function is aberration free, then $T_\phi(\mathbf{k})$ vanishes when $R(\mathbf{k})$ is an even function of $\mathbf{k}$, and $T_a(\mathbf{k})$ vanishes when $R(\mathbf{k})$ is an odd function. In principle, this means that it is possible to get a complete separation of the amplitude and phase information from the specimen by forming images with two different detector configurations, one with an even response, and one with an odd response. In the latter case the image is said to be in differential phase contrast (DPC), since the signal is proportional to the phase gradient of the specimen transmittance.

The ideal form of DPC detector is one that provides the first moment of the intensity distribution landing on it. This can be shown to have an unconditionally linear response to phase gradients [7]. The split detector, first proposed for use in the scanning transmission electron microscope [8], and opposite quadrant (OQ) detector configurations are easier DPC configurations to realise in practice, since they do not require a continuously varying detector response. Both configurations are available when the usual brightfield detector is dissected into four quadrants centred on the optical axis. Figure 1 shows a set of images illustrating a number of possible detector response functions, while Fig. 2 shows the corresponding contrast transfer functions.

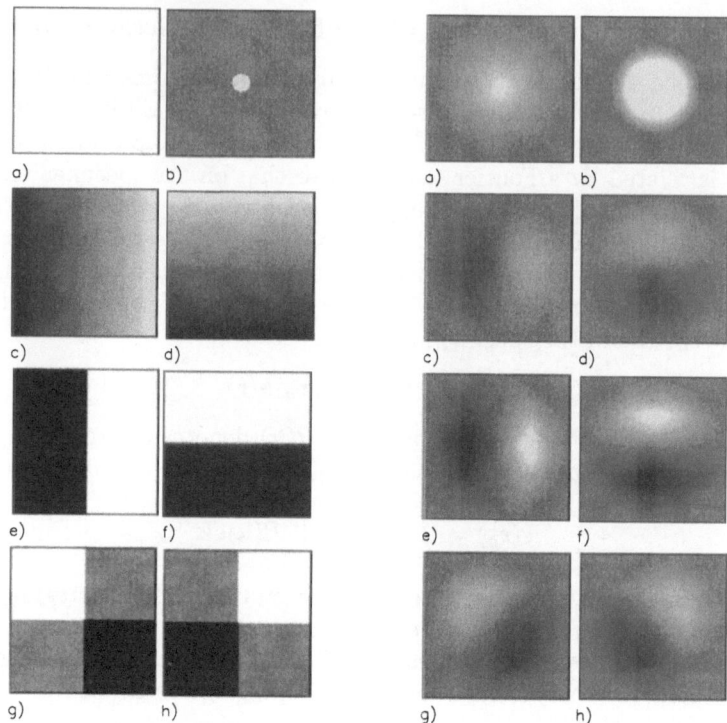

**Fig. 1.** *Left:* Images showing the response functions for a number of different detector configurations. Mid grey corresponds to a zero value, white to a value of +1 and black to a value of −1. a) Uniform detector for incoherent brightfield imaging. b) Uniform axial disk detector for partially coherent brightfield imaging. c) First moment detector sensitive to horizontal phase gradients. d) First moment detector sensitive to vertical phase gradients. e) Split detector sensitive to horizontal phase gradients. f) Split detector sensitive to vertical phase gradients. g) OQ detector sensitive to diagonal phase gradients. h) OQ detector sensitive to diagonal phase gradients (orthogonal to (g)).

**Fig. 2.** *Right:* Images showing the contrast transfer functions for the detector response functions shown in Fig. 1. The key to figures a) to h) is given in the caption for Fig. 1.

## 2.3    Practical Implementation of Configured Detector Systems

The usual mode of operating the STXM is for there to be a single detector downstream of the sample that collects all of the X-ray flux transmitted by the sample, producing an incoherent brightfield image where the contrast represents an X-ray absorption map of the sample. The basic requirement for a configured detector is that it can select only a fraction of the transmitted flux, but the advantage of using a multi-element detector in the STXM is that different detector response functions can be simulated after the image has been collected, simply by forming appropriate combinations of the signals from the various detectors.

This means that a number of different detector configurations are available simultaneously, and that a single scan of the specimen can provide sufficient data to analyse both the amplitude and phase terms of its complex transmittance.

**Darkfield Imaging.** A darkfield image is formed when only radiation scattered outside the brightfield cone is allowed to fall on the detector. This results in an image which has the maximum possible contrast, although the contrast is related non-linearly to the object transmittance. The first darkfield images were recorded simply by placing a stop in front of the usual STXM detector to occlude the brightfield cone [9], but more recent examples of the technique have also made use of a multi-element charge-coupled device (CCD) detector [10].

**Quadrant Detector Imaging.** The quadrant detector has been the form of configured detector most commonly used to produce differential phase contrast images in visible-light and electron microscopy. A practical system for X-ray microscopy requires that each quadrant is capable of high-speed photon counting, and that there be both the minimum dead space and the minimum of cross-talk between the four channels, while still maintaining the maximum uniformity of response. Solid-state avalanche photodiode detectors have been shown to be suitable for use with soft X-rays [11], but the quadrant version has been found to suffer from some problems with cross-talk that requires special signal processing electronics to deal with [12].

**Configurable CCD Detector.** The most flexible form of configurable detector is one that is divided into a large number of small X-ray-sensitive elements. Back-face thinned CCD detectors have now replaced photographic film as the recording medium of choice in almost all the current generation of TXM's, as they provide high detective quantum efficiency with the convenience of on-line electronic recording and visualisation of the image data [13]. In some senses such a CCD is also an ideal form of configurable detector for the STXM, but there are two main problems: the first is that the CCD arrays used in the TXM have typically $512^2$ or $1024^2$ pixels and the readout time for a full CCD frame is usually measured in seconds, the second is that a huge volume of data is generated by even a single image scan if a full CCD frame is recorded for each pixel position in the STXM image scan. Both problems can be alleviated if the individual cells are combined electronically at the readout stage to produce a smaller effective array size, but the CCD frame readout times are still too long to allow large raster sizes to be used for the STXM scan. Ideally, the CCD frame readout time should be no more than a few milliseconds.

Despite the present difficulties with using a CCD system as a detector in the STXM, one has been used on the STXM at the NSLS Brookhaven [10], initially as an aid to the alignment of the zone plate optics, and as a means of recording X-ray microdiffraction patterns. It has subsequently been used to record microdiffraction patterns for each pixel in the raster scan of STXM images, and these

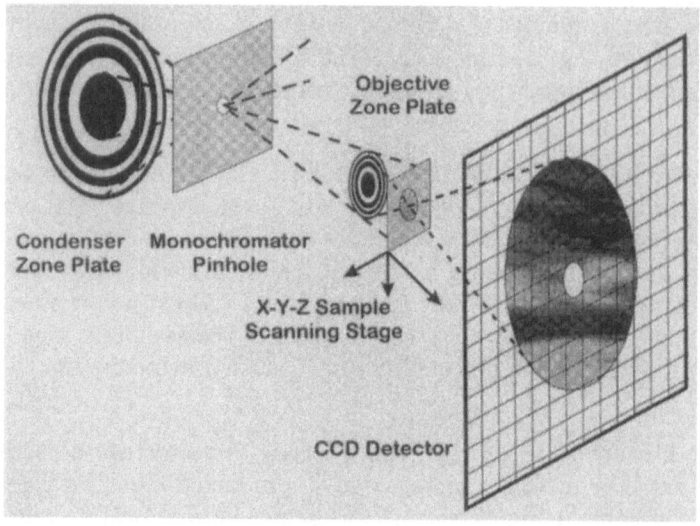

**Fig. 3.** Schematic diagram of the BESSY TXM converted to operate as a configured detector STXM

data have been used to reconstruct complex amplitude and phase maps of the specimen [14] using the method of Wigner-distribution deconvolution [15].

## 3    Configured Detector Imaging at BESSY

The Göttingen TXM at BESSY was the first to be fitted with a back-face thinned CCD detector to record full-field image data. Despite the relatively slow readout times offered by the CCD, it was decided to build a demountable scanning system that could be inserted into the existing TXM beamline to convert it into a prototype STXM. In particular, the aim was to assess the performance of the CCD system as a configurable detector, and to compare experimentally the image signals generated by various antisymmetric detector configurations.

### 3.1    Operation in STXM Mode

To accommodate the change to a STXM mode of operation the usual objective zone plate of the TXM was removed, and a scanning system, allowing 3-axis motion of the sample under stepper-motor control, was mounted inside the vacuum of the CCD camera chamber, as shown schematically in Fig. 3. A new, long focal length objective zone plate, with 200 nm resolution, was placed about 7 mm upstream of the sample. The objective zone plate produced a demagnified image of the monochromator pinhole on the sample plane and, in the absence of a sample, the diverging brightfield cone illuminated approximately half the available area of the CCD detector. Data collection was controlled by the existing acquisition software for the CCD (PMI software from Photomatrix), using

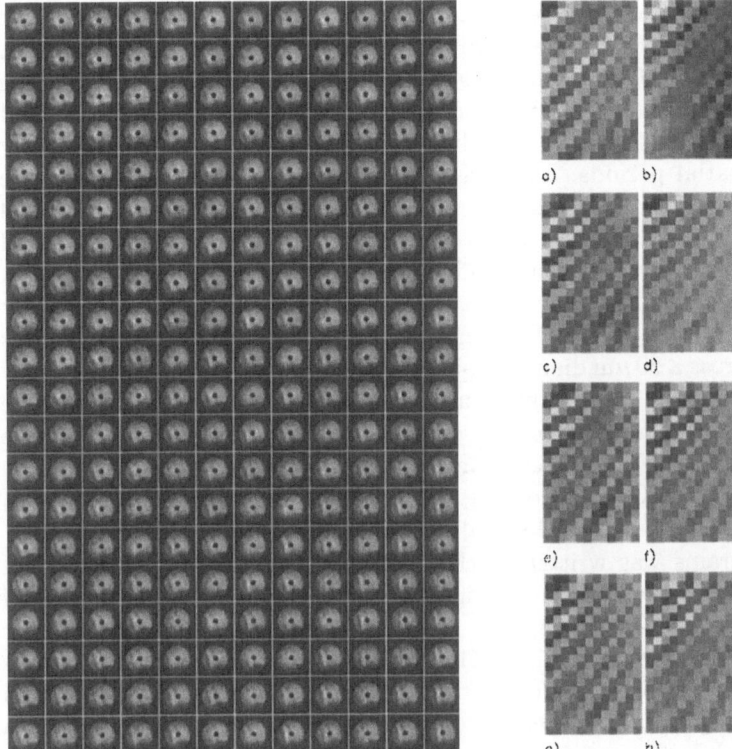

**Fig. 4.** *Left:* Each frame in this 12 by 20 raster shows the signal on the CCD detector for a single position in the STXM scan of a Simens star test pattern. The CCD dark background signal has been subtracted and the zero-order light from the objective zone plate has been masked off.

**Fig. 5.** *Right:* A set of configured detector images calculated from the data in Fig. 4. Each frame in Fig. 4 yields one pixel value for each image in Fig. 5a) to h), using the corresponding detector response functions shown in Fig. 1a) to h). The scan step between pixels was approximately 200 nm.

command macros to collect sequences of CCD frames at user-definable intervals. The microprocessor system that moved the scanning stage was triggered by the closure of a mechanical shutter in front of the detector, so that all sample motion took place during the readout phase of the CCD. The minimum interval between pixels was about 2s. To keep the data sets to a manageable size and to keep the overall acquisition times to a reasonable duration, the CCD frame was rebinned during the readout phase from a $1024^2$ to a $64^2$ array, and the STXM image scans were generally restricted to rasters of no more than 20 by 20 pixels.

An example of a STXM mode scan is shown in Fig. 4, which shows the full CCD frame recorded for each pixel position in a 12 by 20 scan of the sample. The scan step size was approximately 200 nm in both axes, and the specimen imaged was part of a 19 $\mu$m diameter Simens star test pattern fabricated in 150 nm thick germanium on a 100 nm thick silicon nitride membrane by P.S. Charalambous at King's College London. Germanium has a relatively high ratio of the refractive index to the absorption index at the operating wavelength of around 2.4 nm, and the pattern consists of a series of 48 radial spokes, so it provides a continuously varying set of spatial periods that cover the expected resolution range of the imaging system. Although it is usual to operate an STXM with both a central stop on the zone plate and a order-selecting aperture between the zone plate and the sample, to eliminate undiffracted radiation and to prevent higher diffraction orders from reaching the detector plane, neither of these features was necessary when operating with the geometry shown in Fig. 3. In this case, the specimen was itself mounted across a 20 $\mu$m diameter pinhole which eliminated all but a narrow pencil of zero-order light coming through the centre of the zone plate. The active area of the CCD was 25 mm by 25 mm, so the zero-order pencil would illuminate only a few central cells of the detector, and it was straightforward to mask off the signal from these few CCD cells during the subsequent data processing phase, without significantly affecting the signal of interest from the brightfield cone.

A set of programs was written in the IDL programming language to read in image data such as that shown in Fig. 4, and then to calculate a number of possible image signals by weighting each of the CCD frames by different detector response functions, as shown in Fig. 1. These calculations take only a few seconds on a 120 MHz Pentium PC, and the resulting set of images could be displayed on the beamline immediately after the data acquisition. An example of such a set of configured detector images in shown in Fig. 5.

## 3.2    Interpretation of DPC Images

Each frame in Fig. 4 is effectively a convergent beam diffraction pattern from a different point in the test object, and a careful examination of the frames shows a rich variation in the intensity distribution across the detector plane that was very sensitive to the position of the focused probe. Indeed, it was noted during the alignment phase that a simple real-time observation of the CCD signal as the sample position was adjusted was remarkably effective in identifying when the sample was close to focus. The usual brightfield image signal, formed from the integrated intensity in each CCD frame, is insensitive to much of this information. A smaller brightfield detector produces partially coherent imaging conditions, but there will be no phase contrast in the absence of lens aberrations, and the spatial resolution is poorer because the contrast transfer function is narrower. Antisymmetric detector configurations produce phase contrast from the information they give on the redistribution of intensity across the detector plane, and they can do this without sacrificing resolution. However, the DPC signal has a directional dependence, so that there will be little or no information transfer for some features in the object. At X-ray wavelengths, strong phase gradients, are usually accompanied by strong absorption gradients that can give image contrast if the sample is out of focus. Clearly, at least two independent

DPC images must always be examined, while the incoherent brightfield image can provide complementary information on the absorption contrast.

## 4    Conclusions

The use of a CCD detector provides a completely configurable form of detector that can be used in the STXM. As such it can provide a wealth of specimen information that is not available when a single brightfield detector is used. The additional information can be exploited by sophisticated analyses such the Wigner-distribution deconvolution method, but the degree of processing such techniques require means that they cannot presently be applied in real time. However, the use of a few simple antisymmetric detector configurations in addition to the usual brightfield combination allows real-time observation of the sample in both amplitude and phase contrast. Significant reductions in the volume of data generated during routine operation could be achieved if only the processed image data (for example, brightfield and two independent DPC signals) is stored, although the likely future availability of high-capacity, low-cost storage media may make this an unnecessary economy. The main technical challenge that needs to be addressed is the speed at which each frame of data is read from the CCD detector. Devices with $1024^2$ pixels are excellent for imaging applications, but the requirement for STXM operation is for a much smaller array ($\approx 64^2$ is quite adequate) that can be read out at frame rates $\approx 1\,\mathrm{kHz}$. Fortunately, there is good reason to believe that such devices are soon to become readily available.

## Acknowledgments

It is a pleasure to acknowledge the support provided by the research groups at King's College London and at Georg-August Universität Göttingen. Particular thanks are due to P. Guttmann, M. Peuker, T. Wilhein, M.T. Browne, and P.S. Charalambous.

## References

1. Rudolph, D., Schmahl, G., Niemann, B.: Amplitude and phase contrast in X-ray microscopy. In: Duke, P.J., Michette, A.G. (eds.): Modern Microscopies. Plenum, New York 1990 pp.59–67
2. Schmahl, G., Rudolph, D., Schneider, G., Guttmann, P., Niemann, B.: Phase contrast X-ray microscopy studies. Optik **97** (1994) 181–182
3. McNulty, I.: The future of X-ray holography. Nucl. Inst. Meth. Phys. Res. A **347** (1994) 170–176
4. Snigirev, A., Snigireva, I., Kohn, V., Kuznetsov, A., Schelokov, I.: On the possibilities of X-ray phase-contrast microimaging by coherent high-energy synchrotron radiation. Rev. Sci. Instrum. **66** (1995) 5486–5492

5. Zeitler, E., Thomson, M.G.R.: Scanning transmission electron microscopy I and II. Optik **31** (1970) 258–280, and 359–366
6. Morrison, G.R.: X-ray imaging with a configured detector. In: Aristov, V.V., Erko, A.I. (eds.): X-ray Microscopy IV. Bogorodski Pechatnik, Chernogolovka 1994 pp.479–484
7. Waddell, E.M., Chapman, J.N.: Linear imaging of strong phase objects using asymmetrical detectors in STEM. Optik **54** (1979) 83–96
8. Dekkers,, N.H.de Lang, H.: Differential phase contrast in a STEM. Optik **41** (1974) 452–456
9. Morrison, G.R., Browne, M.T.: Darkfield imaging with the scanning transmission X-ray microscope. Rev. Sci. Instrum. **63** (1992) 611–614
10. Chapman, H.N., Jacobsen, C.J., Williams, S.: A characterisation of dark-field imaging of colloidal gold labels in a scanning transmission X-ray microscope. Ultramicroscopy **62** (1996) 191–213
11. Palmer, J.R., Morrison, G.R.: The use of avalanche photodiodes for the detection of soft X-rays. Rev. Sci. Instrum. **63** (1992) 828–831
12. Browne, M.T., Morrison, G.R.: A 100ns anti-coincidence circuit for multi-channel counting systems. Measurement Science and Technology **6** (1995) 1487–1491
13. Wilhein, T., Rothweiler, D., Tusche, A., Scholze, F., Meyer-Ilse, W.: Thinned, back-illuminated CCDs for X-ray microscopy. In: Aristov, V.V., Erko, A.I. (eds.): X-ray Microscopy IV. Bogorodski Pechatnik, Chernogolovka 1994 pp.470–474
14. Chapman, H.N.: Phase-retrieval X-ray microscopy by Wigner distribution deconvolution. Ultramicroscopy **66** (1996) 153–172
15. Rodenburg, J.M., Bates, R.H.T.: The theory of super-resolution electron microscopy via Wigner-distribution deconvolution. Phil. Trans. R. Soc. Lond. A **339** (1992) 521–553

This article was processed using the LaTeX macro package with LLNCS style

# A New Near-Field Scanning Transmission X-Ray Microscopy with 10-nm Resolution

R. E. Burge, X.-C. Yuan, J. N. Knauer

Cavendish Laboratory, University of Cambridge, Cambridge, CB3 0HE, UK
and
Physics Department, King's College London, Strand, London WC2R 2LS, UK

**Abstract.** We show by computer modelling that scanning soft X-ray microscopy using a 10 nm to 20 nm aperture in near field conditions can provide a point to point resolution in transmission of 10 nm. Implementation of this new proposal will radically move forward the present best resolution in scanned X-ray imaging of about 35 nm. The method is realistic in terms of exposure time per point with third generation synchrotron sources of soft X-rays.

## 1 Introduction

The development of soft X-ray transmission microscopes and their applications in biology have been reviewed recently [1]. The concentration on biological applications has been due in large part to the design of the synchrotron sources of soft X-rays currently in mature operation which are near optimum for studies in the "water-window" at wavelengths between the C and O K-absorption edges.

The present study considers the imaging of "biological" material by scanning X-ray microscopy in the near-field and in the water-window, but 10 nm resolution is also expected to be of value in the solution of structural problems in material science and in spectromicroscopy. The optical set-up assumes the collection of all the photons transmitted by the specimen by a detector in the far-field. A wavelength of 2.4 nm, within the water window (X-ray energy about 500 eV), is selected for the calculations below and the material to be imaged is taken to be the organic photoresist PMMA. Detailed specimen composition is not critical to the computer modelling to be presented.

The point to point resolution so far achieved in scanning (and imaging) X-ray microscopes is limited by the problems of fabricating nanodimensional Fresnel zoneplates with high zone positional accuracy [2]–[4]. Current point to point resolutions in practice in scanning X-ray microscopy [1] are at best about 35 nm, even after deconvolution for the probe size [5]. Very considerable efforts are being made to increase the zoneplate resolution, or to work with a zoneplate focus higher than the first order, in pursuit of the goal of 10 nm resolution [6].

**X-Ray Microscopy and Spectromicroscopy**
Eds.: J. Thieme, G. Schmahl, D. Rudolph, E. Umbach
© Springer-Verlag Berlin Heidelberg 1998

There has been recent intense activity in the study of scanning near-field optical systems(SNOM, e.g. [7], [8]) where the scanning optical probe is defined by the light passing through a sub-wavelength diameter aperture or an optical fibre forming a nanoscale dielectric tip. The best lateral resolution so far achieved in SNOM is about 20 nm with a scanning aperture of the same size. However, because the probe retains the width of the scanning aperture only as far downstream as about one-half the diameter of the aperture itself before very strong divergence of the probe occurs, there is a strong limitation on the specimen thickness for high-resolution transmission imaging.

We propose imaging in near field for the soft X-ray case also using a small aperture to provide improved spatial resolution but for much thicker specimens than are possible with SNOM. To improve on the resolution so far achieved by X-ray microscopy, we consider a range of apertures with widths between 5 nm and 20 nm at the primary focus of a zoneplate in the schematic optical arrangement of Fig. 1. The zone plate acts as a condenser and calculations have been made for monochromatic X-rays consistent with its use in first order.

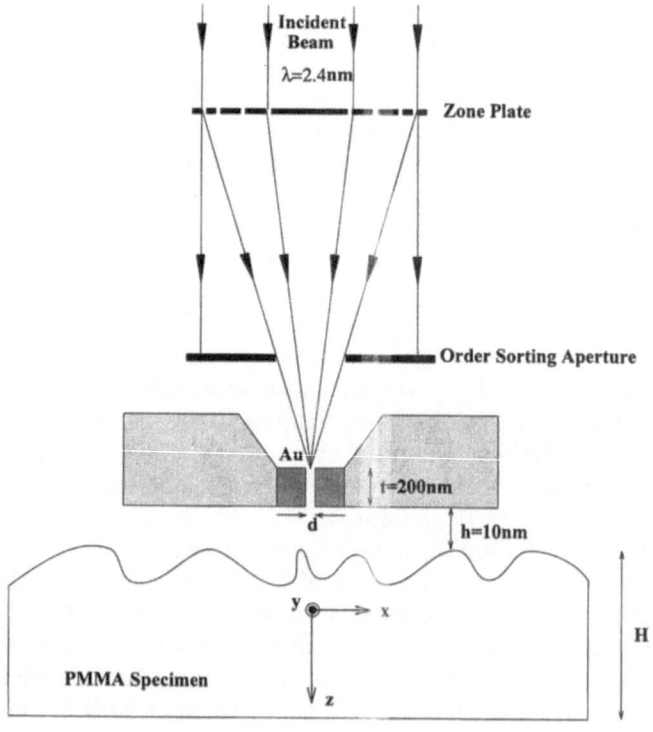

**Fig. 1.** Schematic diagram of the proposed optical arrangement.

## 2   Probe Modelling

Calculations of the X-ray flux transmitted through X-ray masks with sub-micron feature sizes have been made [9]–[11]. We adopt here the full vector theory of Maxwell's equations as the aperture size is close to the wavelength dimension and we require to establish the effects of polarisation.

Methods of calculating SNOM images [12], [13] are available and may be supplemented by the various accurate methods to calculate the near field diffraction from a dielectric grating structure. We have chosen the coupled wave theory [14] for the transmission of the X-ray probe through the aperture and specimen regions and Rayleigh-Sommerfeld theory to propagate the wave motions through vacuum from the specimen towards the detector. The coupled wave method requires a periodic structure in order to expand the electromagnetic fields in

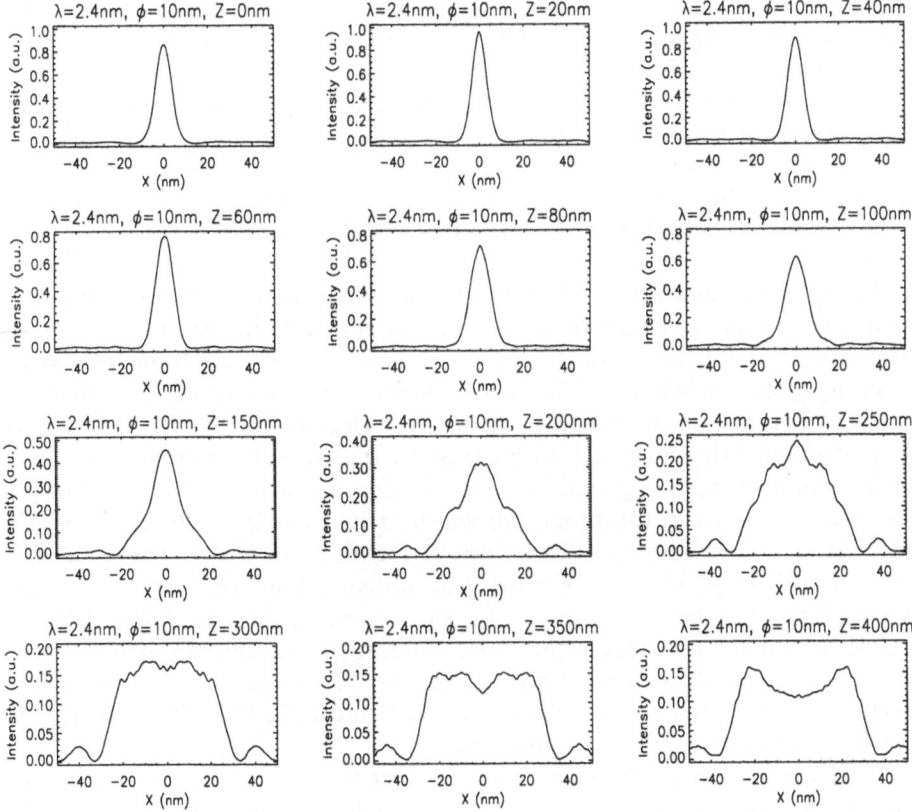

**Fig. 2.** Probe propagation through vacuum: intensity distributions $z = 0$nm to $z = 400$nm TE tangential component. Thickness of gold aperture defining probe $=200$ nm, $n_{\text{gold}} = 1 - 0.4274 \times 10^{-2} + i0.4295 \times 10^{-2}$ for $\lambda = 2.4$nm.

terms of spatial harmonics matched at the boundary interface. A grating is chosen which locally approximates to a pinhole. For example a gold grating with period 100 nm and slit width 10 nm represents a 10 nm pinhole. Calculations for large period gratings require many spatial harmonic components in the coupled wave expansion of the field. We use 131 spatial harmonics here. In the soft X-ray region the refractive indices of materials are close to unity (see Figure legends for Henke values of refractive indices for 2.4 nm wavelength X-rays).

For computation in a realistic time with available resources the probe defining aperture has to be taken as two-dimensional i.e. a slit with given width along the $x$ dimension, and infinite in $y$; the probe propagates along $z$, (see Fig. 1). Similarly the specimen structure is taken to be constant along the $y$ dimension. Thus TE polarisation is polarised parallel and TM polarised perpendicular to the length of the slit. It needs to be made clear that the quantitative aspects of the calculations for two-dimensional imaging will be modified in detail when undertaken with a square or circular aperture rather than a long slit. However there is good evidence from optical calculations [7], [8] that the general pattern of image features seen in one dimension is repeated in two dimensions but obviously with reduced fractional transmission. Calculations were made for both TE and TM polarisation.

The material of the aperture was selected as gold 200 nm thick for the given X-ray energy. It is instructive [10] to consider the aperture as a waveguide and, according to the slit width, a series of guided TE and TM modes will be transmitted by the slit, with $TE_0$ (and $TM_0$) modes dominant for the narrow slits to be considered here. However small the slit width the zero order modes never reach cut off.

In Fig. 2 are shown the calculated probe shapes and normalised intensities for the TE (radial) polarisation of the probe emerging from a slit of width 10 nm into vacuum at different $z$ distances downstream in the range from zero, at the probe aperture, to 400 nm. The probe shows rapid divergence as $z$ increases beyond 200 nm, but the width is maintained close to the aperture dimension for many times the slit width in contrast to the effect for visible light. These results and corresponding ones for 5 nm, 15 nm, 20 nm wide slits are shown in Fig. 3 as variations in the probe full width at half maximum height (FWHM) with distance downstream. The curves in Figure 3 are closely similar despite the four-fold range of aperture width and are consistent with the transmission of the single $TE_0$ and $TM_0$ modes for such narrow slits. The FWHM for the four slits increase smoothly from about 10 nm near the slit to about 20 nm at $z = 200$nm. The efficiencies of photon transfer of the flux from the condenser zoneplate intercepted by the slits are respectively $17\%, 29\%, 32\%$ and $35\%$ for $z = 10$nm and the 5, 10, 15, 20 nm slits.

Considering TM polarisation, account must be taken of both the tangential and the normal components of the electric field. Calculations show that the radial TE and TM results are virtually identical for all the slit widths studied and the normal TM field is negligible in comparison.

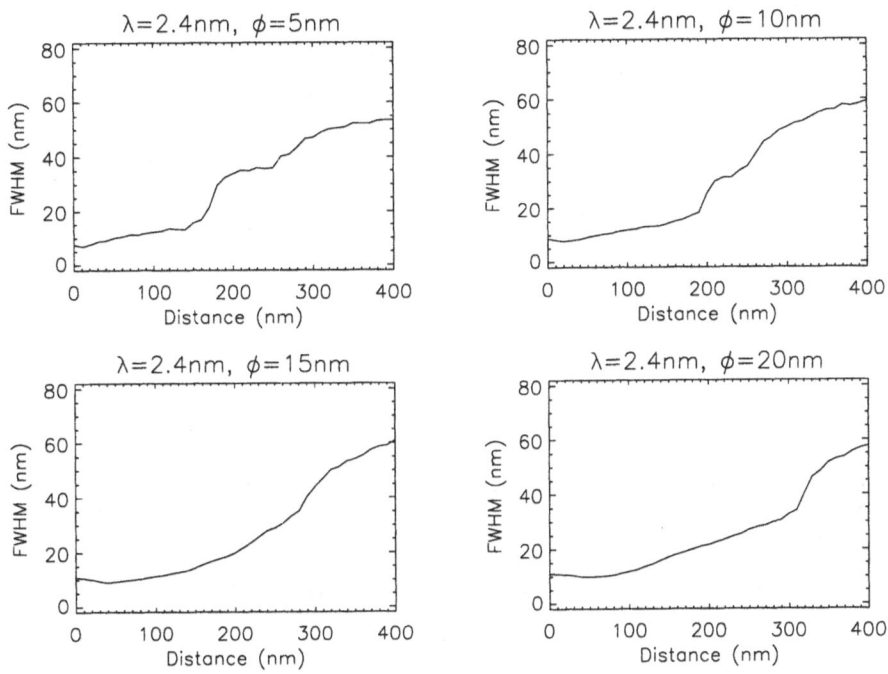

**Fig. 3.** Probe propagation through vacuum: FWHM of intensity distributions on axial propagation, for slit widths 5, 10, 15, 20 nm.

# 3    Image Calculations and Image Resolution

The calculations to be discussed, based on TE polarisation, refer to the high-resolution potential of the new method. Because of the limited range in $z$ over which the probe retains the slit width then, as in the SNOM, the specimen surface must be kept within a few nm from the slit aperture. Calculations are shown in Fig. 4, for a wavelength of 2.4 nm, of the intensities in the images of a hypothetical thin film support (400 nm thick PMMA) with a superposed rectangular block of calcium, 50 nm wide (and hence near in dimension to the present resolution of the zoneplate microscope) and 40 nm thick; in one image the calcium block is on top of the thin film and it is underneath the film in the second. In effect the block is imaged, as shown in Fig. 2, with probes that are respectively 10 nm and 60 nm in width. The definition of the image with the block on top of the film near the slit is greatly superior to that of the second image. The dotted lines in Fig. 4 show for comparison the absorption images for both object locations determined directly from the Henke data; at sufficiently long distances from the edges of the specimen the Henke data become valid.

The image edge spread functions derived from the block boundaries (calcium to PMMA boundaries) are consistent with a resolution of 10 nm, but fur-

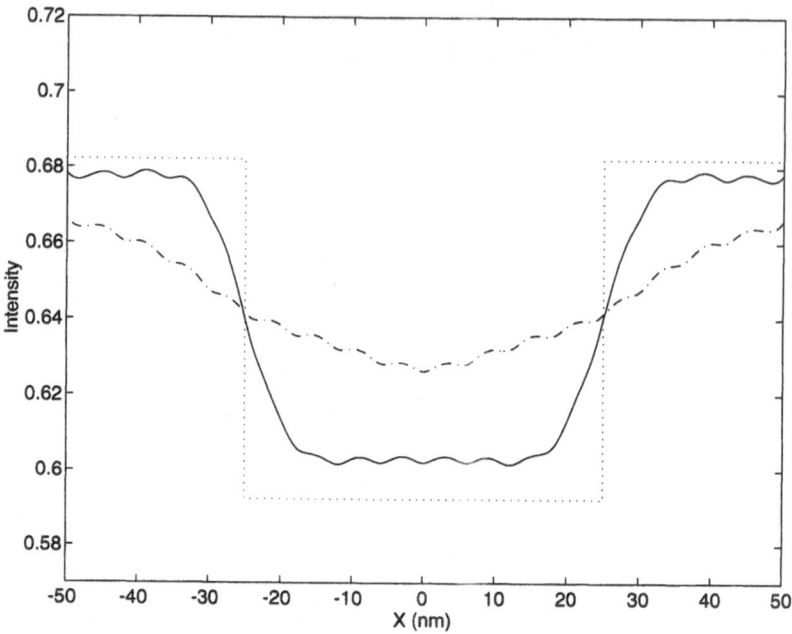

**Fig. 4.** Comparison of images of 50 nm wide, 40 nm thick particle of calcium on 400 nm thick PMMA support. Solid curve: particle on top of film near probe; Dashed curve: particle on bottom of film further from probe; Dotted line: Henke Data. $n_{PMMA} = 1 - 0.8731 \times 10^{-3} + i0.1826 \times 10^{-3}$, $n_{Ca} = 1 - 0.9396 \times 10^{-3} + i0.6747 \times 10^{-3}$ at $\lambda = 2.4$nm.

ther consideration was undertaken with similar materials (PMMA) at the edge, and the results were again compared with the Henke data. Calculations of the edge spread functions were made for a 50 nm PMMA film, with a perpendicular edge partly covering a uniform PMMA film also 50 nm thick and for a similar 100 nm/100 nm film combination. These calculations are shown in Fig. 5.

A final computer experiment was conducted by deconvolving for the effects of the slit (or probe) width on the edge spread functions. First deconvolution of the edge spread functions was carried out using the correct probe as in Fig. 2 with noise-free data following a Fourier transform approach. Then deconvolution of noisy data was carried out using the Wiener filter to simulate experimental conditions after adding Poisson noise with the aid of a random number generator. Additive noise of 2% was assumed on the thinner PMMA specimen; the corresponding additive noise for the same number of incident photons per unit area of the thicker sample is 6.6%. The estimated resolution derived from the edge spread functions before and after deblurring for the probe are given in Table 1. In accord with the different definitions for resolution in common use, we

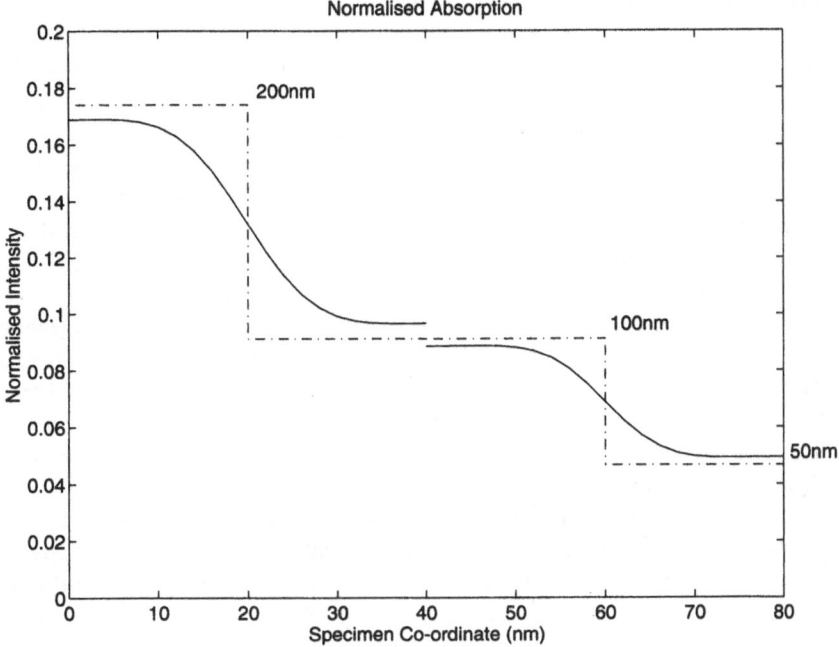

**Fig. 5.** Normalised absorption at density steps from 50 nm to 400 nm. Solid curves: full vector calculation; Dashed curves: Henke Data. $\lambda = 2.4$nm.

**Table 1.** Image Resolution Estimated from Edge Spread Function

|  | 20% − 80% criteria | | 10% − 90% criteria | |
|---|---|---|---|---|
|  | 50 nm step | 100 nm step | 50 nm step | 100 nm step |
| Original Noise Free Image | 9 nm | 10 nm | 11 nm | 13 nm |
| Noise Free Deconvolution | 6 nm | 6 nm | 9 nm | 9 nm |
| Deconvolution of Noisy Data | 6 nm | 6 nm | 9 nm | 9 nm |

show figures for resolution based on both the "20% to 80%" and the "10% to 90%" edge criteria.

These calculated results, following deconvolution for the probe, show that image resolution at a level better than 10 nm is predicted for specimens with thickness up to about 200 nm, which is consistent with our motivation.

# 4   Experimental Arrangements

By analogy with SNOM, and among several possibilities, the control of the height of the probe relative to the moving specimen might be accomplished by combining the near-field scanning X-ray microscope with an atomic force microscope (AFM, e.g. [15]). We note the use [15] of a combined AFM and optical transmission scanning tip (in the given example of 20 nm diameter) and the successful work to pull micropipettes [16] also to an outer diameter of about 20 nm. For a circular 20 nm aperture a conservative value for the transfer of X-ray energy through a 100 nm thick specimen from the focus of the zoneplate to the detector is about 5%. Synchrotron sources of high brightness are appearing which are ideally suited for near field X-ray microscope operation. The third generation ELETTRA storage ring at Trieste, Italy, produces $2 \times 10^9$ photons/s [1] at 600 eV energy into the $0.2\mu m$ diameter focus of a zoneplate which in the given example, for good counting statistics, corresponds to the satisfactory pixel dwell time of order 1ms.

Deconvolution for the known probe shape, shown experimentally, and consistent with Table 1, to improve resolution in scanning X-ray microscopy by a third [5], will improve the resolution for thin specimens below 10 nm and extend the thickness range for 10 nm resolution to a thickness of at least 200 nm. For much thicker specimens a larger probe size and poorer resolution is consistent with the averaging of specimen structure through the depth.

# 5   Discussion

We are moving towards experimental implementation of our proposal, which implies the possibility of macromolecular resolution with soft X-rays, subject to the reduction of radiation damage possibly by cryo-specimen techniques. Further consideration will be made of the experimental requirements of the near-field approach, particularly for higher energy X-rays than the range of the water window. As to the apparent limitation of a 200 nm specimen thickness for high resolution imaging, we note that many studies in zone plate X-ray microscopy and spectromicroscopy are routinely carried out in the 100 nm to 200 nm range of specimen thickness. Unlike for visible light, the TE and TM plane polarised components for soft X-rays behave in similar ways relative to image formation.

# Acknowledgements

This work was supported by the Leverhulme Trust and the Royal Society and has benefitted over a number of years from discussions with Dr Michael Browne and Dr Pambos Charalambos.

# References

1. J. Kirz, C. Jacobsen, and M. Howells, Quat. Rev. Biophys. **28**, 33 (1995).
2. E. Anderson and D. Kern, in Springer Series in Optical Sciences, Berlin, Springer-Verlag, **67**, 75 (1992).
3. T. Matthies, C. David and J. Thieme, J. Vac. Sci. Technol. **B 11**, 1873 (1993).
4. P. Charalambous, P. Anastasi, R.E. Burge, and K. Popova, Procs SPIE 1995 Symposium on Optical Science, Engineering and Instrumentation, San Diego, **2516**, 2 (1995).
5. C. Jacobsen, S. Williams, E. Anderson, M.T. Browne, C.J. Buckley, D. Kern, J. Kirz, and X. Zhang, Opt. Comm. **86**, 351 (1991).
6. R.E. Burge, Royal Institution Proceedings, **64**, 137 (1992).
7. L. Novotny, D.W. Pohl, and P. Regli, J. Opt. Soc. Am. **A 11**, 1768 (1994).
8. J.L. Kahn, T.D. Milster, F.F. Froehlich. R.W. Ziolkowski, and J.B. Judkins, J. Opt. Soc. Am. **A 12**, 1677 (1995).
9. S.D. Hector, H.I. Smith, and M.L. Schattenburg, J. Vac. Sci. Technol. **B 11**, 2981 (1993).
10. R.K. Kupka, Y. Chen, F. Rousseaux, A.M. Haghiri-Gosnet and H. Launois, J. Vac. Sci. Technol. **B 11**, 667 (1993).
11. R.E.Burge and X.Yuan, Procs SPIE 1995 Symposium on Optical Science, Engineering and Instrumentation, San Diego, **2516**, 108 (1995).
12. D. Van Labeke and P. Barchiesi, in Near field optics, Eds. D.W. Pohl and D. Courjon, Kluwer Academic Publishers, Netherlands, NATO ASI series, Series E: Applied Sciences, **242**, 157 (1993).
13. Ch. Girard, A. Dereux, D. Andre, A. Castiaux and J.P. Vigneron, Phys. Low-Dim. Struct., **4/5**, 53 (1995).
14. M.G. Moharam and T.K. Gaylord, J. Opt. Soc. Am. **71**, 818 (1981).
15. F. Baida, D. Courjon and G. Trebillon, in Near field optics, Eds. D.W. Pohl and D. Courjon, Kluwer Academic Publishers, Netherlands, NATO ASI Series, Series E: Applied Sciences, **242**, 71 (1993).
16. E. Betzig, J.K. Trautman, T.D. Harris, J.S. Weiner and R.L. Kostelak, Science, **251**, 1468 (1991).
17. B.L. Henke, J.C. Davis, E.M. Gullikson, and R.C.C. Perera, A Preliminary Report on the X-ray Photoabsorption Coefficient and Scattering Factors for 92 elements in the 10–10,000 eV Region. (University of California, Berkeley, Calif., 1988), Rep. LBL-26259.

This article was processed using the LaTeX macro package with LLNCS style

# X-Ray Microscopy Applications

X-Ray Microscopy Applications

# Visualization of Soil Colloids by X-Ray Microscopy

Jürgen Niemeyer[1], Jürgen Thieme[2], Galina Machulla[3]

[1] FB VI Soil Science/Soil Chemistry, University of Trier, D-54286 Trier, Germany
[2] Institute for X-ray Physics, University of Goettingen, Geiststrasse 11,
D-37073 Goettingen, Germany
[3] Institute of Soil Science and Plant Nutrition, University of Halle,
Am Weidenplan 14, D-06108 Halle/Saale, Germany

**Abstract.** X-ray microscopy allows the direct visualization of soil colloids in water. This method is very well suited for the investigation of aggregation processes and for the visualization of inner-aggregate pore systems.

## 1 Introduction

It has been shown in the last years, that it is possible to image colloids i.e. particles with a characteristic length $\leq 1\mu m$ directly in water by X-ray microscopy [1]. In soils this particle fraction consists mainly of clay minerals, iron aquoxides, humus particles and organo-mineral complexes. Due to their large specific surface area (up to 800 $m^2/g$ ) and their small size many physical, chemical and biological processes in soils are influenced by theses particles. The movement of water and gases or the sorption and desorption of nutrients and toxicants are typical examples of theses processes. In this context, it has to be pointed out, that on earth a large variety of very different soils is existing. In the following, the use of X-ray microscopy and the interpretation of the images obtained will be demonstrated.

## 2 Results and Discussion

### 2.1 Clay Dispersions

Very often, the main fraction of the soil colloids are clays. Under appropriate chemical and physical conditions, these sheet-like minerals aggregate. Theses aggregates strongly influence the flow of water and dissolved substances, so that very often the growth of plants is influenced. To prepare theses small and very fragile aggregates for electron-microscopic investigations is very difficult and time consuming. Furthermore the formation of artifacts is very likely [2].

The figures 1 and 2 show typical X-ray microscopic images of particles found in aqueous dispersions of different montmorillonites [3]. Montmorillonite is a three-layer clay minerals and very often found in most soils. This sheet-like structure is

X-Ray Microscopy and Spectromicroscopy
Eds.: J. Thieme, G. Schmahl, D. Rudolph, E. Umbach
© Springer-Verlag Berlin Heidelberg 1998

clearly visible. The internal pore system of theses aggregates has a very open structure, resulting in a nearly unhindered flow of water and solutes. In addition, theses
figures show, that the size of the montmorillonite particles differs for deposit to
deposit.

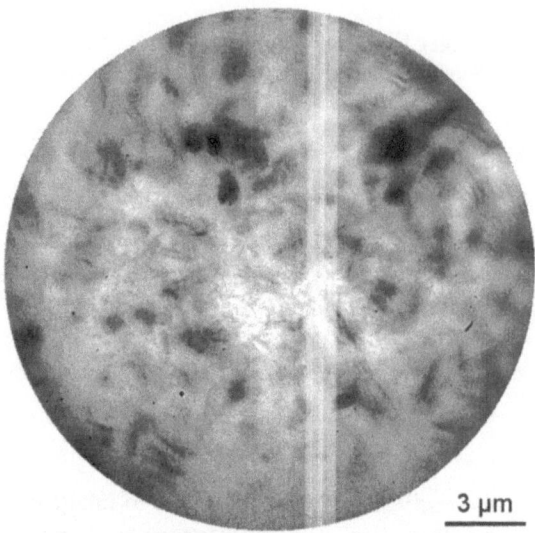

**Fig. 1.** X-ray microscopic image of a dispersion of a montmorillonite clay from Wyoming

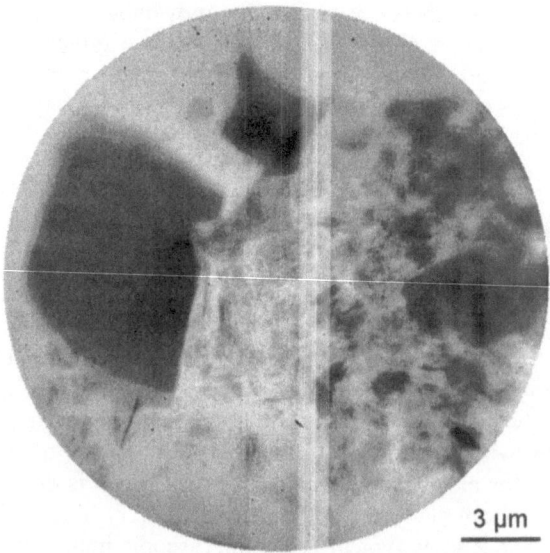

**Fig. 2.** X-ray microscopic image of a dispersion of a montmorillonite clay from Bavaria

## 2.2 Clay Dispersions and Cationic Surfactants

In the figures 3 and 4 the large influence of the composition of the dispersing aqueous medium on the aggregation behavior of clay particles in shown.

In model experiments a cationic surfactant ( CTB ) was added to montmorillonite dispersions. The reason for theses experiments was that in some technologies for soil decontamination surfactants are used in order to increase the solubility of the pollutant to be removed.

Figure 3 shows the open sponge like structure of the original montmorillonite aggregates, whereas in figure 4 dense, compact aggregates are imaged. This compacting effect is due to the attraction of  hydrophobic hulls around adjacent clay particles by the so called hydrophobic interaction. These hulls are formed by adsorption of the positively charged surfactant. It is obvious, that this coagulation drastically reduces the decontamination capacity of this soil detergent.

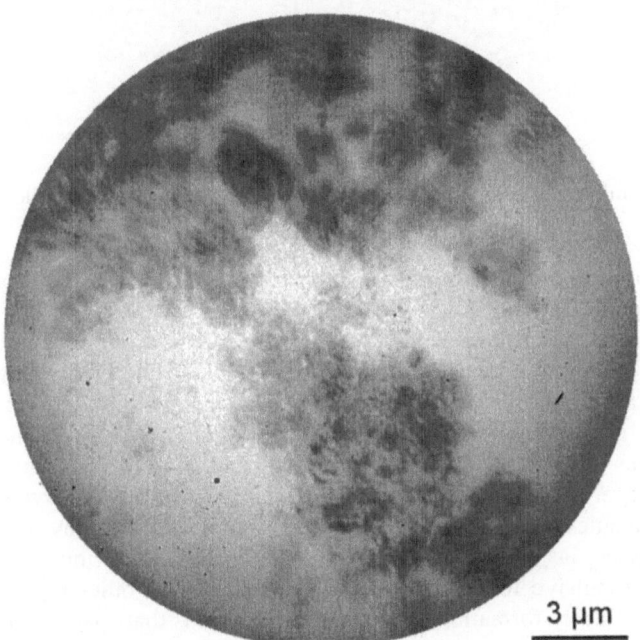

3 µm

**Fig. 3.** X-ray microscopic image of the open sponge like structure of the original montmorillonite aggregates

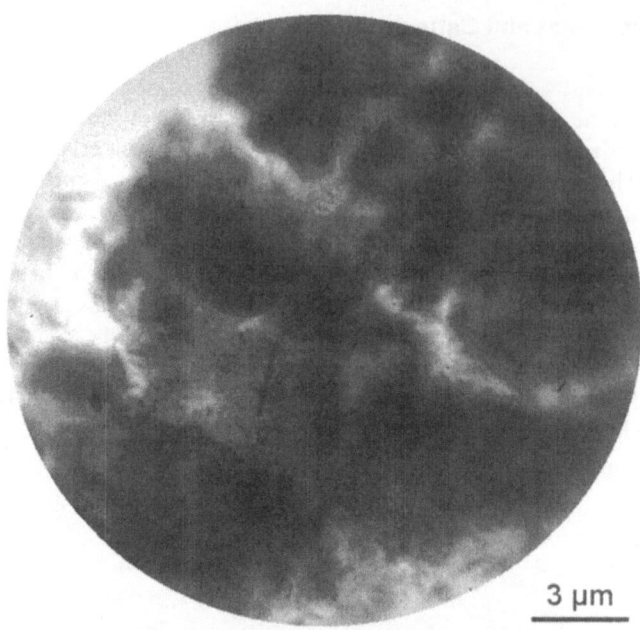

3 μm

**Fig. 4.** X-ray microscopic image of the compact structure of the montmorillonite aggregates formed by the addition of CTB

On the other hand, the same surfactant can act both as a compacting and as a dispersing agent [4]. This is shown in the figures 5, 6 and 7. For this experiments, nontronite, an iron-rich three layer clay was used. This high amount of iron in the clay lattice resulted in a high X-ray contrast. In figure 5 the highly frayed outermost parts of the nontronite particles are shown. The sheetlike structure is again clearly imaged. The addition of the cationic surfactant leads ( i ) to the formation of many small particles ( figure 6 ) and ( ii ) to the compactation of the edge regions of the remaining larger particles ( figure 7 ). An explanation of this observation is, that the cationic molecules are attached to the outer regions of the clay stack. If their is sufficient space, a bilayer of surfactant molecules is formed resulting in an electrostatic repulsive force between adjacent sheets. In other regions of the clay stack, this bilayer formation is not possible and that part is contracted by hydrophobic interactions. As a consequence of this repulsion and attraction, the outermost parts of the thin sheets break off.

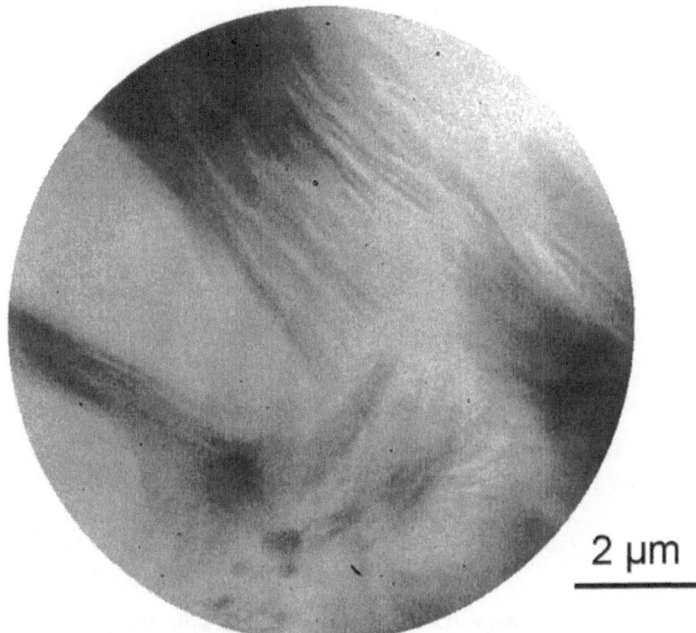

**Fig. 5.** X-ray microscopic image of the highly frayed outermost parts of the original nontronite particles

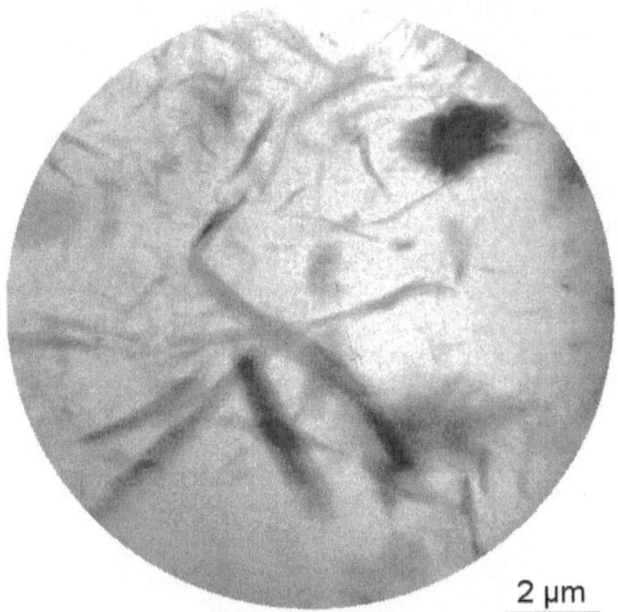

**Fig. 6.** X-ray microscopic image of small nontronite particles formed by the addition of CTB

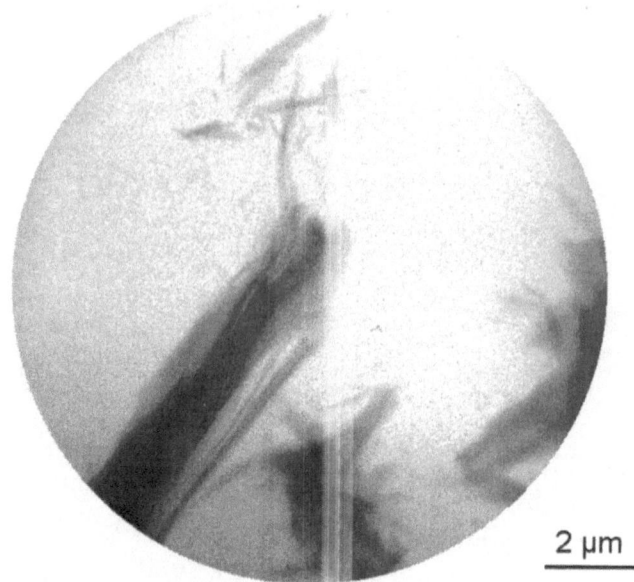

2 µm

**Fig. 7.** X-ray microscopic image of a compact edge region with attached small particles of nontonite after addtion of CTB

## 2. 3  Clay Dispersions and Biomolecules

Besides the structure-changing activities of synthetic organic molecules, the structural activities of bio- or natural molecules were investigated by X-ray microscopy too. For the results obtained on the interaction of micro-organisms with clay particles the reader is referred to [5].

Very often whole organisms or at least parts of larger organisms in soils are surrounded by clay or other small particles. These attached agglomerates are very loosely structured, as shown in figure 8. In these pictures, parts of a fungal hypha is imaged. To this biological structure an aggregate, consisting of mainly clay particles is attached [6].

As already mentioned, in this case the structure is easily permeable by water. Based on other findings, mainly model experiments, other authors assume that organisms excrete some substances into the surrounding medium which cleave former dense structures [7]. The interaction of bacteria and clay particles with this respect is discussed in detail in [5].

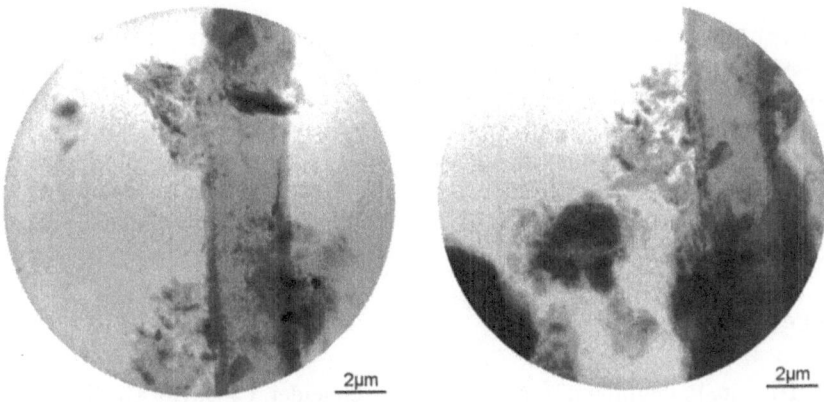

**Fig. 8.** X-ray microscopic image of a part of a fungal hypha with attached clay aggregate

By the use of X-ray microscopy it was possible to image single organo-mineral complexes directly in water. Figure 9 shows such a complex with a very extended veil attached to the mineral core. This organic material reaches very deeply into the surrounding medium; a fact that has not been considered before.

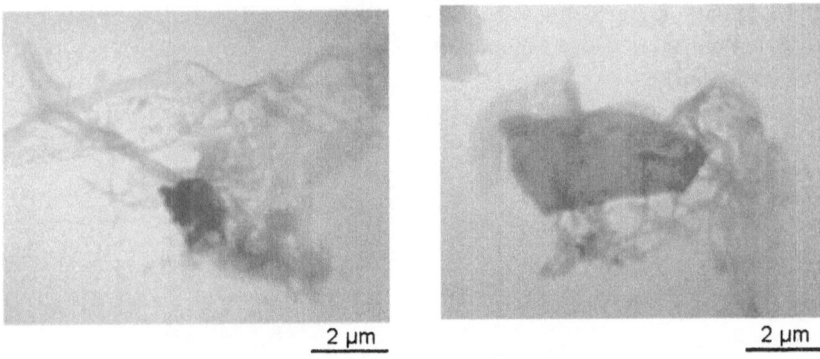

**Fig. 9.** X-ray microscopic image of organo-mineral complexes

# 3 Conclusions

The application of X-ray-microscopy allows the direct investigation of different structure forming processes in soils and other natural materials. This method is a very useful completition of already established microscopic techniques. It can expected, that X-ray microscopy will become in the near future an established technique for the investigation of many other natural materials.

## Acknowledgements

This paper represents publication no. 19 of the Priority Program 546 "Geochemical processes with long-term effects in anthropogenically-affected seepage- and groundwater". Financial support was provided by Deutsche Forschungs-gemeinschaft. In addition, this work has been supported partly by the Federal Ministry of Education, Science, Research and Technology, BMBF, Bonn, under contract number 05 644 MAG, and by the Deutsche Bundesstiftung Umwelt, Osnabrück, under contract number 03149. We thank the BESSY staff for excellent working conditions.

## References

1    J. Thieme, P. Guttmann, J. Niemeyer, G. Schneider, C. David, B. Niemann, D. Rudolph, G. Schmahl, Nachr. Chem. Tech. Lab. 40 (1992) 562-563.
2    R.C. Foster in *Modification of Soil Structure*, W.W. Emerson, R.D. Bond, A.R. Dexter, (Eds.) (John Wiley, Chichester, 1978) 103.
3    J. Niemeyer, J. Thieme, P. Guttmann, T. Wilhein, D. Rudolph, G. Schmahl, Progr. Coll. & Polym. Sci. 95 (1994) 139-142.
4    J. Niemeyer, J. Thieme, J. Plant Nutr. Soil Sci. 160 (1997) 93-95.
5    G. Machulla, J. Thieme, J. Niemeyer, this volume.
6    J. Niemeyer, G. Machulla, J. Thieme, in preparation.
7    C. Chenu in *Environmental impact of soil component interactions,* Vol.1; P. M. Huang, J. Berthelin, J.-M. Bollag, W.B. McGill, A.L. Page, (Eds.) (Lewis Publishers, Boca Raton, 1995) 217.

# Aggregation of Colloids Observed by X-Ray Microscopy

J. Thieme[1], J. Niemeyer[2], G. Machulla[3], U. Schulte-Ebbert[4]

[1]Forschungseinrichtung Röntgenphysik, Georg-August-Universität Göttingen,
Geiststraße 11, D-37073 Göttingen, Germany
E-mail: jthieme@gwdg.de

[2]Fachbereich VI – Geowissenschaften, Abteilung Bodenkunde, Universität Trier,
D-54286 Trier, Germany
E-mail: niemeyer@uni-trier.de

[3]Institut für Bodenkunde und Pflanzenernährung, Martin-Luther-Universität,
Weidenplan 14, D-06108 Halle, Germany
E-Mail: laoec@mlucom2.urz.uni-halle.de

[4]Institut für Wasserforschung GmbH Dortmund,
Zum Kellerbach 46, D-58239 Schwerte, Germany
E-Mail: ifw_mail@compuserve.com

**Abstract.** Many aggregation processes of colloidal particles take place in an aqueous phase. Thus, to ensure a detailed visualisation of the aggregation processes it is necessary to image the aggregates within this environment. Due to the size of the primary particles many of these processes can not be observed directly in light microscopy, as the resolution is too low. The aim of these studies is to show that by X-ray microscopy aggregation phenomena in aqueous phase can be observed directly.

## 1 Introduction

X-rays within the wavelength range between the K-absorption edges of oxygen at $\lambda = 2.34$ nm and carbon at $\lambda = 4.38$ nm are very well suited for X-ray microscopy studies of aqueous colloidal systems [1]. Here, photoelectric absorption and phase shift are the two dominating processes of interaction of X-rays with matter. The radiation is weakly absorbed by water but strongly absorbed by iron oxides, silicates, organic matter, etc. resulting in a good amplitude contrast of objects in aqueous environments. These differences are even larger when looking at the phase shift of X-rays penetrating water or other materials [2]. The graph in Fig. 1 shows the linear absorption coefficient of three substances, i.e. water, the phyllosilicate smectite, and the organic molecule phenol, leading to amplitude contrast in X-ray images. Thus, it is possible with an X-ray microscope to image objects in aqueous media directly and without preparational steps as drying or staining.

**X-Ray Microscopy and Spectromicroscopy**
Eds.: J. Thieme, G. Schmahl, D. Rudolph, E. Umbach
© Springer-Verlag Berlin Heidelberg 1998

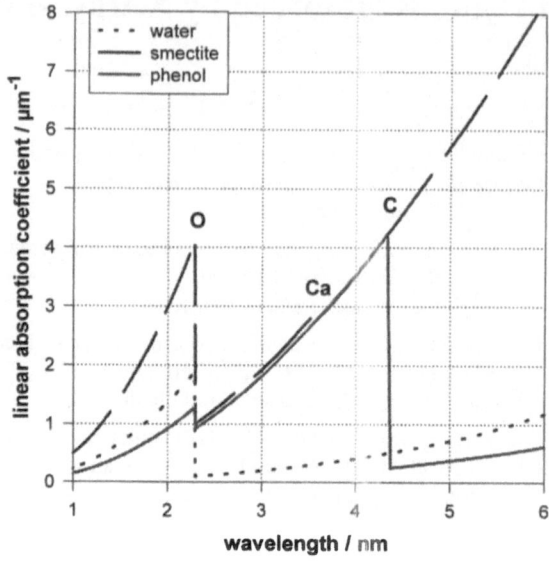

**Fig. 1.** Linear absorption coefficient of water, smectite and phenol as function of wavelength

# 2 Aggregation Phenomena

## 2.1 Aggregation of Hematite Particles

X-ray microscopy allows to visualise directly aggregation phenomena in colloidal dispersions. This has been demonstrated exemplary with a hematite dispersion as a model system [3]. Increasing amounts of $Na_2SO_4$ were added to a stable hematite dispersion to induce coagulation of hematite particles to larger structures which are called aggregates. Figure 2a shows an X-ray image of a stable dispersion comprising hematite particles with a radius of 80 nm approximately. The hematite particles were synthesised following the method described in [4]. Figure 2b was made after adding 7.5 µl of a 1% solution of $Na_2SO_4$ to a 1 ml aliquot of the dispersion. The critical coagulation concentration (ccc) [5], i.e. the concentration above which the dispersion collapses, was determined to be reached adding 8 µl of the $Na_2SO_4$ solution. Figure 2c was taken after adding 10 µl.

The single aggregates were measured with the box counting method. The fractal dimension showed an increase from $D_F = 1.36$ after the addition of 7.5 µl $Na_2SO_4$, i.e. below the ccc, to $D_F = 1.77$ after the addition of 10 µl $Na_2SO_4$, above the ccc, as can be seen in Fig. 3. This result of increasing fractal dimension with increasing $Na_2SO_4$ concentration up to now is not in accordance with the values produced by light scattering experiments [6] or by numerical approaches [7].

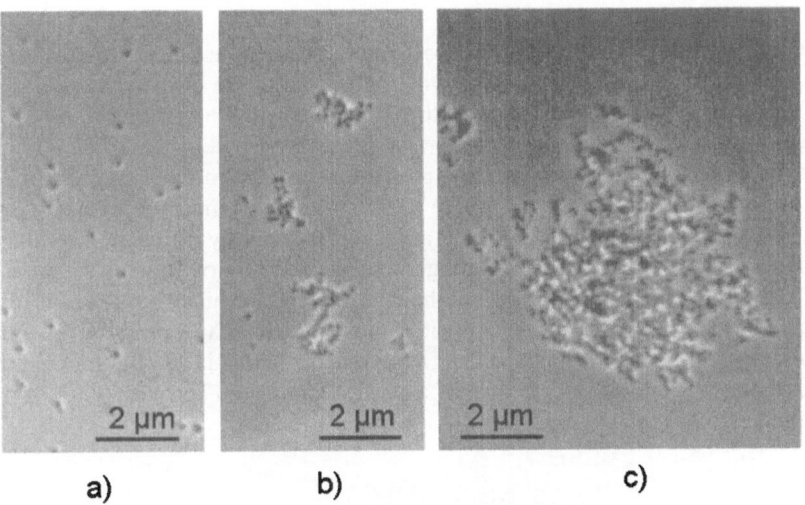

**Fig. 2.** X-ray images of the stable hematite dispersion (a), and of single hematite aggregates after the addition of 7.5 µl (b) and 10 µl (c) of $Na_2SO_4$

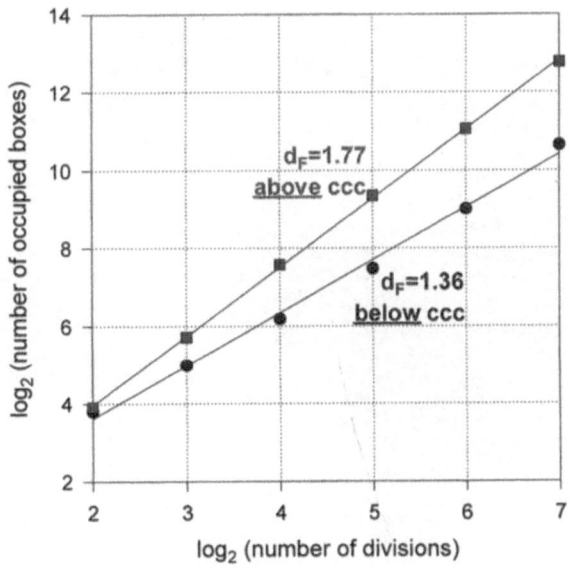

**Fig. 3.** Box-counting plot, derived below the critical coagulation concentration (ccc) from the aggregate at the bottom of Fig. 2b. and above the ccc from the aggregate in Fig. 2c.

## 2.2 The Micro Pore System of Soils

Soils are penetrated by atmosphere, hydrosphere and geosphere [8]. This penetration is expressed in the pore system of soils, where the distribution of the pore radii shows a wide range. This pore system and with it the form of the inner surface of soils determines to a great extend the transport of substances within the soils. Transport processes are extremely important, examples are the water movement and diffusive transport of nutrients and toxicants. In the range down to 10 µm pore radius the inner structure of soils can be well determined and characterised by porosimetric methods. These methods fail in the colloidal range where the radii of the micro pores are < 1 µm. Indirect model supported methods are used, which base on diffusion measurements. With X-ray microscopy it is possible to image directly the porous inner structure of soils in the colloidal range and to study it [9,10]. For example, Figs. 4 and 5 show the microstructure of a dystric cambisol. The very open form of the structure can be seen clearly.

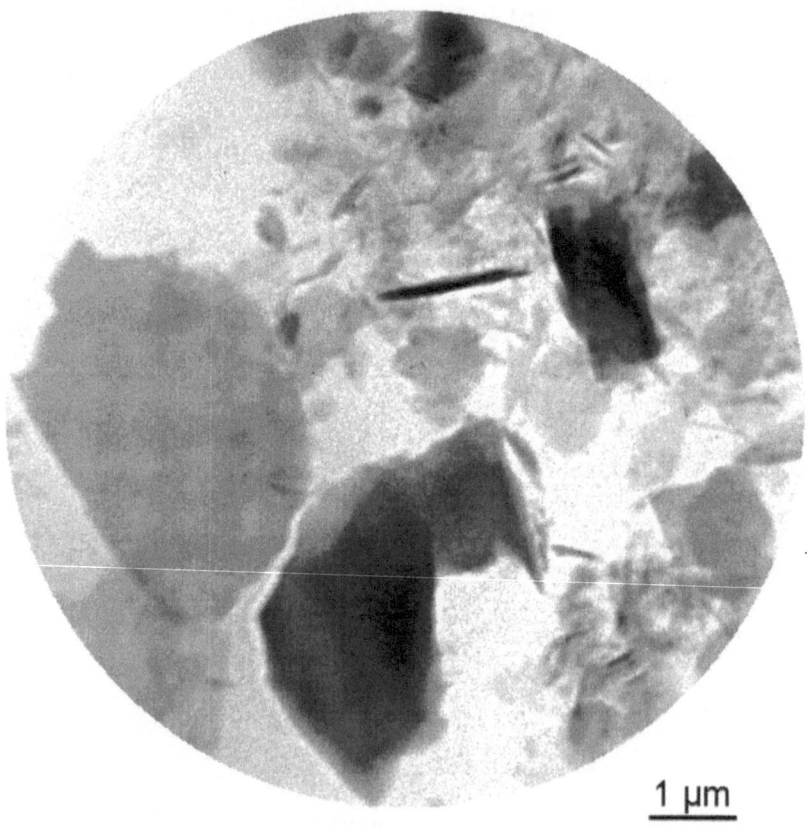

1 µm

**Fig. 4.** X-ray microscopic image of the microstructure formed by colloidal particles within a dystric cambisol.

1 μm

**Fig. 5.** X-ray microscopic image of the microstructure formed by colloidal particles within a dystric cambisol.

## 2.3 Interaction of Humic Substances with Soil Colloids

In the upper part of soils the influence of biological activities is especially prominent. Humic substances, humins, are one result of these activities in soils [11]. Humins are anionic polyelectrolytes. Reactions with cations occur within the aqueous environment because of their negative charge. These reactions influence the microstructure of soils and may even alter it. Humins interact with soil particles among others by the formation of network-like structures. These structures aggregate and, in addition, entangle existing aggregates of other soil colloids. Important parameters of soils can be substantially influenced, as for instance the water flow or the transport of matter by diffusion. The top image in Fig. 6 shows an X-ray image of colloidal aggregate within a 1% dispersion of a chernozem. The bottom image shows an aggregate of this chernozem after the addition of 5% humins (weight-to-weight to chernozem). The network-like structure between the soil particles is clearly visible.

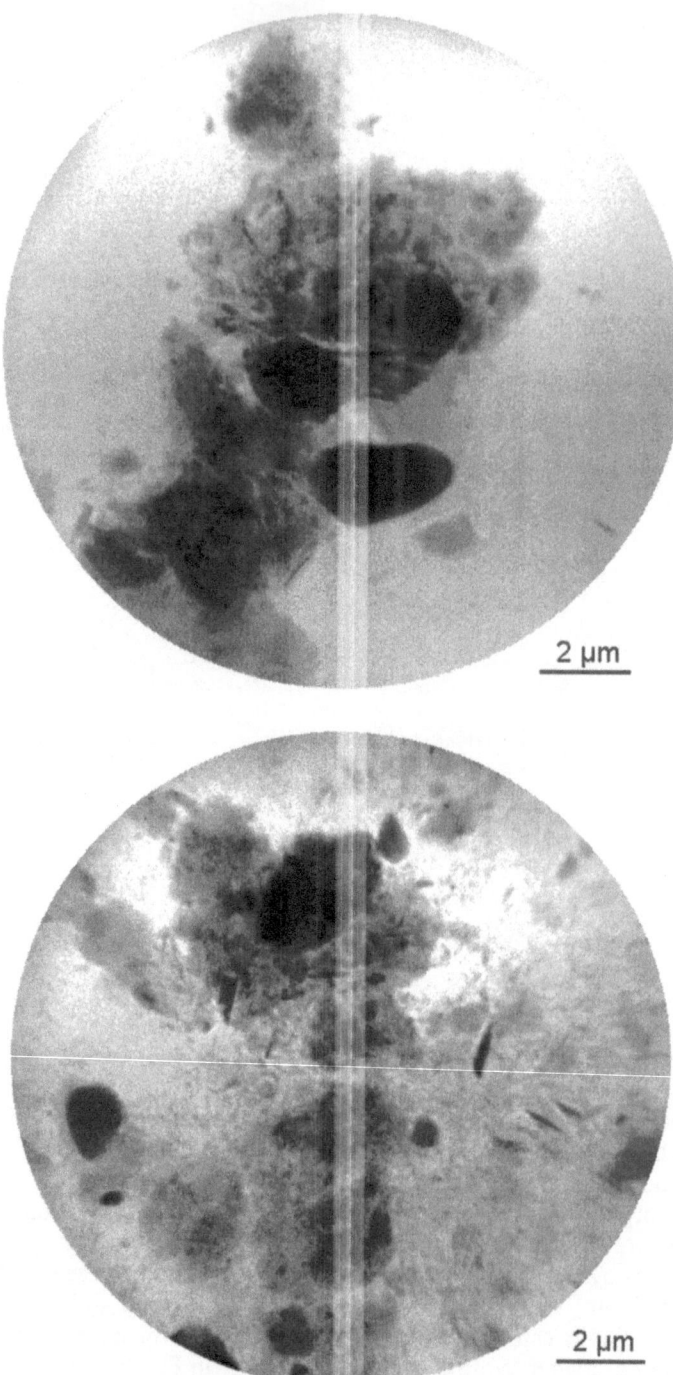

**Fig. 6.** Microstructure formed by colloidal particles within a 1% dispersion of chernozem before (top image) and after (bottom) the addiion of humins.

## 2.4 Formation of Colloidal Particles Due to the Interaction of Humic Substances with Detergents

Humic substances can be extracted from soils by alkaline solutions. As mentioned in 2.3. they are able to influence many reactions in soils, e.g. mass transport and water flow. Detergents are able to reach the ecosystem and hence soils nearly unchanged. Therefore, it is important to study the interaction of both substances in the soil solution. The cationic detergent dodecyltrimethylammoniumbromide (DTB) was added to a 0.05% dispersion of humins to study these interactions. Figure 7a shows small spheres which resulted after the addition of 1 µl of a 1% DTB solution to a 1 ml aliquot. The radius of the spheres is 100 nm within a small limit. Figure 7c shows larger spheres with more different radii after the addition of 7 µl. In addition, aggregates occur. By adding larger amounts of DTB spheres do not occur anymore. Instead, a network like structure appears as can be seen in a very extended form in Fig. 7c, where 50 µl of a 1% DTB solution was added to a 1ml aliquot.

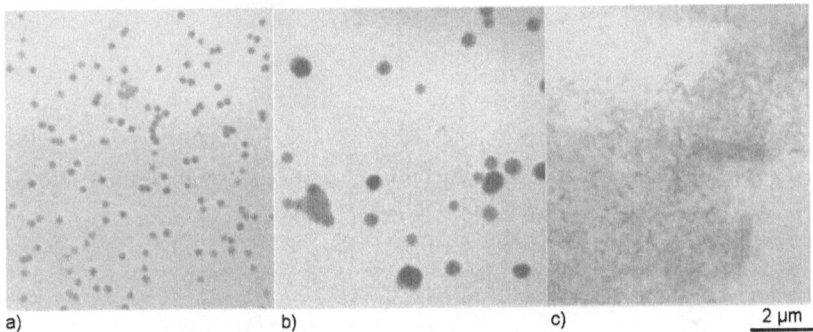

a)                              b)                              c)                    2 µm

**Fig. 7.** Spheres an network like structures as a result of the interaction of humins with a cationic detergent.

# 3 Particle Formation in Ground Water

Hydrochemical changes caused by the degradation of organic matter and reduction of electron acceptors or by mixing of different groundwater types may result in a redox gradient in the aquifer [12]. This gradient can induce the formation of particles by precipitation or the remobilisation of particles which were fixed in mineral coatings on the aquifer material. Iron, as an example, is an abundant cation in groundwater. In anaerobic groundwater aquifers it is present in a reduced form as a bivalent cation [13]. At the groundwater surface or by mixing of anaerobic bankfiltrate and aerobic water the groundwater may get in contact with oxygen. The bivalent iron cation is oxidised to a trivalent state, insoluble compounds with iron are formed in consequence [14]. This gives rise to the formation of new colloidal particles at the transition from anaerobic to aerobic groundwater. Figures 8 and 9 show X-ray images of aggregates of such particles in originally anaerobic ground water after oxidisation. Figure 8 shows two such structures, the larger one looking like an oak leaf, both attached to a much denser colloidal particle. These gel-like structures may contract to form dense iron containing particles in the end. The formation of larger aggregates consisting of these particles is among other things influenced by the microbial activity in the groundwater. In Fig. 9 single particles can be seen in open and loose aggregates, revealing micro organisms and fibrous structures, presumably of organic origin, on which iron containing particles accumulate preferably.

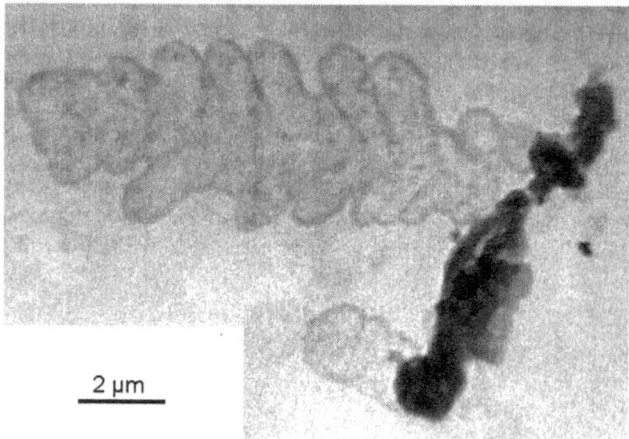

**Fig. 8.** Iron containing structure with a gel-like appearance found in oxidised, formerly anaerobic groundwater.

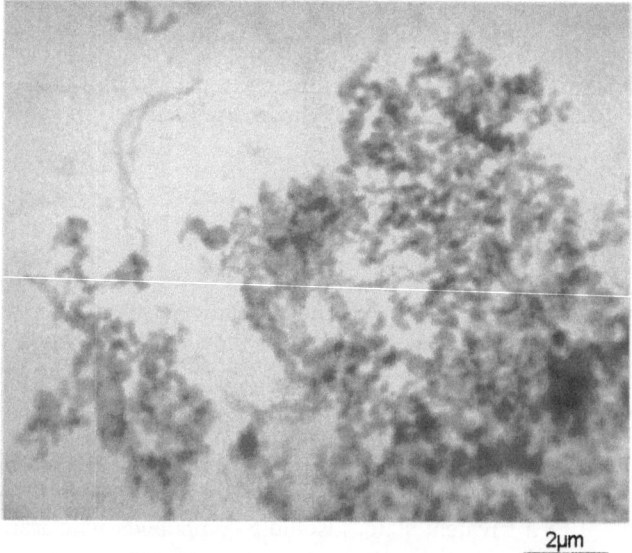

**Fig. 9.** Microbial influenced aggregation of iron containing colloidal particles in oxidised, formerly anaerobic groundwater

## Acknowledgements

This paper represents publication no. 20 of the Priority Program 546 "Geochemical processes with long-term effects in anthropogenically-affected seepage- and groundwater". Financial support was provided by Deutsche Forschungsgemeinschaft. In addition, this work has been supported by the Federal Ministry of Education, Science, and Technology, BMBF, under contract number 05 644 MAG, and by the Deutsche Bundesstiftung Umwelt under contract number 03149. We would like to thank the staff of BESSY for providing excellent working conditions.

## References

1    Schmahl G., Rudolph D., Niemann B., Guttmann P., Thieme J., Schneider G.:
     Röntgenmikroskopie, Naturwissenschaften **83** (1996) 61-70.
2    Schmahl G., Rudolph D., Guttmann P., Schneider G., Thieme J., Niemann B.,
     Wilhein T.: Phase-contrast X-ray microscopy, Synchrotron Radiation News **7** (4)
     (1994) 19-22.
3    Thieme J., Niemeyer J.: Fractal characterization of hematite aggregates by X-ray
     microscopy, Geol. Rundsch. **85** (1996) 852-856.
4    Schwertmann U., Cornell R.: Iron oxides in the laboratory, VCH Weinheim
     (1991) .
5    Brezesinski G., Mögel H.-J.: Grenzflächen und Kolloide, Spektrum Akade-
     mischer Verlag Heidelberg (1993).
6    Amal R., Raper J., Waite T.: Effect of fulvic acid adsorption on the aggregation
     kinetics and structure of hematite particles, J Colloid Interface Sci **151** (1992)
     244–257.
7    Viscek T.: Fractal growth phenomena, Word Scientific Singapore (1989).
8    Scheffer F., Schachtschabel P.: Lehrbuch der Bodenkunde, 13. Aufl.,
     Enke Verlag Stuttgart (1992).
9    Thieme J., Guttmann P., Niemeyer J., Schneider G., David C., Niemann B.,
     Rudolph D., Schmahl G.: Röntgenmikroskopie zur Untersuchung von wäßrigen
     biologischen und kolloidchemischen Systemen, Nachr. Chem. Tech. Lab. **40**
     (1992) 562-563.
10   Niemeyer J., Thieme J.: Visualization of soil colloids by X.-ray microscopy,
     this volume.
11   Sparks D.: Environmental soil chemistry, Academic Press San Diego (1995).
12   Hölting B.: Hydrogeologie, Enke Verlag Stuttgart (1996).
13   Appelo C., Postma D.: Geochemistry, Groundwater and Pollution, A.A.Balkema
     Rotterdam (1994).
14   Cornell R., Schwertmann U.: The Iron Oxides, VCH Weinheim (1996).

# Interaction of Microorganisms with Soil Colloids Observed by X-Ray Microscopy

Galina Machulla[1], Jürgen Thieme[2], Jürgen Niemeyer[3]

[1] Institut für Bodenkunde und Pflanzenernährung, Martin-Luther-Universität, Weidenplan 14, D-06108 Halle, Germany

[2] Forschungseinrichtung Röntgenphysik, Georg-August-Universität, Geiststrasse 11, D-37073 Göttingen, Germany

[3] Fachbereich VI - Geowissenschaften, Abteilung Bodenkunde, Universität Trier, D-54286 Trier, Germany

**Abstract.** In an X-ray laboratory study the interaction of bacteria with a sterile montmorillonite suspension was studied. It was found that the inoculation of the sterile montmorillonite suspension with a culture of soil bacteria resulted in adhesion and aggregation of montmorillonite platelets on the surface of bacterium cells. The platelets appear to be stuck in bacterial slime in a typical edge-to-face association, parallel to each other. The adhesion is due to the extracellular polysaccharides produced by the soil bacteria.

## 1 Introduction

Soils build complex environments that generally contain large amounts of microorganisms. The viability, activity and mobility of bacteria and other microbes in soil depend strongly on the extent to which they are attached to the surfaces of organic and inorganic soil particles. Adhesion of microorganisms to mineral soil colloids may lead to aggregation of soil mineral components, which improves soil structure and its stability [1]. This, in turn, may increase again the biological activity and soil productivity.

In view of the fact that microorganisms excrete extracellular polymer substances, the production of microbial substances is assumed to be the major mechanism by which bacteria and fungi contribute to aggregation processes. This is the material that first makes contact between a cell and a surface and which can cause an irreversible adhesion [2].

It is hypothesized that there are four main types of adhesion between microorganisms and soil solids such as mineral particles. These particles can either be larger than microorganisms, of equal size, or smaller, as in the case of clay particles [3]. The mechanisms involved in this adhesion are of a physicochemical nature. They include van der Waals, electrostatic, hydrogen-bonding as well as hydrophobic interactions [4]. Many polysaccharides can adsorb several particles simultaneously, and thus bind and flocculate them.

Phenomena of building mineral colloidal flocculates, as well as the microbial biodegradation of some xenobiotics, can be considered as basic features in the process of soil and water decontamination [5]. Therefore, studies about the interaction of microorganisms with mineral colloids are fundamental for the science of microbial ecology and of great practical importance for the disciplines of agronomy and plant pathology.

The development of a new technique – X-ray microscopy – has made it possible to study visually the role of bacteria in building colloidal flocculates as well as the mode of their association. Due to a much shorter wavelength, X-ray microscopy provides higher resolution than optical microscopy. Most importantly, X-ray microscopy has the potential for imaging hydrated specimen with high resolution. Moreover, the preparation of samples is simple and the biotic-abiotical system can be studied without distortion.

## 2 Methods and Materials

The method most generally used in laboratory experiments to observe the interaction of microorganisms with mineral colloids applies liquid systems. This method was also used for the present study. The microbes observed in our investigations represent a species of bacterium *Bacillus megatherium* on one hand and a mixed culture of several soil bacteria which are common in most native German soils on the other. The cell length of *Bacillus megatherium* is about 4 µm and in pure cultures bacterial chains could be observed. The culture of soil bacteria includes cells of different size and shape.

To obtain the bacterial suspension and to study the aggregating ability of soil bacteria, we cultured *Bac. megatherium* in a standard nutrient medium consisting of 2 g glycerine, 0.25 g peptone, 0.1 g $K_2HPO_4$, 0.004 g $CaCO_3$, 0.3 g NaCl and 0.25 g $MgSO_4$ per liter of distilled water. The culture was grown for 72 h at a temperature of 28 °C, and after the incubation a 0.1% Na-montmorillonite (Wyoming montmorillonite) suspension was inoculated with this microbial culture at a 1 : 1 - ratio. Subsequently this suspension was studied by X-ray microscopy.

The mixed culture of soil bacteria was maintained in a diluted (0.1% or 1.0% of the original medium) and in the original (100%) nutrient medium after the addition of the Na-montmorillonite (1g per liter). Test tubes containing 2 ml of this bacteria-montmorillonite mixture were incubated at 28 °C for 48 h. At the end of the incubation period the mode of interaction between microorganisms and clay particles was determined with X-ray microscopy.

## 3 Results

In Figures 1 and 2 the mode of cell-montmorillonite association and the influence of the medium concentration on the delimitation of Wyoming montmorillonite in a liquid system is shown. It can be seen (Fig. 1, *top*) that the swollen montmorillonite aggregates disperse into tactoids (or quasicrystals), which consist of several packs of crystallites. In general, the crystallites are associated in a subparallel manner (face-to-face) and have to be considered as interactions of clay particles with microorganisms. After the culture of *Bac. Megatherium* was added to the clay suspension, a close look was taken at the  clay platelets (or crystallites, Fig. 1, *bottom*) and at the very small clay particles (Fig. 2, *top*). It was observed (see Fig. 1, *bottom*) that the crystallites are in an edge-to-face association with the bacterial cells. The very small particles, however, are located on the surface of the microbial cells and in between (Fig. 2, *top*). In these location the concentration of microbial slime is higher then in the surroundings. Large quantities of polysaccharide secretion are seen as several microns thick "shadows" around the microbial bodies. The extracellular polysaccharide slime and its ability to bind clay particles has also been observed by means of ultra-thin section and low-temperature scanning electron microscopy [1, 2].

Clay particles in soil are a main source of nutrients for microbes. The microbes obtain these nutrients through biological weathering processes. These processes are of a biochemical nature and result from the secretion of acid metabolites. For pH-values smaller than 5, it has been reported that mineral destruction by complexation or dissolution takes place [6]. This biochemical weathering leads to a destruction of clay quasicrystals, which fall apart into thinner subparticles (Fig. 2, *bottom*). In the X-ray micrograph, fully dispersed montmorillonite tactoids in the 1.0% nutrient solution can be observed. This 1.0% solution appears to be the optimal microhabitat for an active mixed culture, since it destroys clay mineral tactoids completely. Soil bacteria cultured in the 0.1% solution, however, appear to be inactive (Fig. 3, *top*), because of the extremely low nutrient concentration, whereas in the case of the 100% solution (Fig. 3, *bottom*) they find sufficient nutrients in the solution. In both nutrient solutions (0.1% and 100%) the montmorillonite suspension keeps therefore its original appearance. It thus seems possible that mineral destruction by soil microbes occurs in particular, whenever the microbe population begins to starve after there is a rapid drop of nutrient concentration in the soil solution.

## 4 Conclusions

1. X-ray microscopy allows the direct visualization of bacteria in soil, their extra-cellular polymer substances in microsystems, as well as the affected mineral aggregates and particles.
2. The polysaccharide secretion is noticed as a white „shadow" that surrounds the cells.
3. The inoculation of the sterile montmorillonite suspension with bacteria resulted in the adhesion and aggregation of montmorillonite platelets on the surface of the cells. The platelets appear to be attached in the bacterial slime in a typical edge-to-face association with a parallel orientation to each other. This result seems to be due to the extra-cellular polysaccharides produced by the soil bacteria.
4. By secretion of organic polymers and by physicochemical action, microorganisms can change the organization and physical characteristics of the media in which they live.

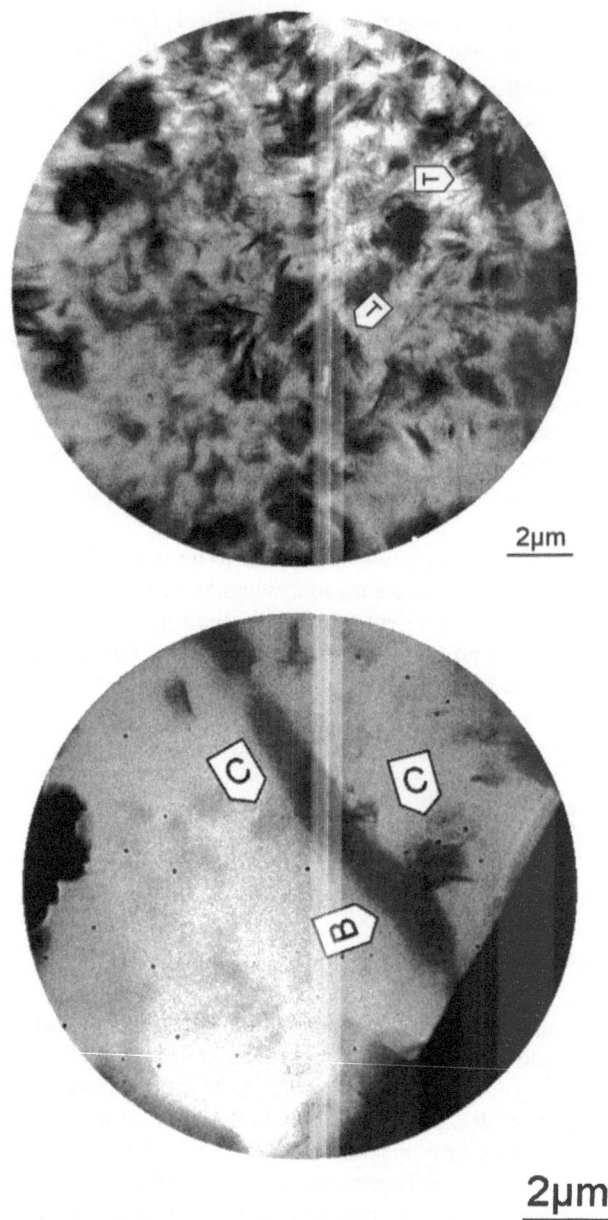

2µm

2µm

**Fig. 1.** The interaction of soil microbes with colloidal particles: (*top*) montmorillonite aggregate dispersion; (*bottom*) polysaccharide secretion and crystallite adhesion caused by *Bacillus megatherium*.
T - tactoid, C - crystallite, P - small clay particles, B - bacterial cell associated with clay crystallites in an edge- to-face manner.

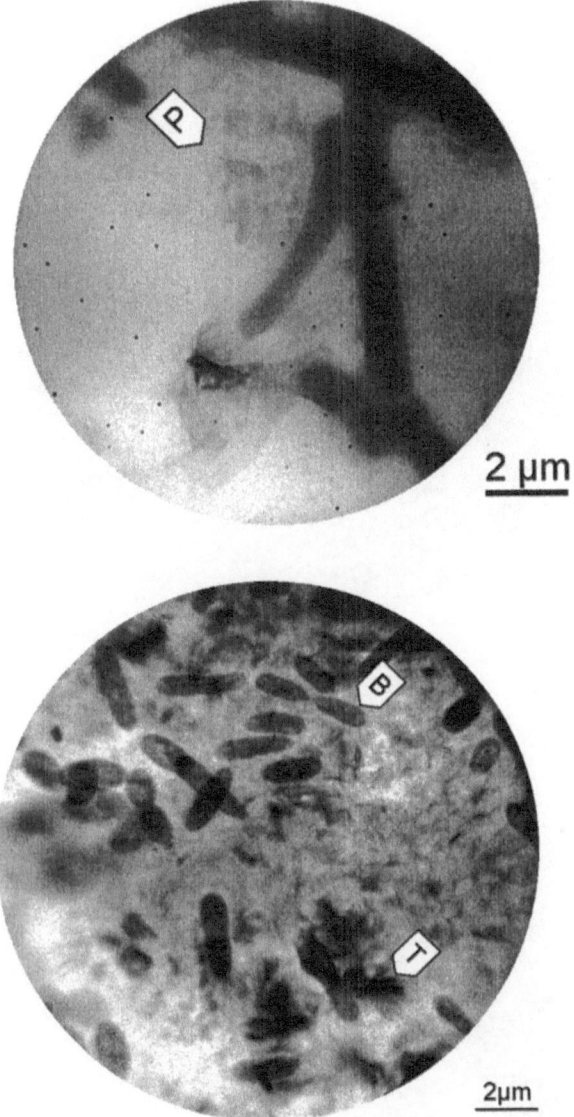

**Fig. 2.** The interaction of soil microbes with colloidal particles: (*top*) small clay particles adsorption in polysaccharide slime; (*bottom*) tactoids destruction in a 1.0% nutrient solution. T - tactoid, C - crystallite, P - small clay particles, B - bacterial cell associated with clay crystallites in an edge- to-face manner.

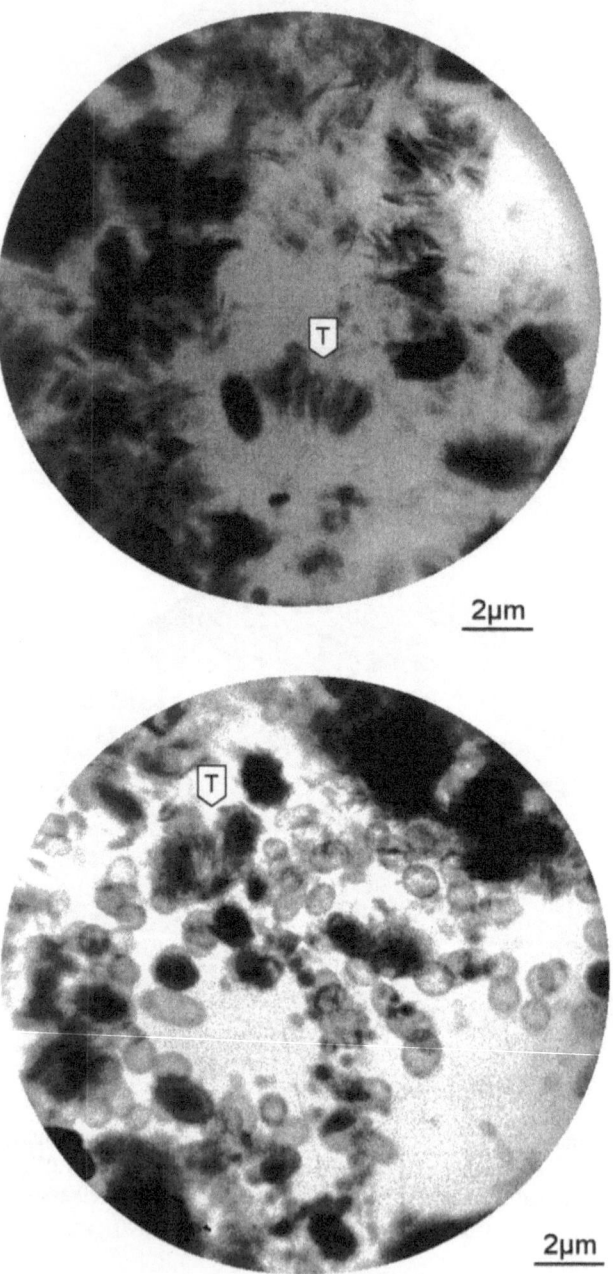

**Fig. 3.** Microbe - clay mineral system in the 0.1% (*top*) and 100% (*bottom*) nutrient solutions.

# References

1    Emerson, W.W., R.C. Foster, and J.M. Oades in *Interactions of Soil Minerals with Natural Organics and Microbes* (Huang and Schnitzer; SSSA, Inc., Madison, Wisconsin, 1986), pp. 521 - 548.

2    Robert, M. and C. Chenu in *Soil Biochemistry, Vol.7* (Stotzky and Bollag; Marcel Dekker, Inc., NewYork, Basel, Hong Kong, 1992) pp. 333-360.

3    Hattori, T., *Microbial Life in the Soil* (Marcel Dekker, Inc.; New York, 1973), 235 p.

4    Zvyagintsev, D.G., *Pochva i microorganizmy/Soil and Microorganisms* (University of Moscow; Moscow, 1987) p. 35.

5    Capone, D. G. and J. E. Bauer in *Environmental Microbiology* (Mitchell; Wiley-Liss, Inc., New York,1992), pp. 191-238.

6    Robert, M. and J Berthelin in *Interactions of Soil Minerals with Natural Organics and Microbes* (Huang and Schnitzer; SSSA, Inc., Madison , Wisconsin, 1986), pp. 453-495.

# Applications of X-Ray Microscopy to the Analysis of Sperm Chromatin

R. Balhorn[1], R. E. Braun[2], B. Breed[3], J. T. Brown[4], D. Evenson[5], J. M. Heck[4],
J. Kirz[6], I. McNulty[7], W. Meyer-Ilse[4], X. Zhang[6]

[1]Biology and Biotechnology Research Program, Lawrence Livermore
National Laboratory, Livermore, CA 94550, USA, E-Mail: balhorn2@llnl.gov
[2]University of Washington, Department of Genetics, Seattle, WA 98175, USA
[3]The University of Adelaide, Department of Obstetrics and Gynecology,
Adelaide, South Australia
[4]Center for X-ray Optics, Lawrence Berkeley National Laboratory,
Berkeley, CA 94720, USA
[5]South Dakota State, Department of Chemistry, Brookings, SD 57007, USA
[6]State University of New York at Stony Brook, Department of Physics,
Stony Brook, NY 11794, USA
[7]Argonne National Laboratory, Chicago, IL 60439, USA

**Abstract.** Chromatin structure has been particularly difficult to study in mammalian sperm cells because the DNA molecules are so tightly packed inside the nucleus that fluorescent probes cannot access the interior of the nucleus and electrons can only penetrate thin sections of the sperm head. X-rays can readily be used to interrogate the interior of the intact sperm head without sectioning or decondensing it. This has made it possible for us to determine the extent of hydration of sperm chromatin (working at a wavelength inside the water window), identify the composition of chromatin in the sperm of several different mammals (using X-ray absorption near edge spectroscopy), visualize vacuoles located inside the intact sperm head, obtain biochemical information about the structure of the equatorial segment and perinuclear theca surrounding sperm chromatin, and examine the nature and uniformity of chromatin compaction in the marsupial mouse, Sminthopsis crassicaudata, and several different lines of transgenic mice. These studies have shown that the sperm cell is a particularly good target for X-ray microscopy studies.

## 1 X-Ray Microscopy Can be Used to Examine DNA Packing Inside the Sperm Nucleus

Light, electron and atomic force microscopy have all been used to study how DNA is organized inside the nucleus of mammalian sperm cells. Conventional light microscopy has been limited by the small size of the sperm nucleus and the limited resolution of the technique to providing information about the general morphology of the sperm head and the uniformity of staining of sperm chromatin using fluorescent probes. Electron microscopy (EM) has revealed that DNA is packed so densely inside the nucleus of the mature sperm that electrons cannot penetrate it (Dooher and

X-Ray Microscopy and Spectromicroscopy
Eds.: J. Thieme, G. Schmahl, D. Rudolph, E. Umbach
© Springer-Verlag Berlin Heidelberg 1998

Bennett, 1973; Roosen-Runge, 1962). The process of DNA compaction within the maturing spermatid has been followed by transmission EM of thin sections of the spermatid nucleus. These studies have shown that the process of condensation is initiated at the apical end of the nucleus and progresses toward the tail as the nucleus elongates and takes shape. During this process, the diffuse chromatin characteristic of all somatic cells is completely reorganized. A family of very arginine-rich proteins, called protamines, are synthesized and bind to DNA, replacing the histones and other chromosomal proteins during mid spermiogenesis (Balhorn, 1989). Upon binding, these small proteins coil the DNA into toroidal structures that contain up to 50Kb of DNA (Hud et al., 1993). Once the process is completed, each sperm nucleus contains approximately 50,000 of these structural subunits. The coiling of DNA into toroidal structures in vitro has been examined by EM (Hud et al., 1993; Hud et al., 1995) and their existence in vivo has been confirmed by atomic force microscopy (AFM) (Balhorn et al., 1993).

While X-ray microscopy cannot achieve a resolution comparable to EM or AFM, the ability of X-rays to penetrate the sperm nucleus makes it possible to probe the interior of the sperm nucleus and obtain structural and compositional information that reflects the entire nucleus. We have used this technique to obtain structural information about sperm chromatin organization that could not be obtained using other forms of microscopy. These studies have provided new information in three different areas: 1) the extent of chromatin hydration, 2) the biochemical composition of sperm chromatin and associated structures, and 3) the uniformity of chromatin packing inside the nucleus.

## 2 Extent of Sperm Chromatin Hydration

Sperm cells are extremely unusual in that they, as a terminal differentiation product of an organ (the testis), are not destined to undergo apoptosis and cell death after performing their function. Each sperm cell instead carries the complete genomic blueprint of the individual that produced it, and the process of sperm development has been designed to temporarily "deprogram" the genome it carries so when it is combined with the genome of the egg (following fertilization) and reactivated, it can be reprogrammed to function as an embryonic cell, not a testis cell. This temporary inactivation of an entire genome and the attendant condensation of the sperm's DNA into a biochemically inert, highly compacted state has not been observed to occur in any other type of cell. In an effort to estimate the extent of this compaction, data obtained from earlier biochemical studies performed in our laboratory and estimates of sperm nuclear volumes obtained by serial section EM were used to calculate the concentration of DNA and the density of its packing inside the sperm nucleus (Pogany et al., 1981). The results indicated that the volume of the sperm nucleus, and the physical volume of the DNA molecule packed inside it, were essentially identical. Thus the DNA, packed at a concentration of approximately 750mg/ml, appeared to fill the entire volume of the sperm nucleus.

These calculations suggested that the sperm chromatin must contain very little water. Only the minor groove of the helix appeared to contain possible sites for water binding, because the protamine molecules fill the major groove (Hud et al., 1994, Prieto et al., 1997). While the data were convincing, the results seemed inconsistent with the way the sperm chromatin was known to decondense both *in vitro* and *in vivo* after fertilization. In both cases, the highly compacted sperm head are observed to decondense rapidly, with the entire mass of sperm chromatin swelling relatively uniformly throughout. Since this decondensation requires reduction of a series of intermolecular disulfide bonds that interlock protamine molecules around the DNA helix, and this reduction could only be achieved if the protamine is hydrated and accessible to the reducing agent, the two findings appeared inconsistent.

Working inside the water window at wavelengths (4.483 nm) where protein and DNA absorb strongly but water does not (Fig. 1), we were able to use X-ray microscopy, combined with atomic force microscopy, to obtain reasonably accurate estimates for the water content of air-dried rat sperm chromatin (DaSilva et al., 1992).

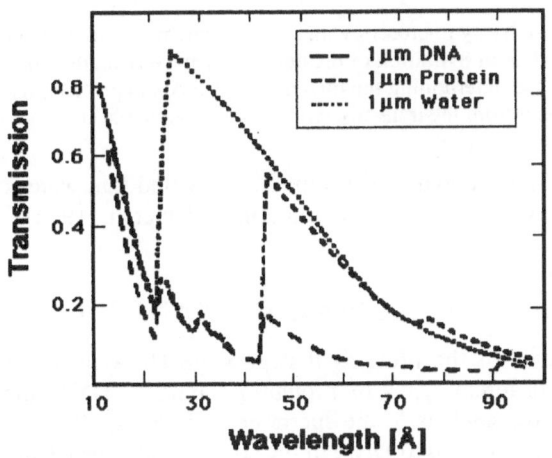

**Fig. 1.** X-ray transmission through 1 micron of DNA, protein and water.

Accurate thickness measurements were obtained for individual rat sperm nuclei air-dried onto silicon nitride windows using the atomic force microscope (Fig. 2A). Transmission images were subsequently taken of the same sperm nuclei (Fig. 2B) using LLNL's pulsed X-ray laser microscope. Using the known composition of the protamine-DNA complex, the density of the complex, and the transmission of 4.483nm X-rays through a particular region of the nucleus, we were able to calculate the thickness of the DNA-protamine complex (470nm) inside the rat sperm nucleus. Since the actual thickness of the nucleus in this region was determined to be 700nm by AFM, the results indicated that ~33% of the volume of the dried rat sperm nucleus must be occupied by water. Subsequent studies using the AFM to monitor changes in mouse sperm nuclear volume upon dehydration provided similar results. In these studies, the volume of the air-dried nucleus was found to be 26–36% greater than the volume of the completely dehydrated nucleus (Allen et al., 1996).

While the pulsed-X-ray laser studies could not provide information about the amount of water inside the nucleus of fully hydrated sperm, they did provide the first, clear evidence that sperm chromatin must be extensively hydrated, even in its highly compacted state. The subsequent AFM studies that confirmed the result extended the

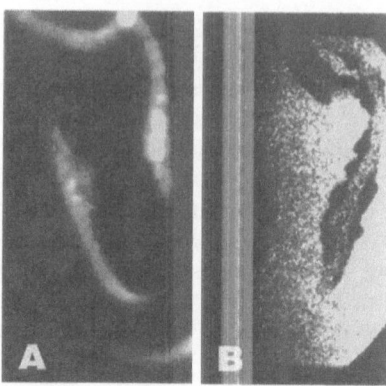

**Fig. 2.** AFM and X-ray microscopy image of rat sperm. Similar images were used to determine the extent of hydration of the nucleus by measuring the thickness of the head with the AFM and determining the thickness of DNA and protein present in the same head from the transmission of X-rays (4.483nm) through it.

analysis to include fully hydrated sperm and revealed that water comprises as much as 64–69% of the volume of sperm chromatin (Allen et al., 1996).

## 3 Composition of Sperm Chromatin (XANES Imaging)

Biochemical studies of the proteins that package DNA in mammalian sperm have shown that two different types of protamines bind to DNA and work together to package it inside the nucleus of the sperm cell (Balhorn, 1989). The smaller protein, protamine 1, is found bound to DNA in the sperm of all species of mammals. A larger histidine-rich protamine 2 molecule is only present in the sperm of selected species (predominantly rodents and primates). Unlike the histone proteins that package DNA in all other cells, which are always present in the same proportion, electrophoretic analyses of the isolated protamines have shown that the relative proportion of the two protamines bound to sperm DNA differs dramatically among species (Balhorn, 1989). These studies were not, however, able to provide both DNA and protamine content information for the same sperm cell. Consequently, the absolute amount of protamine 1 and 2 bound to DNA in sperm chromatin could not be determined accurately.

Even though the amount of protamine 2 is highly variable among species, several studies have indicated that its presence is critical for male fertility (Balhorn et al., 1988; Bach et al., 1990; Belokopytova et al., 1993; de Yebra et al., 1993). To understand the significance of this variation, we must first understand how the two proteins package DNA. A first step in this process requires that we know the mass ratio of protamine to DNA in sperm chromatin. Because semen contain normal, abnormal and immature sperm cells as well as cells from supporting tissues, accurate determinations of the protamine to DNA mass in sperm can only be obtained by analyzing individual cells. This allows the investigator to select normal, fully matured sperm cells for analysis and avoid including data obtained from defective, immature or supporting cells.

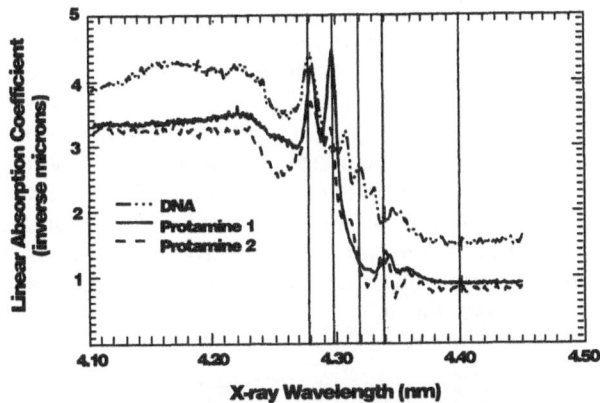

**Fig. 3.** X-ray absorption spectra of DNA, protamine 1 and protamine 2 at the carbon edge. Images of sperm heads were obtained at the wavelengths indicated by vertical lines and the spectral differences between DNA and protamines were used to map out these components separately.

Previous studies have shown that X-ray absorption near edge spectroscopy (XANES) can be used in combination with scanning transmission X-ray microscopy to discriminate between the protein and DNA components of individual Chinese hamster ovary cells (Kirz et al., 1994). X-ray absorption spectra obtained for DNA, protamine 1 and protamine 2 (Fig. 3) revealed spectral differences between DNA and the protamines that could be used to map (and quantitate) these components separately inside the sperm nucleus of four different species of mammals. To accomplish this, individual dried sperm were imaged at six different wavelengths (4.100, 4.279, 4.297, 4.318, 4.339, and 4.400nm) that represent specific peaks in the DNA or protein absorption spectra using the Scanning Transmission X-ray Microscope at Brookhaven National Laboratory. The optical density of the sperm nucleus at each wavelength was obtained from these images and the data were used to calculate the mass of DNA and protamine present in the nucleus using the Singular Value Decomposition method (Zhang et al., 1996). Sperm from four species were chosen as representatives of the range of protamine 2 variation that is known to occur among different mammalian species. Bull sperm contain only protamine 1. Stallion, hamster and mouse sperm contain increasing amounts of protamine 2 (14%, 34% and 67% respectively). Attempts were also made to obtain data for human sperm (~50% protamine 2), but the nuclei proved to be too thick for quantitative analysis.

Using this method, DNA and protein maps were obtained for the sperm nuclei of all five species (Zhang et al., 1996). In each case, as shown in Fig. 4 for hamster, the DNA was found to be confined to the nucleus, as expected. Occasionally the midpiece of the tail appeared in the DNA images, suggesting the analysis picks up the small amount of mitochondrial DNA located in this region of the tail. Protein maps indicate the protein is distributed fairly uniformly throughout the majority of the head. The extra protein present in the acrosome is also apparent as additional material surrounding the anterior end of the nucleus. Quantitative analyses were performed on sperm nuclei from each species treated to remove the acrosome and tails, leaving only the sperm chromatin. The results show that the mass of protamine in the sperm nucleus, relative to DNA, is constant for all four species irrespective of the protamine 2 content. This has allowed us to discriminate between two possible scenarios for DNA packing by protamine in mammals (Fig. 5). One possibility is that a protamine 1 molecule binds to each turn of DNA in each species, and those species that contain protamine 2 have an increase in protamine above that found in bull sperm (a species that contains only protamine 1). A second is that all the DNA is covered uniformly by protamine, and when protamine 2 is present, it replaces protamine 1. The XANES data show that the total protamine content of the sperm nucleus is constant in each species. If the amount of protamine 2 used to package DNA increases, the amount of protamine 1 decreases proportionately.

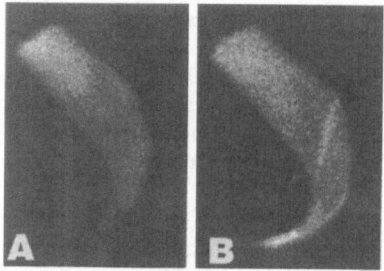

**Fig. 4.** XANES images were used to produce DNA (A) and protein (B) maps of hamster sperm heads. This particular head was not attached to a tail.

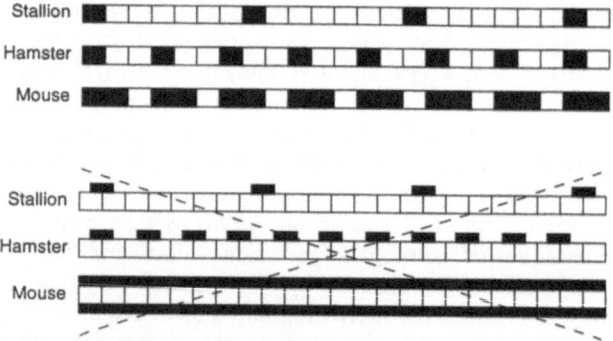

**Fig. 5.** Possible scenarios for protamine 1 and protamine 2 binding to DNA. Each protein is assumed to cover approximately one turn of DNA (rectangle represents protein aligned along the DNA molecule). White rectangles are protamine 1, black are protamine 2. The XANES studies indicate the total mass of protamine bound to DNA in the species is constant, ruling out the possibility that protamine 2 is present in addition to protamine 1.

## 3.1 Analysis of Vacuoles in Human Sperm Chromatin

Qualitative analyses of DNA and protein maps of human sperm nuclei have provided additional information about defects or inhomogeneities in the packing of DNA in human sperm. A structural feature often observed in human sperm chromatin are small voids or vacuoles (Fig. 6). These vacuoles appear to be present at a higher frequency in the sperm of infertile individuals. EM studies have suggested that these regions are simply voids in the chromatin, regions of the nucleus that appear to be empty. These vacuoles are visible in DNA maps of the human sperm nucleus obtained by XANES imaging (Fig. 7), but they appear to be obscured or missing in the protein maps of most nuclei (Fig. 8). This observation has provided the first evidence that the vacuoles are not really empty, and suggests that while they do not contain DNA, they do contain significant amounts of protein.

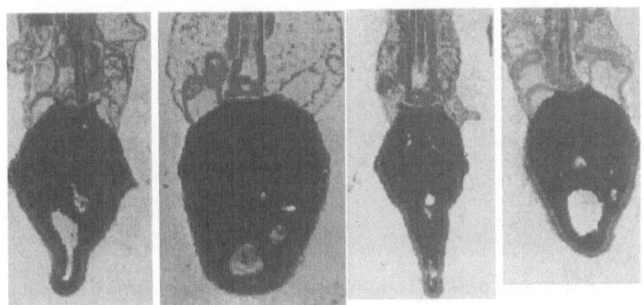

**Fig. 6.** The densely packed chromatin that fills the human sperm head occasionally contains voids or vacuoles, as shown here by electron microscopy.

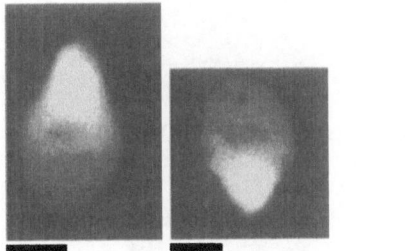

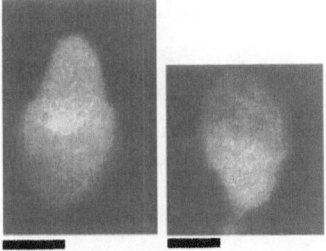

**Fig. 7.** DNA maps of human sperm nuclei obtained by XANES imaging. These two sperm heads have vacuoles (holes) that do not contain DNA.

**Fig. 8**. Protein maps of human sperm nuclei obtained by XANES imaging. The holes that were visible in the DNA maps of these same sperm heads do not appear to be empty, but contain protein.

## 3.2 Biochemical Composition of the Equatorial Segment

In certain species, a structure called the equatorial segment becomes visible when the acrosome is removed from the head (Allen et al., 1995). While the structure and function of this component of the nucleus remain a mystery, it is known to be the first part of the nucleus that comes in contact with the egg upon fertilization (Bedford et al., 1979). The equatorial segment shows up clearly in AFM images of bull sperm heads (Fig. 9) as a triangular belt wrapped around the nucleus. This structure is not visible in DNA maps of sperm chromatin (Fig. 10). But it stands out clearly in protein maps of the sperm heads obtained by XANES imaging (Fig. 11), providing evidence that this structure contains predominantly protein.

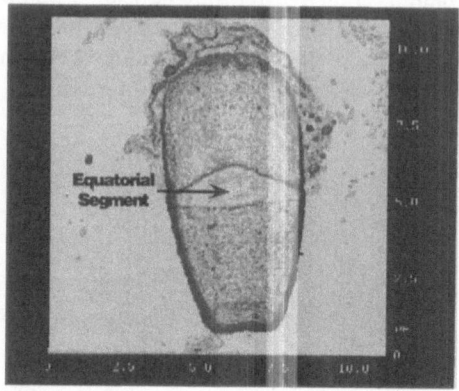

**Fig. 9.** AFM image of a bull sperm head with the acrosome disrupted. The equatorial segment appears as a triangular belt wrapped around the nucleus.

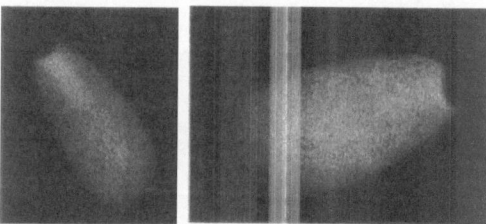

**Fig. 10.** DNA maps of bull sperm heads obtained by XANES imaging. The equatorial segment is not visible in these images.

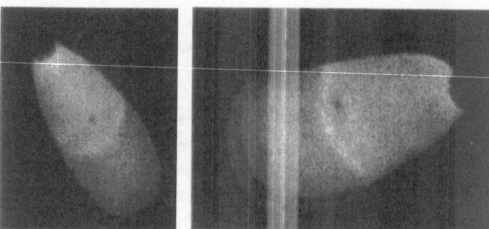

**Fig. 11.** Protein maps of bull sperm heads obtained by XANES imaging. The equatorial segment is clearly visible in these images.

Scanning transmission X-ray microscopy images of the sperm chromatin stained with a maleimide derivative of nanogold (Fig. 12) show the perimeter of the equatorial segment is stained intensely, at a level that is well above the background staining achieved for the rest of the nucleus. This suggests that at least the edges of the equatorial segment are extremely rich in cysteine. Cysteine is the only amino acid present in proteins that reacts with maleimide under the conditions used to prepare the nuclei for analysis.

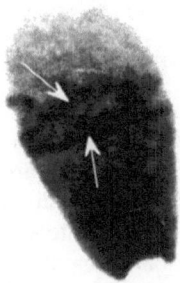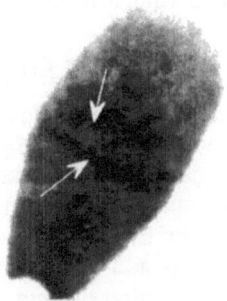

**Fig. 12.** Scanning transmission X-ray microscopy images of two amembraneous bull sperm nuclei stained with a maleimide derivative of nanogold. The edges of the equatorial segment are stained more densely than the rest of the nucleus, indicating that the equatorial segment may contain cysteine rich proteins.

### 3.3 Perinuclear Theca

Both EM and biochemical studies have indicated that a thin layer of proteinaceous material covers the surface of sperm chromatin, lying immediately underneath the plasma membrane (Longo and Cook, 1991; Bellve, 1992; Oko and Maravei, 1994).

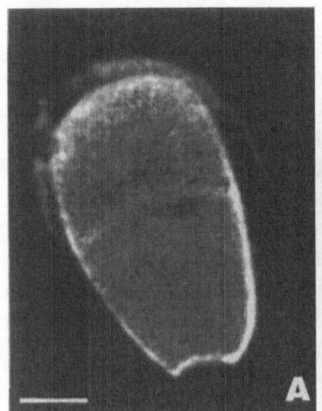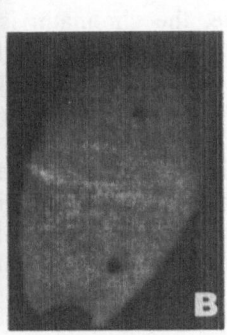

**Fig. 13.** XANES images of bull sperm heads were used to plot the protein to DNA ratio of the head. A. Ratio for an intact head showing the acrosome and a protein rich ring around the head. B. Bull sperm head treated with the detergent mixed alkyltrimethyl ammonium bromide, which removes the acrosome and perinuclear theca, a protein rich layer surrounding the chromatin.

The function of this material, referred to as the perinuclear theca, is not known. But it appears to be present in all mammalian sperm. Treatments of the sperm heads with certain detergents, such as mixed alkyltrimethylammonium bromide (MTAB), in the presence of a reducing agent dissolve the membranes that surround the chromatin as well as the proteins that make up the perinuclear theca (Balhorn et al., 1977). Two dimensional plots of the XANES data obtained for bull sperm, as the ratio of protein to DNA (Fig.13), show the presence of a very protein rich, DNA deficient layer surrounding the chromatin (Zhang et al., 1996). This layer, which is not present in MTAB treated nuclei, is too wide (~200nm) to be the plasma and nuclear membranes and appears to be the perinuclear theca.

# 4 Uniformity of Chromatin Organization

Because the mammalian sperm nucleus is not much more than a micron thick, the transmission of X-rays through the densely packed DNA-protein complex that makes up sperm chromatin can be used to map the uniformity of chromatin condensation throughout the entire nucleus without having to examine individual thin sections of the head as is required by EM. This allows the investigator to examine large numbers of individual sperm cells and obtain information on chromatin organization in each cell in a relatively short period of time. We have used this capability to examine how alterations in protamine synthesis in transgenic mice affect the uniformity of DNA compaction inside the maturing sperm head. The method has also been used to examine the chromatin of the marsupial mouse Sminthopsis crassicaudata and confirm the existence of two different regions inside the nucleus that appear to differ in the nature of their organization.

### 4.1 Early Expression of the Protamine 1 Gene in Mice

The final stage of DNA compaction in differentiating spermatids occurs when the histones and transition proteins bound to DNA in spermatid chromatin are displaced by the two arginine and cysteine rich protamines, protamine 1 and protamine 2. The synthesis and deposition of these two protamines onto DNA occurs during step 11 in mouse spermatids, approximately the same time the nucleus begins to develop its characteristic hook-like shape (Balhorn et al., 1984). Protamine deposition onto DNA and chromatin compaction proceeds in a specific, highly ordered fashion, being initiated at the apical end of the sperm nucleus and progressing inward and toward the implantation fossa. Once complete, the chromatin is so tightly packed that the individual DNA molecules are separated by only 5-7Å (Hud et al., 1994).

As part of a study designed to examine the DNA sequence domains that control the timing of expression of the protamine 1 gene, several lines of transgenic mice were produced by Lee et al (1995) that express the protamine 1 gene beginning in step 7 spermatids, several days earlier than normal. Sperm produced by two of these lines were examined by X-ray microscopy using the XM-1 microscope at the Advanced Light Source, Lawrence Berkeley to determine if early expression of the protamine 1 gene disrupts the process and the uniformity of chromatin compaction that normally occurs in the mouse sperm nucleus (Fig. 14 and 15).

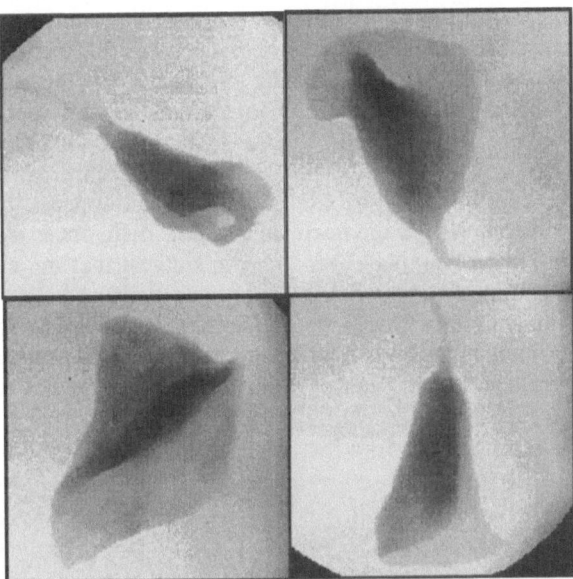

**Fig. 14.** X-ray microscopy images of sperm produced by a line of transgenic mice (Line 6) that express the protamine 1 gene beginning in step 7 spermatids, several days earlier than normal. These images show the chromatin is not uniformly dense and the edges of the nuclei appear to be folded, taking on the appearance of a flower.

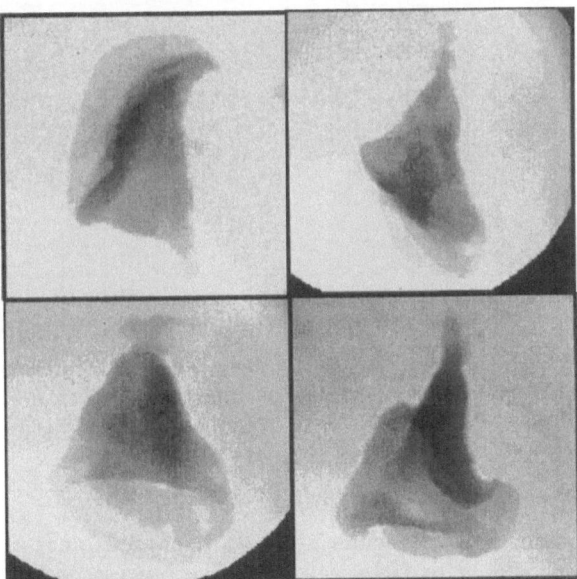

**Fig. 15.** X-ray microscopy images of sperm produced by a second line of transgenic mice (Line 13) that also express the protamine 1 gene beginning in step 7 spermatids. The chromatin is not uniformly dense in these nuclei either, and the edges of these nuclei also appear to be folded.

While heterozygotes from both of these transgenic lines appear to be fertile, X-ray microscopy of the sperm heads show that the majority of the sperm produced by these animals (Lines 6 and 13) exhibit dramatic differences in the uniformity of DNA compaction and sperm head morphology. In contrast to the sperm heads obtained from control mice (Fig. 16), the head shapes of sperm produced by both Line 6 and Line 13 animals are grossly distorted. In most cases, the chromatin in the transgenic lines appear to be convoluted and folded, and the head appears to be shaped like a flower (Fig. 14 and 15). Within the nucleus, dramatic differences are also observed in the density of chromatin packing. These results suggest that the early synthesis and deposition of protamine 1 onto DNA has a dramatic effect both the shaping of the sperm head and the pattern of chromatin condensation. Other than the general flower-like shape of the nucleus and apparent folding observed in thinner regions of most nuclei, the specific shape and pattern of condensation appeared to be different for each nucleus.

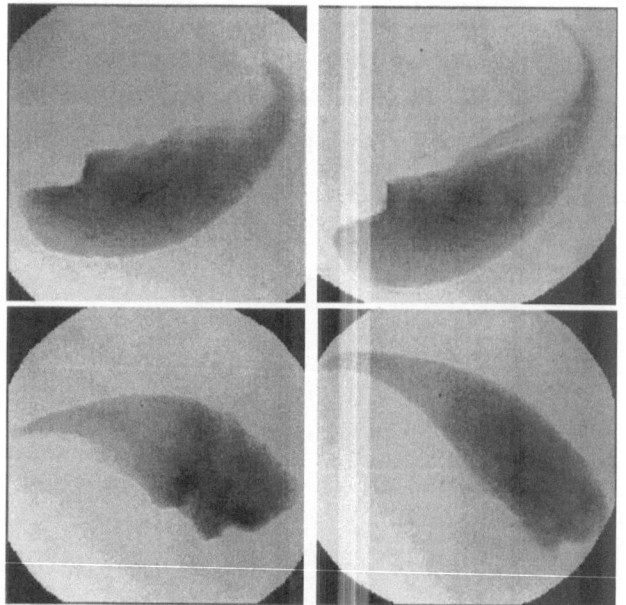

**Fig. 16.** X-ray microscopy images of sperm produced control mice that express the protamine 1 gene at the proper time (step 11).

## 4.2 Co-expression of Mouse and Chicken Protamine Genes

In an effort to examine how the synthesis of abnormal protamines in transgenic mice might compete for binding to DNA and affect the process of DNA condensation, sperm maturation, and male fertility, Rhim et al. (1995) generated several lines of transgenic mice expressing the chicken protamine gene in addition to the normal mouse protamine 1 and protamine 2 genes. The chicken protamine molecule was chosen because it is nearly twice as large as mouse protamine 1 and because it does not contain any cysteine residues, the amino acids that form the disulfide bonds that crosslink neighboring protamine molecules together during the final stages of sperm maturation in mammals. Biochemical and immunological studies of the sperm produced by transgenic animals expressing the chicken protamine gene revealed that the chicken protamine was incorporated into sperm chromatin. Staining of sperm nuclei with antibodies to the chicken protein suggested that every sperm cell contains a detectable amount of chicken protamine. Preliminary EM studies also indicated that the chromatin in a number of mature sperm contained regions of the chromatin that are less condensed than normal. This suggested that the DNA complexed with the chicken protamine was not packed as tightly as the DNA packaged by the mouse protamines. Based on the combination of immunological and EM data, the investigators concluded that all sperm produced by the transgenic males contained regions of chromatin that are not properly packaged. And yet at least a subpopulation of the sperm produced by these transgenic males were fully functional (the males were fertile).

The extent of chromatin condensation observed in certain regions of the sperm head by EM was significantly less than normal, and our previous studies with mouse

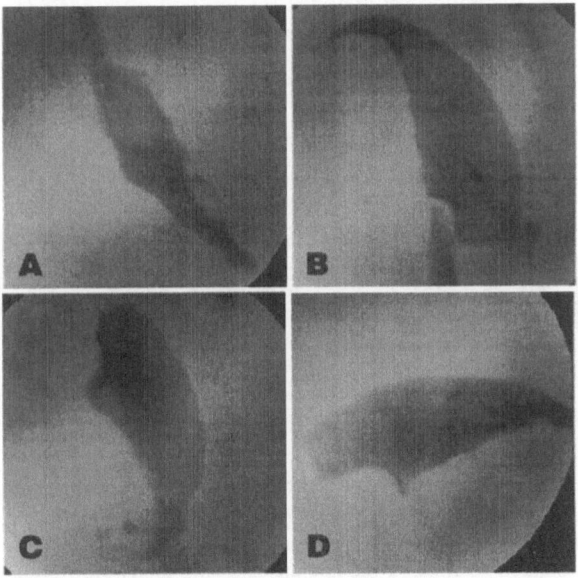

**Fig. 17.** X-ray images of sperm produced by transgenic mice expressing both the mouse and chicken protamine genes. A small percentage of the sperm heads contain less densely compacted regions of chromatin as shown in A. Electron microscope images of similar sperm show these regions to have less densely packed chromatin similar to that found in chicken sperm. B-D. The majority of the sperm appear to exhibit normal patterns of chromatin condensation.

sperm indicated that sperm containing these regions could be easily detected by X-ray microscopy without having to resort to sectioning and analyzing multiple sections through each nucleus. Using the transmission X-ray microscope at the Advanced Light Source, Lawrence Berkeley Laboratory, we could also examine relatively large numbers of intact fully hydrated sperm to determine what percentage of sperm in the population actually contained these "pockets" of less condensed chromatin. Although the analysis of sperm from these transgenic animals is not yet complete, the preliminary results suggest that the majority of the sperm produced by transgenic males expressing the chicken protamine gene contain normally condensed chromatin. Only a very small percentage of the sperm (Fig. 17A) appear to contain pockets of lesser condensed chromatin similar to those observed by EM.

### 4.3 Heterogeneity of Chromatin Organization inside the Nucleus of Marsupial Mouse Sperm

The normal fertile sperm produced by most mammalian species contain chromatin that is, for the most part, uniformly condensed throughout the nucleus. Alterations in this uniformity usually signal that the process of DNA repackaging that occurs during spermatid maturation is defective. Even in human sperm, where as much as 10-15% of the DNA remains packaged by histones (Gatewood et al., 1987), EM analyses of sections through the nucleus of a normal sperm cell show it to be uniform in compaction.

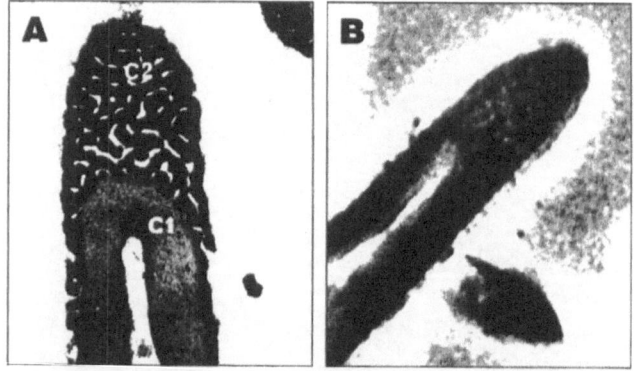

**Fig. 18.** Sperm chromatin organization in heads of the marsupial mouse *Sminthopsis crassicaudata*. A. Transmission electron microscopy (TEM) image of stained sections of the nucleus showing the two different types of chromatin organization in regions C1 and C2. B. Transmission X-ray microscopy images suggest that the unusual chord-like organization in region C1 may be real, and not an artifact caused by tissue dehydration and imbedding for TEM.

Transmission EM studies performed by Breed et al. (1994) have suggested that the marsupial mouse Sminthopsis crassicaudata may be an exception to the rule. Sections of the nucleus stained with uranyl nitrate and lead citrate revealed what appeared to be two different types of chromatin. One region located at the apical end of the nucleus under the acrosome contains chromatin that appears to be more electron dense that the remainder of the chromatin, which is more granular and appears less condensed. While both regions have been shown to contain DNA by their staining with fluorescent DNA binding dyes (Soon and Breed, 1996), it has not been possible to confirm that the two regions of chromatin are condensed to different degrees by EM. These regions can only be observed after staining. Consequently, the observed differences might be attributed to biochemical differences in the chromatin that affect their intensity of staining with uranyl nitrate and lead citrate.

In an effort to attempt to confirm the existence of two distinct regions of chromatin in Sminthopsis sperm that differ in their condensation state, transmission X-ray microscopy images were obtained of air-dried sperm using XM-1 at the Advanced Light Source. While the results are very preliminary, and only a few nuclei have been examined to date, the images (Fig. 18) do suggest that two different types of chromatin are actually present in the sperm of these mice. Additional experiments will be conducted to confirm this result by imaging cells in fluid and at the oxygen edge. If the crevices in the apical chromatin actually exist, the additional water present should help make the crevices stand out when imaged at the oxygen edge.

## 5 Future Applications of X-Ray Microscopy to Sperm

Because some of the results we have presented are preliminary, we must focus our initial efforts on completing the studies we have just described. However, the intriguing successes we have had in combining x-ray microscopy with XANES and applying it to the analysis of individual sperm cells indicates this technique may prove to be extremely useful for examining the content and distribution of particular proteins within the sperm nucleus. Consequently, we hope to focus a significant portion of our future efforts on the analysis of a variety of proteins in mammalian sperm, including the distribution of protamine 2 precursors in mouse spermatids, the localization of histones in human and marsupial mouse sperm, and the co-localization of chicken protamine and the lesser condensed chromatin domain in the sperm of transgenic mice. Our ultimate goal will be to apply the various X-ray microscopy methods we have described to the analysis of sperm from infertile men. The purpose of these studies will be to determine if these males produce a population of normal sperm that can be identified and distinguished from defective sperm based on 2D hydration maps, histone and protamine 2 precursor contents and distributions, and the extent of vacuolization. Such studies will help us identify the physical or biochemical causes for certain types of male infertility, as well as provide new information that can be used to help clinicians select normal, fully functional sperm cells produced by infertile individuals for use in vitro fertilization or related techniques of fertility intervention.

## 6 Conclusions

Perhaps one of the most obvious conclusions to be drawn from this work is that the sperm cell appears to be an ideal target for study by X-ray microscopy. Because its DNA is so densely packed inside the nucleus, most other techniques cannot obtain information about its organization without resorting to sectioning or decondensing the sperm head prior to analysis. The ability to obtain structural or compositional information on large numbers of individual cells also makes it possible to examine variation within the population and obtain reasonable statistical data. The availability of fluid cells for obtaining images in water and the development of cryo-techniques will allow us to investigate structure in its fully hydrated state, eventually under conditions that minimize X-ray damage.

The examples we have described clearly demonstrate that X-ray microscopy can be used to obtain new information about biological structure without having to push the resolution beyond its current limit. In certain cases, data obtained by X-ray microscopy may need to be combined with data obtained by other techniques to provide the results we need. It is also clear that the biological systems we study should be picked carefully so they can actually provide new structural information. In the early stages of X-ray microscope development, which we have experienced in the last few years, this has not been as important. Groups needed objects to study that had been well characterized by light and electron microscopy so they could compare the quality and resolution of their X-ray images with the state-of-the-art offered by other methods. But if future projects using X-ray imaging or analysis are to be funded, we must direct our energies toward studies that put more emphasis on the attainment of new information about biological structure, not only microscope development. It is not necessary, however, that we identify and pose questions that only X-ray microscopy can answer. In biology, as in the other sciences, it is critical that any important finding be confirmed using more than one technique. Two of the studies we have just described are good examples. The results we've obtained on sperm chromatin hydration, as determined initially by X-ray microscopy, were later confirmed using a very different approach and technique, atomic force microscopy. Our determination of the protamine and DNA contents of sperm chromatin from different species, and the observation that the mass ratio of protamine to DNA is constant irrespective of the cell's protamine 2 content, was achieved both by XANES and particle induced X-ray emission spectroscopy (PIXE). In this latter case, the two techniques provided corroborating as well as complementary information; XANES identified the total protein content of the nucleus, while PIXE provided information about the protamine 1 and protamine 2 content of sperm chromatin (Bench et al., 1996).

## Acknowledgments

We thank all the unnamed individuals that helped make these studies possible, either by providing materials for analysis, assistance in sample preparation, or various other types of support. This work was supported by the United States Department of Energy, Office of Basic Energy Sciences and the Office of Health and Environmental Research under contracts W-7405-ENG-48, FG02-89ER60858, and DE-AC 03-76SF00098 and the National Science Foundation grant BIR-9316594.

# References

1   Allen, M.J., Bradbury, E.M., Balhorn, R. The Natural Subcellular Surface Structure of the Bovine Sperm Cell. J. Struct. Biol. 114 (1995), 197-208.

2   Allen, M.J., Lee, J.D. IV, Lee, C., Balhorn, R. Extent of Sperm Chromatin Hydration Determined by Atomic Force Microscopy. Mol. Reprod. Develop 45 (1996), 87-92.

3   Bach, O., Glander, H.-J., Sholz, G., Schwarz, J. Electrophoretic Patterns of Spermatozoal Nucleoproteins (NP) in Fertile Men and Infertility Patients and Comparison with Somatic Cells. Andrologia 22 (1990), 217-224.

4   Balhorn, R., Gledhill, B.L., Wyrobek, A.J. Mouse Sperm Chromatin Proteins: Quantitative Isolation and Partial Characterization. Biochem. 16 (1977), 4074-4080.

5   Balhorn, R., Weston, S., Thomas, C., Wyrobek, A.J. DNA Packaging in Mouse Spermatids. Synthesis of Protamine Variants and Four Transition Proteins. Exp. Cell Res. 150 (1984), 298-308.

6   Balhorn, R., Reed, S., Tanphaichitr, N. Aberrant Protamine 2 Ratios in Sperm of Infertile Human Males. Experientia 44 (1988), 52-55.

7   Balhorn, R. Mammalian Protamines: Structure and Molecular Interactions. In: Molecular Biology of Chromosome Function, K.W. Adolph, ed. Springer-Verlag, New York. 1989. p366-395.

8   Balhorn, R., Lee IV, J.D., Allen, M.J. Atomic Force Microscopy of Human Sperm Chromatin. Mol. Biol. Cell 4 (Suppl) (1993): 401A.

9   Bedford, J.M., Moore, H.D.M., Franklin, L.E. Significance of the Equatorial Segment of the Acrosome of the Spermatozoon in Eutherian Mammals. Exp. Cell Res. 119 (1979), 119-126.

10  Bellve, A.R., Chandrika, R., Martinova, Y.S., Barth, A.H. The Perinuclear Matrix as a Structural Element of the Mouse Sperm Nucleus. Biol. Reprod. 47 (1992), 451-465.

11  Belokopytova, I.A., Kostyleva, E.I., Tomilin, A.N., Vorobev, V.I. Human Male Infertility May be due to a Decrease of the Protamine 2 Content in Sperm Chromatin. Mol. Reprod. Develop. 34 (1993), 53-57.

12  Bench, G.S., Friz, A.M., Corzett, M.H., Morse, D.H., Balhorn, R. DNA and Total Protamine Masses in Individual Sperm from Fertile Mammalian Sperm. Cytometry 23 (1996), 263-271.

13  Breed, W.G., Leigh, C.M., Washington, J.M., Soon, L.L. Unusual Nuclear Structure of the Spermatozoon in a Marsupial, Sminthopsis crassicaudata. Mol. Reprod. Develop. 37 (1994), 78-86.

14  DaSilva, L.B., Trebes, J.E., Balhorn, R., Mrowka, S., Anderson, E., Attwood, D.T., Barbee Jr., T.W., Brase, J., Corzett, M., Gray, J., Koch, J.A., Lee, C., Kern, D., London, R.A., MacGowan, B.J., Matthews, D.L., Stone, G. X-ray Laser Microscopy of Rat Sperm Nuclei. Science 258 (1992), 269-271.

15  deYebra, L., Ballesca, J.L., Vanrell, J.A., Bassas, L., Oliva, R. Complete Selective Absence of Protamine P2 in Humans. J. Biol. Chem. 268 (1993), 10553-10557.

16  Dooher, G.B., Bennett, D. Fine Structural Observations on the Development of the Sperm Head in the Mouse. Am. J. Anat. 136 (1973), 339-361.

17  Gatewood, J.M., Cook, G.R., Balhorn, R., Bradbury, E.M., Schmid, C.W. Sequence-Specific Packing of DNA in Human Sperm Chromatin. Science 236 (1987), 962-964.

18  Hud, N.V., Allen, M.J., Downing, K.H., Lee, J., Balhorn, R. Identification of the Elemental Packing Unit of DNA in Mammalian Sperm Cells by Atomic Force Microscopy. Biochem. Biophy. Res. Commun. 193 (1993), 1347-1354.

19  Hud, N.V., Milanovich, F.P., Balhorn, R. Evidence of a Novel Secondary Structure in DNA-Bound Protamine is Revealed by Raman Spectroscopy. Biochemistry 33 (1994), 7528-7535.

20  Hud, N.V., Downing, K.H., Balhorn, R. A Constant Radius of Curvature Model for DNA in Toroidal Condensates. Proc. Natl. Acad. Sci. 92 (1995), 3581-3585.

21  Kirz, J., Ade, H., Anderson, H., Buckley, C., Chapman, H., Howells, M., Jacobsen, C., Ko, C.-H., Lindaas, S., Sayre, D., Williams, S., Wirick, S., Zhang, X. New Results in Soft X-ray Microscopy. Nuc. Instrum. Methods Physics Res. B87 (1994) 92-97.

22  Lee, K., Haugen, H.S., Clegg, C.H., Braun, R.E.: Premature translation of protamine 1 mRNA causes precocious nuclear condensation and arrests spermatid differentiation in mice. Proc. Natl. Acad. Sci. USA 92 (1995), 12451-12455.

23  Longo, F.J., Cook, S. Formation of the Perinuclear Theca in Spermatozoa of Diverse Mammalian Species: Relationship of the Manchette and Multiple Band Polypeptides. Mol. Reprod. Develop. 28 (1991), 380-393.

24  Oko, R., Maravei, D. Protein Composition of the Perinuclear Theca of Bull Spermatozoa. Biol. Reprod. 50 (1994), 1000-1014.

25  Pogany, G.C., Corzett, M., Weston, S., Balhorn, R. DNA and Protein Content of Mouse Sperm. Exp. Cell Res. 136 (1981), 127-136.

26  Prieto, M.C., Maki, A.H, Balhorn, R. Analysis of DNA-Protamine Interactions by Optical Detection of Magnetic Resonance. Biochemistry (1997), in press.

27  Rhim, J.A., Connor, W., Dixon, G.H., Harendza, C.J., Evenson, D.P., Palmiter, R.D., Brinster, R.L. Expression of an Avian Protamine in Transgenic Mice Disrupts Chromatin Structure in Spermatozoa. Biol. Reprod. 52 (1995), 20-32.

28  Roosen-Runge, E.C. The Process of Spermatogenesis in Mammals. Biol. Rev. Camb. Philos. Soc. 37 (1962), 343-377.

29  Soon, L.L.L., Breed, W.G. Ultrastructure of Nuclear Condensation and Localization of DNA and Proteins in Spermatozoan of a Dasyurid Marsupial, Sminthopsis crassicaudata. Mol. Reprod. Develop. 43 (1996), 217-227.

30  Zhang, X., Balhorn, R., Mazrimas, J., Kirz, J. Mapping and Measuring DNA to Protein Ratios in Mammalian Sperm Head by XANES Imaging. J. Struct. Biol. 116 (1996), 335-344.

# Mapping the Organic and Inorganic Components of Bone

C. J. Buckley[1], N. Khaleque[1], S. J. Bellamy[1], M. Robins[2], X. Zhang[3]

[1]Department of Physics, King's College London, Strand, London WC2R 2LS, UK
[2]Department of Physiology, King's College London, Strand,
London WC2R 2LS, UK
[3]Department of Physics, State University of New York,
Stony Brook, NY 11794, USA

**Abstract.** A mapping technique which uses a scanning transmission soft X-ray microscope (STXM) is described. The technique has been developed and used to quantitatively map the calcium mineral and protein mass thicknesses in undemineralised, unstained, thin bone sections. Near complete femoral-neck sections of sibling normal and ovariectomised mice have been mapped. The results show the quantitative relationship between calcium and protein on the macro and microscopic scales for both tissues.

## 1 Introduction

A number of methods exists for measuring and mapping bone mass [1]. On the macroscopic and cellular scale, electron probe microanalysis (EPMA) [2], electron energy loss spectroscopy (EELS) [3], light [4] and infrared [5] microscopies are available. While these techniques have a number of positive attributes, they are not well suited to the quantitative mapping of mineralised tissue due to either very long pixel times, or specimen damage. EPMA is primarily sensitive to elemental composition, whereas EELS and IR imaging offer chemical state contrast. In the case of EELS, its use is generally limited to inorganic atoms and molecules as significant specimen damage usually results with organic specimens. IR micro-analysis does not produce significant specimen damage, but it is impractically slow in imaging mode, and has poor spatial resolution.

With the advent of bright X-ray sources and improved optics, microscopy with energy-tunable X-ray probes [6] has opened up some remarkable mapping possibilities on mineralised tissues [7, 8]. Low energy, mono-energetic X-rays can be used in transmission mode to quantitatively map both the distribution of the organic and inorganic molecules. This is achieved by utilising the absorption differences obtained via spectral features at the carbon and calcium K and L edges respectively. In the study of osteoporotic bone, it is the distribution and relative amounts of calcium based mineral and collagen (protein) which are of interest, as collagen forms the template for mineralisation. In the study reported here, we have quantitatively mapped and compared the calcium and protein content of femoral neck section in normal and ovariectomised mice. The purpose of ovariectomisation was to produce a mouse with low oestrogen levels and cause an early onset of osteoporosis.

**X-Ray Microscopy and Spectromicroscopy**
Eds.: J. Thieme, G. Schmahl, D. Rudolph, E. Umbach
© Springer-Verlag Berlin Heidelberg 1998

## 2 Materials and Methods

One of a pair of female sibling mice was ovariectomised at 100 days old to suppress the production of oestrogen, and the pair were sacrificed at 209 days. The mouse femurs were removed and fixed in formal saline for ten days and dehydrated by graded ethanol solutions at room temperature. These were then impregnated with LR White methacrylate resin for a period of four weeks at 4° C, after which they were cured at 60°C for 18 hours. The blocks were trimmed to the femoral neck surface, and cut to the centre of the neck. Sections were cut using a microtome equipped with a diamond knife set at an angle of 4° to a thickness of 200 to 400 nm. The whole-neck sections were transferred to a silicon nitride support membrane and imaged in the STXM.

## 3 Scanning Transmission X-Ray Microscopy

The STXM at Brookhaven National Laboratory [9] was used for the work reported here. The undulator source and spherical-grating monochromator provided mono-energetic photons for the microscope. A zone plate X-ray lens demagnified the mono-energetic source forming an X-ray probe of 50 nm in diameter. The specimen was scanned, and the transmitted intensity was detected by a gas-flow X-ray counter. The spatial resolution was controlled by the step size of the sample stages, and the defocus of the X-ray beam. The pixel times for the images shown here varied from 5 ms to 40 ms. The transmission ($\tau$) of the mono-energetic X-rays by a given point in a sample composed of n elements, having a thickness t, is given by equation (1), where $\mu_i$ is the mass absorption constant of the ith element, and $\rho_i$ its density. Note that $\rho_i t$ is the mass of the element per unit area sampled by the X-ray probe and is referred to as the mass thickness.

$$\tau = \exp(-\sum_{i=1}^{n} \mu_i \rho_i t) \tag{1}$$

## 4 Calculation of the Calcium Maps

The creation of quantitative calcium maps is based on the technique of absorption differences between images formed using a number of X-ray energies close to the calcium L absorption edge. An absorption spectrum is shown in Fig. 1. In its simplest form, the technique uses two images either side of the calcium L absorption edge, and the mass thickness of calcium is found by subtracting the optical densities of these two images and dividing through by the difference in absorption coefficients, i.e.

$$[\rho t]_{Ca} \approx \frac{\ln(\tau_1) - \ln(\tau_2)}{\mu_2 - \mu_1} \tag{2}$$

where the subscripts refer to the two energies used.

A more accurate measurement of the calcium mass thickness is obtained if the difference in absorption due to other elements in the specimen at the pre and post calcium L edge energies is taken into account. The absorption changes produced by the calcium-salt mineral elements can be accounted for by calculation, however the changes associated with the organic matrix can only be determined by mapping the carbon mass thickness ($[\rho t]_C$) at the carbon K-edge. The details of how this is achieved has been reported [10], and the mass thickness of the calcium is calculated from

$$[\rho t]_{Ca} = \frac{\ln(I_1 I_{02} / I_2 I_{01}) - (\mu_{C2} - \mu_{C1})[\rho t]_C}{(\mu_{Ca2} - \mu_{Ca1}) + \sum_{j=1}^{m} [\mu_{mj2} - \mu_{mj1}] k_j} \tag{3}$$

where $\mu_{C1}$ and $\mu_{C2}$ are the mass absorption coefficients of carbon at the pre- and post calcium L edge energies, $\mu_{mj2}$ and $\mu_{mj1}$ are those for the elements associated with the mineral at the pre- and post calcium L edge energies, while $k_j$ is the fractional density of the jth element in the mineral with respect to the calcium density.

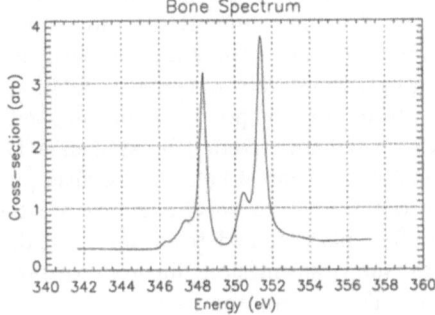
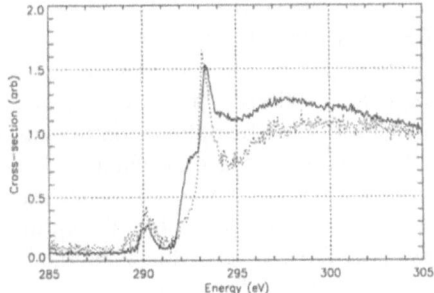

**Fig. 1.** An X-ray absorption spectrum taken through a mineralised area of a normal bone section. The energies used for mapping were 342.5, 350.6 and 359.4 eV.

**Fig. 2.** X-ray absorption spectra taken though L R White embedding medium (solid) and BSA protein (dotted). The energies used for mapping were: 281.8, 292.4, 293.05 and 302.4 eV.

The aim of the work reported here was to map both calcium and the protein in the mineralised tissue sections. The change in absorption cross-section at the pre and postedge energies of the calcium L edge are similar in magnitude to those for carbon at the carbon K edge. However, carbon has a considerably greater atomic abundance than calcium in the specimens, and the thickness of the specimens needs to be 300nm or less to provide sufficient transmission in order to map the organic components with a good signal to noise ratio. In order to obtain sufficient absorption by calcium on this thickness, the NEXAFS peak at 350.6 eV in figure 1 was used. The method by which quantitative calcium maps using NEXAFS peaks can be obtained is detailed in Buckley [10]. The use of the NEXAFS peak in combination with the analysis of equation (3) results in a high signal to noise quantitative calcium map with an accuracy limit of about 3% imposed by photon statistics. The lower detection limit is close to $1 \times 10^{-7}$ g/cm$^2$ [9] I.e. for a 200nm thick section this is equivalent to 5 mg/cm$^3$.

## 5 Calculation of the Protein Maps

There are two major carbon based components in the sample: the embedding material and the biochemical organic components. By imaging the specimen at energies below and above the carbon K absorption edge, the total carbon mass thickness can be measured . This is similar to the calcium mass thickness measurement described in the previous section. To map and measure the protein mass thickness alone, the fine structures in the carbon absorption spectra were also used. These fine structures near the absorption edge are due to the covalent bonds the carbon makes with adjacent atoms. The energy position of these peaks is influenced by the different chemical environment surrounding these bonds. Different organic components of the specimen can therefore be distinguished using the size and position of these peaks.

The embedding material is primarily bisphenol A dimethyl acrylate and methyl methacrylate, while the organic component in the tissue is mainly collagen. Fortunately, the differences in their chemical structures give rise to the slightly different carbon edge spectra as shown in figure 2. The differences allowed us to separate collagen from the embedding media by imaging at the energies detailed in the spectra captions. If the chemical formula of the compound to be mapped is known, then the absorption coefficients can be obtained from tabulated data [11] which are valid at energies which are not close to the fine structure. The absorption coefficients for the NEXAFS peaks can be obtained from spectra which are then normalised by the cross-edge absorption coefficients obtained from tabulated data. These coefficients can be combined with the transmission images to obtain quantitative maps of the principal components. The mineralised tissue sections were treated as being composed of three components: protein (collagen), embedding medium and mineral. Protein and embedding medium maps were obtained by using images of the same area taken at four different energies.

There are a number of possibilities for determining the mass-thicknesses of the protein and embedding media. These are, direct method (such as for the calcium map), square matrix inversion [12], and singular value decomposition [13] (SVD). There are several energies at which a difference is observed in absorption coefficient between the protein and the embedding medium. To optimise the separation between the protein and embedding medium, more than two energies were used. An advantage of SVD is that it can be used on an over-determined data set where there are more optical density equations than unknown mass thickness values. SVD produces a least squares fit to the mass thickness values produced by the combination of the over-determined data. This method was applied to determine the protein and embedding medium mass thicknesses of the bone sections. The sensitivity of the technique for the protein on the sections used in this study was about $10^{-6}$ g/cm$^2$. which was limited by photon statistics.

## 6 Results

The results shown in Fig. 3 are the raw transmission images required to make both the calcium and the protein maps. The images on the top row of Fig. 3 were formed using the X-ray energies: 281.8, 302.4, 292.4 and 293.05 eV from left to right. These were used to calculate the embedding medium and protein maps shown in Figs. 4a and 4b. The images in the bottom row of Fig. 3  were formed using the X-ray energies 342.5, 359.4 and 350.6 eV and  were used  to make the calcium map of Fig. 4c. The calcium maps can be used to measure the mineral (calcium hydroxy apatite and calcium carbonate) mass thickness by multiplying the map values by a factor of 2.5.

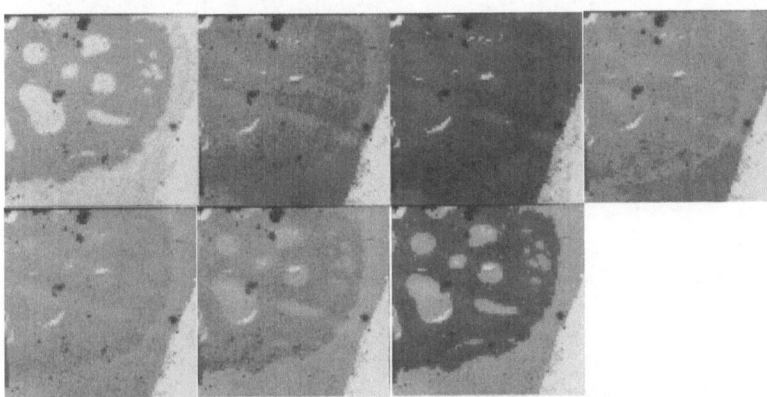

**Fig. 3.** Scanning X-ray transmission images of unstained, undemineralised bone section from the femoral neck of normal mouse sacrificed at 209 days. The transmission  images of the top row (left to right) were formed using the X-ray energies: 281.8, 302.4, 292.4 and 293.05 eV (at the Carbon K absorption edge). While those of the bottom row were made at the X-ray energies: 342.5, 359.4 and 350.6 eV (at the calcium L absorption edge). The set of  even images were used to make quantitative maps of the embedding medium, protein and calcium.

A set of seven images were also taken for each field of view on the ovariectomised mouse sections. The low magnification maps presented in figure 4 are of the majority of the femoral necks. Higher resolution maps were made of  the top right area of the ovariectomised (figure 5). These are the calcium and protein maps together with an embedding medium map which highlights the bone forming cells close to the miner-lising cartilage. The mass thickness scales on figures 4 and 5 can be converted to the concentrations in grams per cubic centimeter by dividing  the mass thickness value by the section thickness. The thicknesses of the sections from the normal mouse and ovariectomised  mice were 205 nm and 375 nm respectively.

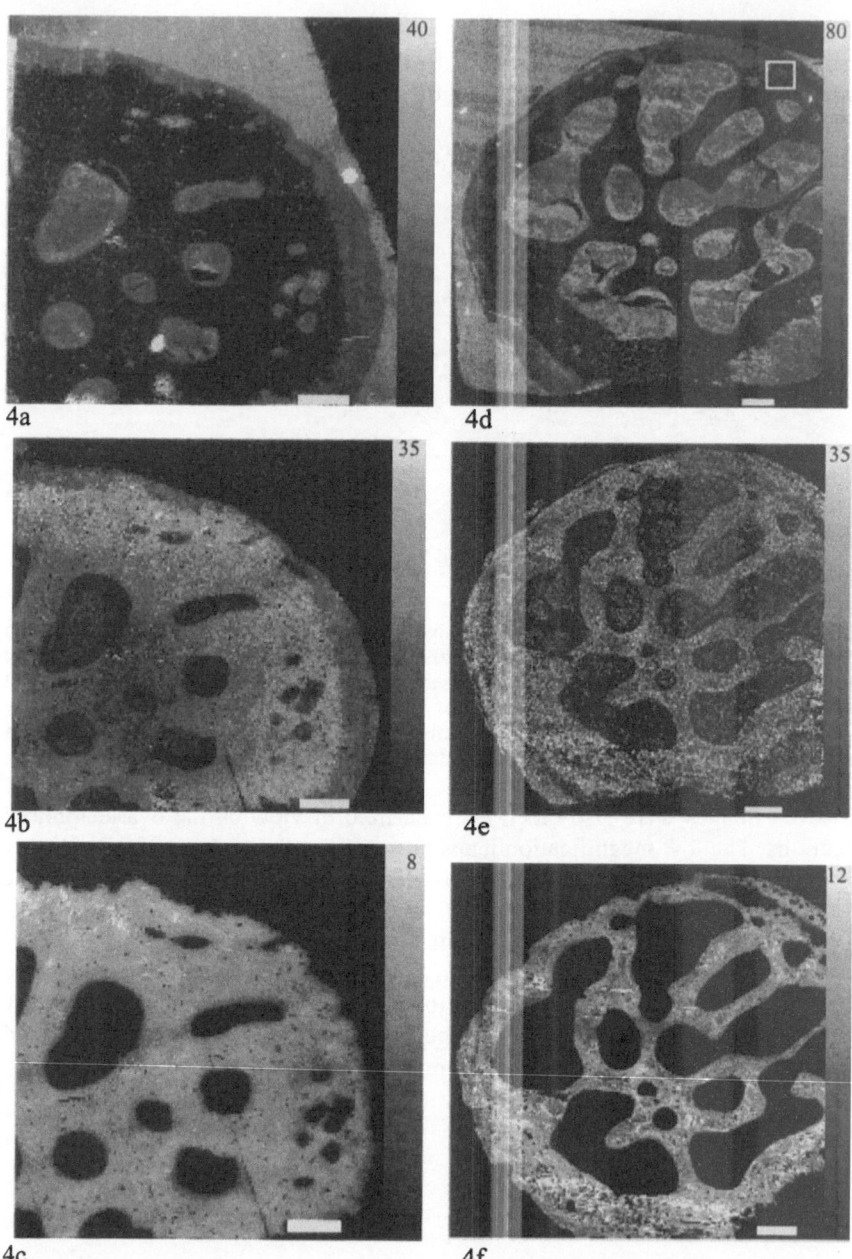

**Fig. 4.** Maps of embedding medium (a & d), protein (b & e) and calcium (c & f) of unstained, undemineralised sections of normal (left) and ovariectomised (right) mouse femoral neck. The mass thickness is indicated by an increasing brightness scales, and the maximum values are indicated in units of g/cm². The scale bars are equivalent to 100 μm.

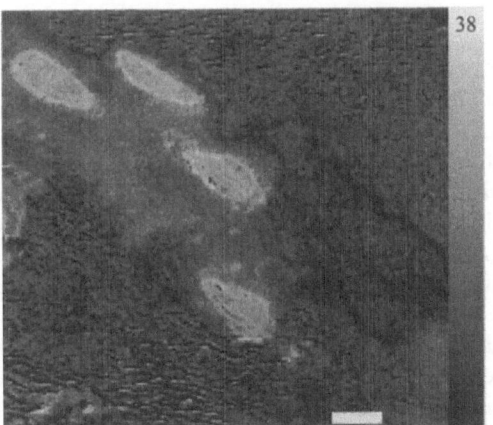

**Fig. 5a.** Embedding medium map of the top right area indicated on figure 4b. This is the area between the cortical bone and the mineralising cartilage. The embedding medium substitutes hydrated volume, and provides a useful means of highlighting cells in relation to the organic and mineralising matrix. The scale bar represents 5 μm and the brightness scales are indicated in units of g/cm$^2$.

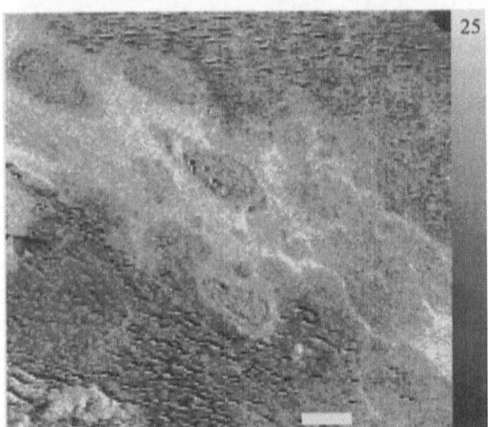

**Fig. 5b.** Protein map. The high concentration of protein surrounding the cells is the collagen matrix exuded by the cells. The collagen forms the template for mineralisation. The scale bar represents 5 μm and the brightness scales are indicated in units of g/cm$^2$.

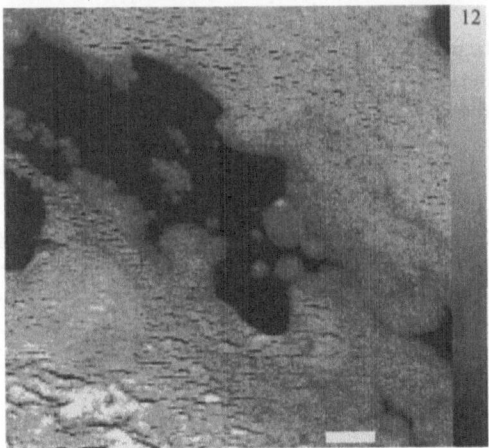

**Fig. 5c.** Calcium map. The more mature collagen matrix is on the upper right and lower left, and shows considerable mineralisation. Initial mineralisation islands can be seen in the vicinity of the cells. The structure is typical of mineralising cartilage. The scale bar represents 5 μm and the brightness scales are indicated in units of g/cm$^2$.

# 7 Discussion

The low resolution images of Figs. 4 show embedding medium, protein and calcium maps of near complete sections of the majority of the femoral neck at equivalent neck positions in the normal (4a, 4b & 4c) and ovariectomised (4d, 4e & 4f) mouse bone sections. Inspection of these maps show the enlarged marrow cavities in the ovariectomised mice. Further, the density of the protein is seen to increase with radius in the normal bone, but not so in the ovariectomised bone. The increase in protein concentration with radius in the normal bone is indicative of the manufacture of new bone, while the relative uniformity of the protein in the ovariectomised bone suggests a less active remodeling caused by the hormone imbalance in the animal.

The ovariectomised sample had a section thickness of nearly double that of the normal mouse sample. This can clearly be seen by inspection of the mass thickness values in the embedding medium periphery of maps 4a and 4d, where 4b shows values which are almost double those of 4a. However, inspection of the protein mass thickness values of figures 4b and 4e show similar maximum values. This means that the density of the protein in the mineralised areas is considerably lower in the ovariectomised sample than in the normal sample. An appreciation of the reduced protein concentration can be gained by comparing the mass thickness values of the protein in the haemopoietic tissue in the cavities with those in the mineralised areas. The protein mass thickness in the mineralised areas is considerably greater than that in the haemopoietic tissue in the normal sample. While the same comparison for the ovarietomised sample shows a smaller difference. Also, the ovariectomised sample shows a higher concentration of embedding medium in the mineralised areas, which indicates a greater micro-porosity in these areas. These findings challenge what has been the prevailing view that the matrix of osteoporotic-like bone always shows a normal composition, and support the findings of others (e.g. Diebold et. al. [14]) who observe significant differences in composition between normal and osteoporotic bone.

To demonstrate the ability of the technique to make quantitative maps in regions of cellular activity, the top right corner of the ovariectomised sample (marked with a rectangle in figure 4d) was mapped. The results are shown in figure 5 where the embedding medium map is shown together with the calcium and protein maps. The embedding medium (figure 5a) primarily fills the hydrated cavities, highlighting the matrix-forming cells. Figure 5b shows the higher concentration of protein (collagen) matrix around the cells while figure 5c shows the calcium distribution in the mineralising matrix. The calcium mass thickness is greater in the more mature matrix in the top right and lower left regions.

This initial study using the technique demonstrates the ability of scanning transmission X-ray microscopy to quantitatively map organic and inorganic constituents of bone over a considerable spatial dynamic range without stain or demineralisation. The analysis presented here has concentrated on calcium, protein and embedding medium. However, it should also be possible to employ the technique to map other components such as calcium carbonate in these specimens.

# 8 Future Work

Bone tissue contains several types of collagen. It has been recognised that some types of collagen are readily mineralised, while others are not. Future work will map and measure the protein and mineral densities on animals which have genetically induced deficiencies of specific types of collagen. The data will be combined with that from other forms of microscopy on the same samples to gain an insight into the mechanisms of mineralisation, its suppression and promotion.

## Acknowledgements

The authors would like thank the Stony Brook X-Ray Group for the use of their microscopy equipment on the X1a beamline of the NSLS at Brookhaven National Lab.
The zone plates used in this work were provided by Erik Anderson of the Centre for X-Ray Optics, and the work was supported in part by a grant from the NSF grant number BIR-9316594 and from funds provided by the department of physics at KCL.

## References

1    Hassager C., Christeiansen C., Calcif. Tissue Int. **57** 1–5 (1995).
2    Kitsugi T., Yamamuro T., Nakamura T., Oka M., Kokubo T., Okunaga K., and Shibuya T., Calcif. Tissue Int. **56(4)**, 331–335 (1995).
3    Jeanguillaume C, Tence M, Zhang L and Ballongue P. Cellular and Molecular Biology, **42(3),** 439–450 (1996).
4    Calder S.J., McCaskie A.W., Belton I.P., Finlay D.B. and Harper W.M., Journal of bone and joint surgery - British volume, **77(4)** 637–63 (1995).
5    Boskey A.I., Pleshko M., Doty S.P. and Mendelsohn R., Cells and Materials **2**, 209–220 (1992).
6    Kirz J., Jacobsen C. and Howells M., Quarterly Reviews of Biophysics, **28(1)**, 33–130 (1992).
7    Buckley C.J., Burge R.E., Foster G.F., Rivers M., Ali S.Y. and Scotchford C.A., Inst. Phys. Conf. Ser. **130**, 621–626 (1992).
8    Buckley C.J., Foster G.F., Burge R.E., Ali S.Y., Scotchford C.A., and Rivers M., Rev. Sci. Instrum. **63**, 588–590 (1992).
9    Jacobsen C., Williams S., Anderson E., Browne M.T., Buckley C.J., Kern D., Kirz J., Rivers M., and Zhang X., Optics Comunications, **86**, 3:0351–36 (1991).
10   Buckley C.J., Rev. Sci. Instrum. **66(2)**, 1318–1321 (1995).
11   Henke B., At. Data Nucl. Data Tables **55** 349 (1993).
12   Cazaux J., Micosc. Microanal. Microstruct. **4,** 513–537, (1993).
13   Zhang X., Balhorn R., Mazrimas J., and Kirz J., Journal of Structural Biology, **116**, 335–344 (1996).
14   Diebold J., Batage B., Stein H., Mulleresch G., Muller P.K. and Lohrs U., Virchows. Archiv A-Pathological Anatomy and Histopathologhy **419(3)**, 209–215 (1991).

# X-Ray Microscopy of Fluid Lipid Membranes

B. Klösgen[1] and P. Guttmann[2]

[1] Physics Dept., Free University Berlin, Arnimallee 14, D-14195 Berlin, Germany
[2] Inst. X-ray Physics, Georg-August-University Göttingen, Geiststr. 11,
D-37073 Göttingen, Germany

**Abstract.** Membranes may be envisaged as two-dimensional liquid crystals. The microscopic molecular arrangement of the involved lipid molecules results in complicated structures in the mesoscopic size. X-ray microscopy is a tool for direct imaging of such appearances. Here we report some new results obtained with the Göttingen X-ray microscope for POPC in pure water. Lipid membranes are observed both with amplitude and phase contrast set-up of the microscope. Cryo experiments complete the observations done under normal and humid conditions. The effect of radiation damage is discussed.

## 1 Introduction

Fluid lipid membranes are the structural basis of biological cell membranes. They form spontaneously by hydrophobic self-organization when amphiphilic molecules like phosphatidylcholines, for example, come into contact with water. Each of these molecules consists of a polar head and two apolar hydrocarbon chains. The respective binary phase diagrams usually exhibit a series of different phases depending on temperature and composition (i.e., the water content), among them lamellar, cubic and hexagonal ones (for a recent review see: [1]). Here we focus on the features of the biologically most relevant lamellar phase that consists of extended bilayers. These membranes themselves can undergo several thermotropic phase transitions of more or less crystalline low temperature states before they end up in the fluid $L_\alpha$-phase. In this lamellar high temperature state, the single lipid molecules are free to diffuse within their sheet of bilayer at rates that are typical for the diffusion in liquids ($\sim 10^{-7}$-$10^{-8}$ cm$^2$/sec$^2$, [2]). Such a membrane may be regarded as a thin plate of a two-dimensional fluid. Herein no shearing forces can be applied, but it can as well be bent and delated with the respective elasticity moduli to describe the contribution of the deformations to the total energy [3].

The lipid molecules cannot arrange to form an infinitely large planar bilayer because this would result in entropically unfavorable open ends at places where the hydrophobic cores of the bilayer were in direct contact with the surrounding aqueous medium. This effect forces the soft fluid bilayer to bend and prefer closed shapes, the so-called lipid vesicles. Their actual shape (as a sphere or a pear, for example) will adjust, in equilibrium, to a minimum in bending energy as given by the bending Hamiltonian $\mathcal{H}$ [4, 5]. The usually low value of the bending elastic modulus of fluid lipid bilayers results in temperature dependent fluctuations of the membranes. These

**X-Ray Microscopy and Spectromicroscopy**
Eds.: J. Thieme, G. Schmahl, D. Rudolph, E. Umbach
© Springer-Verlag Berlin Heidelberg 1998

undulations entropically contribute as a repulsive component to the total interaction potential of adjacent membranes [6].

The complete understanding of the elasticity related effects is thus of principal importance to explain or calculate equilibrium shapes of vesicles [4, 7] and to describe events as shape transformations or pore formation [8, 9] and also the adhesion between membranes [7] up to their fusion.

## 2 Methods

**Samples** were obtained by a variation of our standard method to prepare vesicles for observation with electron microscopy [18, 19]. Our intention here was to prepare vesicle dispersions with object sizes between 0.2 μm and 5μm. Gentle round shaking (100 cycles/min, about 1 h, at 40°C) yield a membrane stock suspension. The resulting vesicle population was very inhomogeneous as judged from optical microscopy. Dense myelin-like tubes coexisted with multilamellar onions and paucilamellar liposomes, either tubular or spherical. The wide variety both of shapes and sizes allowed us to optically select "interesting" objects for a subsequent observation with the X-ray microscope. More homogeneous populations, both as to shapes and appearances, could be obtained by longer shaking times or by applying an additional preparative step ( either extrusion or sonication).

For our purpose to observe the liposome suspensions by the Göttingen transmission X-ray microscope, we applied a small droplet of the suspension onto a commercially available copper grid for electron microscopy that was clamped into the microscope object chamber. By the way we could as well provide defined small liquid cavities, prevent flow within the sample and shield our soft objects from mechanical stress.

**The X-Ray Microscope.** The Göttingen transmission X-ray microscope [20] at BESSY I, Berlin, is a device that uses soft X-rays as available from the electron storage ring for illumination. They are well suitable for our application to a biological sample [21, 22]. In this high frequency spectral range, imaging is no routine procedure and optical components like lenses are not commercially available [23]. The X-ray microscope works at intermediate magnification and resolution. It thus fits into the gap between electron and optical microscopy. Moreover, a special object chamber enables us to observe the sample under normal conditions as to temperature and pressure [24]. We then do not have to bother about artifacts stemming from chemical fixation or freezing.

The X-ray microscope can be driven both in the amplitude [21] and in the phase contrast [25] mode. A more detailed presentation of the technical details is given elsewhere [24]. Very recently, the microscope was equipped with a newly developed version of a cryo-facility which allows operation at cryogenic temperatures [26]. Thus there is no more radiation damage to bother with. It will, however, be seen that one has to account carefully for freezing induced defects, especially with soft material under investigation [19].

Our experiments with such membranes suggested, some years ago, the existence of new properties of some membrane forming lipids.

First, they seemed to produce an additional membrane superstructure of the fluid phase at low tensions. This became evident when experiments on the induced adhesion [10] of adjacent membranes revealed conflicting results for the adhesion energy [11, 12]. The apparent contradiction could not be resolved except if the real membrane surface area was presumed to be much larger than the optically visible projected area. The amount of excess area required to explain the experimental results by far exceeded what could be contributed from the membrane portion (<5%) stored in the membrane undulations. Therefore we postulated the existence of a submicroscopic roughness of membranes [11]. This roughness could as well explain the wide scatter of bending moduli when they were measured in several vesicles and the diverging results for even one vesicle when it was investigated with two different methods [13, 14]. It could also explain stable wiggles in membranes [15] and the slow decomposition of previously ordered lamellar stacks [16].

Second, there seems to be another phase that coexists as a disperse phase with the lamellar membrane phase. This could be concluded from observations on the reversible formation of dark bodies [17]. These fuzzy looking big objects (>10μm) develop from an optically unresolvable source that is contained within the aqueous phase. In order to name this non-micellar component we introduced the term "disperse". The dark bodies are presumably small non-lamellar regions of connected membranes similar to little sponges [1, 9].

Both the formation of membrane superstructures in the fluid phase and of small disperse lipid aggregates are only possible if the membrane is able to bend very sharp and to develop local tips or rims. This cannot be explained by normal bending elastic theory.

The usual description of bending induced effects in membranes bases on the application of a Hamiltonian [3, 5] like

$$\mathcal{H} = \kappa_0 H^2 + \kappa_1 K^2 \tag{1}$$

wherein the two local curvatures $c_1$ and $c_2$ are combined to give either the mean curvature $H = \frac{1}{2}(c_1 + c_2)$ or the Gaussian curvature $K = (c_1 \cdot c_2)$, with their respective bending moduli $\kappa_0$, $\kappa_1$. The mean curvature mainly determines the overall shape of the vesicle (this is the object morphology) and the Gaussian term can drive the formation of holes in surfaces (thus determining the object topology).

This Hamiltonian was successfully applied to explain many of the features that were experimentally observed with lipid vesicles. However, it will yield an energy increase on sharp bends and thus it always works to stabilize smooth membranes. Fine structures as furrows or an arrangement of saddles that we presume to be inherently present on unstressed lipid membranes can only build up and survive if they contribute no or even a negative portion to the total energy. The introduction of higher order bending terms, especially for the Gaussian curvature, seems to offer a solution. However, this requires a reviewed formulation of the elasticity theory of fluid lipid membranes, a project that can only be undertaken on the basis of valid experiments.

Our results from optical microscopy gave first indications that higher order terms might be at work. Direct evidence can only be revealed from techniques of higher resolution. The two main ways of access consist of either scattering or imaging methods. The first yield only information about mean structures. They will be the most promising for crystalline-like structures as, for example, stacks of fluid membranes or lattices of membrane passages under the conditions that the thermal fluctuations are low. We are actually interested in local and singular features that might change in time or that might at least exhibit thermal fluctuations. We thus need high resolution snapshots. Therefore both X-ray and electron microscopy (cryo-TEM and transmission electron microscopy combined with freeze cutting) were combined to investigate our lipid membranes over the whole accessible set of object sizes and resolution ranges. It now seems that the results of the different methods converge and that they confirm the existence at least of the membrane superstructure.

Some of our newer results shall now be presented and discussed in more detail.

## 3 Results

A huge advantage of standard X-ray microscopy as compared to electron microscopy consists in that the samples are investigated under in vitro conditions in their humid surroundings, at normal pressure and at a temperature around 300 K. Image contrast was sufficient to distinguish objects like unilamellar vesicles down to a size of around 300 nm [27].

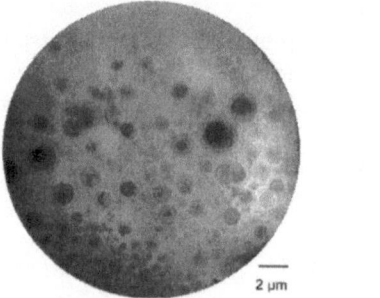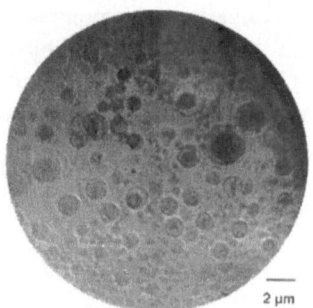

Fig. 1. Two micrographs from different exposures of a population of oligolamellar liposomes of POPC in water. The image on the left side was taken by pure amplitude contrast whereas the right image results from both amplitude (a.c.) and phase contrast (ph.c). Exposure times were 5 sec and 20 sec for a.c. and ph.c., respectively.

Our first experiments were done using the natural amplitude contrast that an organic system exhibits at a X-ray wavelength $\lambda=2.4$ nm, which lies within the so-called water window [21, 28]. The phase contrast setup that is now available widely improves the image contrast for small and light objects as uni- or paucilamellar membranes. The effect of these two principally different contrast mechanisms is demonstrated in Fig. 1. A series of micrographs was taken from a population of oligolamellar liposomes of POPC. Two pictures are shown: the left one was taken using the natural absorption difference of water and organic material for amplitude contrast whereas for the right picture the phase contrast mode was applied. By the way, we have both contrast mechanisms contributing simultaneously to the resulting image. As an improvement, we can now distinguish even small unilamellar particles that were hardly perceptible before. The experimentally determined zone plate resolution limit of about 25 nm by now [29] can thus be fully exploited in that small details became observable with sufficient contrast even in our systems despite their low beam absorbance [28].

We were mostly interested in structural details of single membranes. However, most objects that we found in our samples were oligolamellar as those shown in Fig. 1. This is a consequence of our way of preparation. Multilamellar structures were also abundant in our samples. Their absorption would amount up to 50% for a packed myelin-like structure of 180–200 membranes. As an example for such systems we successively examined the end portion of a multilamellar tube. In the resulting series of pictures taken every 2 min at amplitude contrast and with an exposure time of 5 sec the object under investigation proved to be unstable under the illumination. This is demonstrated in Fig. 2, where the tube end at first is almost black and exhibits no sharp contours even towards the aqueous surroundings. The second micrograph still is very dark around the tube center but the contours are even more fuzzy. The third micrograph of Fig. 2 still reveals the main features with a dark tube center but there are also lots of single membranes perceptible and small vesicles have formed at the edge of the myelin cylinder towards the water.

**Fig. 2.** Pictures from an exposure series taken on a densely packed multilamellar tube. The tube end is shown at the first take (left), at the second one (about 2 min later) and at the tenth (about 20 min after the first shot). Radiation damage leads to an additional swelling of the membranes.

The object in Fig. 2 obviously changes during our observation. The changes started during the first exposure and lasted for some frames. The membranes themselves as the structural basis of the tube seemed to persist. Most remarkable, the image impression became steady as soon as the object density was sufficiently low. We thus have to think about a radiation induced swelling of packed membrane stacks. The underlying reason for this kind of radiation damage [30] is probably due to an Auger mechanism which, as the end of the Auger cascade, first results in an accumulation of positive charges on the membrane surfaces. A strong Coulomb repulsion will at once cause the membranes to drift apart as far as this is possible. The effect is probably strong enough to cause temporary water filled pores to form. Elsehow the initial step of the fast water uptake of the otherwise almost impermeable membranes was impossible. Once the respective distance of adjacent membrane surfaces is wide enough, this process comes to an end. Radiation induced swelling therefore happens only as long as a sample is densely packed.

Subsequent to this initial rapid increase of the intermembrane distance we observed a second process that is much slower. Lipid material seemed to evade from the membranes. This showed up as a slow shrinking of the objects that are observed. An example is given in Fig. 3.

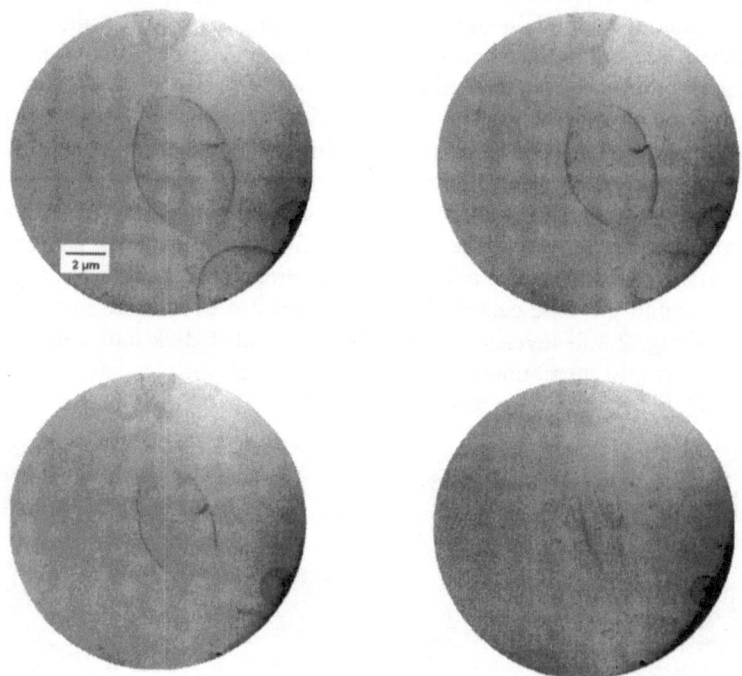

**Fig. 3.** POPC vesicles under numerous illuminations. Three features are to be recognized: first, the central vesicle exhibits a fold that persists all the time. Second, the same vesicle seems to be open at its ends. Third, all vesicles are continuously degraded by radiation damage.

Taking this series of pictures was motivated by the sharp fold that we observed on the surface of the central vesicle. We wanted to check its stability and decided to take many pictures. As one important feature, this linear corrugation persisted all the time. We thus took it as another evidence of the non-smooth structures that unstressed fluid membranes can form [11, 12, 19, 31]. But we also noticed, even from the first image we took, that there was only a faint contour along the upper and lower part of the vesicle. The object seemed to be open at its ends. We therefore expected it to vanish before we could take another image. But it mainly kept its shape over many exposures and only slowly lost its interior water volume. We have at least to conclude that at these sites the local structure is changed. From the soft contrast transition between the membrane structure and the surrounding water we could be tempted to think that there must be some intermediate structural link for the lipid. During all exposures we took the fold never vanished nor did we see the membrane to close its open ends. Instead, we observed a steady degradation of all the objects that were initially present in the picture.

The excited atoms from the initial Auger process will also cause the formation of electron vacancies in the uppermost states of the valence band thus finally inducing the breaking of chemical bonds. The bond scission produces molecular fragments with a physical behavior different from the original lipids [30, 31]. Some of the fragments may tend to destabilize the membrane structure, some may dissolve as micelles or even as monomers into the aqueous volume [1, 17]. The molecular rearrangements caused by C-C or C-O bond scission of the organic molecules and the diffusion of fragments both within the membrane and into the water takes time. This explains the observation of the slow degradation of the membrane structures that we observed in all our samples independent of their initial material density when we exposed them in several illuminations (even of rather low dose).

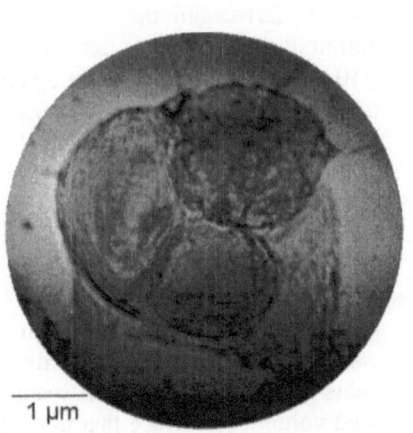
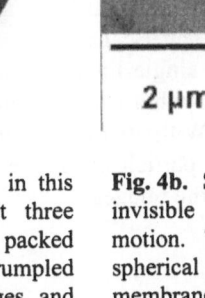
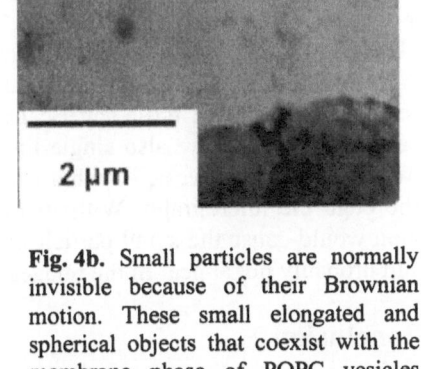

1 µm

2 µm

**Fig. 4a.** The complex body shown in this X-ray micrograph shows at least three different appearances: membranes packed in shells like an onion, crumpled membranes or a lattice of passages and small branches extending into the water.

**Fig. 4b.** Small particles are normally invisible because of their Brownian motion. These small elongated and spherical objects that coexist with the membrane phase of POPC vesicles were present in snapshots obtained by cryo-microscopy.

These results motivated us to reconsider the advantages of working under normal conditions. An alternative experimental possibility is now available by the newly developed cryo-set-up for the X-ray microscope [32]. Freezing the samples would largely reduce the effects of radiation damage. On the other hand, the formation of ice crystals has to be avoided. Cyrstalline ice might destroy or at least modify the local structure especially of soft material as lipid membranes [33]. Even vitrifying the sample may induce mechanical distortions [19]. Two of the first pictures that we took by combining X-ray microscopy with a cryo-technique for fixing the samples are shown in Fig. 4. During the whole time, liquid nitrogen was used to keep the sample at a temperature that was sufficiently low to prevent the recrystallization of ice within the sample.

The features that are distinguishable in Fig. 4a were partially well-know from other experiments. There is a group of vesicles that is tied to each other. Filaments are visible that extend into the outer region. As the ends cannot be seen, they may as well resemble tethers or be part of the disperse phase (Fig. 4b). On the left side, several membranes are packed like the shells of an onion. The intermembrane distance is as wide that single shells can be seen. The center of the complex body contains two discernible objects. Both contain a lot of lipid as can be concluded from the increased gray level. The upper one looks somehow structured. It is not clear whether we have the aspect of a very crumpled surface or a lattice of passages. In the latter case the object might resemble a sponge with a complex membrane surface dividing the aqueous interior into two compartments. The lower object looks smoother with the exception again of some folds. On the low right side, we have again the impression of some membranes with a rough surface. The resolution of the pictures is not sufficient to distinguish the small details in the range of 5-10nm that we observed by electron micrsocopy [19, 31] but the vesicle appearance is definitely different from the aspect of smooth membrane surfaces that we frequently happened to observe, too. Up to now, we could never before take a snapshot of crumpled membrane surfaces with X-ray microscopy at normal temperature and the exposure times of some seconds. This is reasonable because of the fluctuations [34] that tend to smear off fine details into a homogeneous gray level.

The micrograph of Fig. 4b gives another example for the advantage that one has with cryo-X-ray microscopy. Small particles are visible that are contained in the water volume. Some of these spherical particles are connected by extended thin and linear threads. There are also single linear filaments present. One of them seems to grow out from a huge vesicle with a very structured volume or surface that is visible at the edge the micrograph. Without being embedded in vitrified water, Brownian motion would cause the small particles to move all the time. As a consequence, they would probably not appear in the images.

## 4 Conclusion

With cryo-TEM, we observed different types of the superstructure like a membrane covering graininess as one feature and sharp folds or bends as another one [19, 31]. The latter seem to correspond to dark lines that we also observed with X-ray microscopy in large vesicles of some microns of size [28]. We found it both with frozen and unfrozen samples. Thus we can definitely exclude that it can be an artifact stemming from squeezing the sample during the cryo-preparation.

The irregularly packed graininess at low densities that we observed with electron microscopy presumably consists of a saddle structure. It is then distinguishable because the surface portions of increased slopes (where the membranes run along the beam direction) exhibit increased contrast. It seems, as a loosely packed variety, to align in short lines of highs and lows. At sufficient distance, these lines of increased contrast fall into the resolution of X-ray microscopy and become observable. They could, however, previously not be made out because of the fluctuations of the soft material. This used to average off all fine contrast differences during the relatively long exposure times. Now that a temporary state could be fixed by an adapted cryo-preparation [26], we got indications for them also by use of the recently developed X-ray cryo-microscopy.

This method also revealed small particles in the aqueous space. They may be attributed to the disperse phase that we are looking for. First evidence from freeze fracture experiments [35] suggests that the disperse phase might consist of small elongated particles with some knots where they are linked together. This agrees with the few pictures that we got by cryo-X-ray microscopy, too.

In general, the combination of supplementing methods and the comparison of the respective results is of special importance as we have to consider the different preparation induced artifacts. Scale invariance is another problem. The elasticity theory applied is a continuum theory. At the low scale of electron microscopy the appearances could be different. X-ray microscopy now closes the gap between electron and optical microscopy both as to the size of the accessible objects and as to the resolution achieved. The appearance of small X-ray microscopy objects can be controlled by observing similar ones as large electron microscopy objects and vice versa.

Cryo techniques now overcome the motion related old problems that always used to occur with small particles or with soft material as fluctuating lipid membranes. They even allow us to successively observe the same sample with both methods.

Radiation damage is minimized at low temperature. The high viscosity also stabilizes the overall structure even if there is some molecular damage within the sample. Thus radiation induced visible effects are eliminated during the regular illumination time of some seconds. However, freezing times have to be as high as about $10^5$K/sec in order to rule out the formation of ice crystals. Thus only thin samples can be observed.

The application of phase contrast widely improves the image quality and pulls the experimental resolution to its physical limit of some 25-30nm today. The objects could principally be observed with radiation of a different wavelength so that both the absorption and thus the radiation damage can be minimal even at normal temperatures. With the higher brilliance that is expected to be available at BESSY II the exposure time will be reduced and even highly fluctuating small particles may be imaged at normal humid conditions.

## Acknowledgements

This work has been partly funded by the German Federal Minister for Education, Science, Research and Technology (BMBF) under the contract number 05 644 MGA. Special thanks are given to G. Schneider, FE Röntgenphysik, Universität Göttingen, for helping to do the investigations with the cryo X-ray microscope. B. K. is grateful to G. Schmahl, D. Rudolph and their collaborators for their generous support and interest.

# References

1  Seddon, J.M. and R.H.Templer. 1995. Polymorphism of lipid-water systems. In Structure and Dynamics of Membranes, From Cells to Vesicles. R.Lipowskoy, E.Sackmann, editors. Handbook of Biological Physics. vol.1A. A.J.Hoff, series editor. Elsevier Science, Amsterdam.. 97–160.

2  Vaz, W.L.C., R.M.Clegg and D.Hallmann. 1985. Translational diffusion of lipids in liquid crystalline phase phosphatidylcholine multibilayers. A comparison of experiment with theory. Biochemistry 24: 781–786.

3  Helfrich,W. .1973. Elastic properties of lipid bilayers: theory and possible experiments. Z.Naturforsch. 28c: 693–703.

4  Deuling, H.J. and W.Helfrich.1976. The curvature eleasticity of fluid membranes: a catalogue of vesicle shapes. J.Physique 37: 1335–1345.

5  Seifert, U. 1996. Morphology and dynamics of vesicles. 350–357. In Theory of Self-Assembly. 350–357.

6  Helfrich, W. and R.M.Servuss. 1984. Undulations, steric interaction and cohesion of fluid membranes. Il Nuovo Cimento D3: 137–151.

7  Seifert, U., Berndl, K. and R.Lipowsky. 1991. Shape transformations of vesicles: phase diagrams for spontaneous-curvature and bilayer-coupling models. Phys. Rev.A 44: 1181–1202.

8  Jülicher, F., U.Seifert and R.Liposky. Conformal degeneracy and conformal diffusion of vesicles. Phys.Rev.Lett.71,3: 452–455.

9  Charitat, T. and B. Fourcade. 1997. Lattice of passages connecting membranes. J.Phys. II France 7: in press

10 Servuss, R.M. and W.Helfrich. 1989. Mutual adhesion of lecithin membranes at ultralow tensions. J.Phys.France 50: 809–827.

11 Helfrich, W.and B.Klösgen. 1990. Adhesion and roughness of biological model membranes. In Dynamics and Patterns in Complex Fluids. A.Onuki and K.Kawasaki, editors. Springer Proc. in Physics 52: 2–16.

12 Helfrich W. 1995. Tension-induced mutual adhesion and a conjectured superstructure of lipid membranes. In Structure and Dynamics of Membranes. R.Lipowsky, E.Sackmann, editors. In Handbook of Biological Physics, vol.1B. A.J.Hoff, series editor. Elsevier Science,Amsterdam. 691–721.

13 Mutz, M. and W. Helfrich. 1990. Bending rigidities of some biological model membranes as obtained from the Fourier analysis of contour sections. J.Phys.France 51:991–1002.

14 Niggemann, G., M.Kummrow and W.Helfrich. 1995. The bending rigidity of phosphatidylcholine bilayers: dependencies on experimental method, sample cell sealing and temperature. J.Phys.II France 5: 413–425.

15 Beblik, G., R.M.Servuss and W.Helfrich. 1985. Bilayer bending rigidity of some synthetic lecithins. J.Physique 46: 1773–1778.

16 Hartung, J.,W.Helfrich and B.Klösgen. 1994. Transformation of phosphatidyl-choline multilayer systems in excess water. Biophys.Chemistry **49**: 77–81.

17 Harbich, W. and W.Helfrich. 1990. Phases of egg lecithin in an abundance of water. Chem.Phys.Lipids **55:**191–205.

18 Reeves and Dowben. 1969. Formation and properties of thin-walled phospholipid vesicles. J.Cell.Physiol.**73:**49–60.

19 Klösgen, B. and W.Helfrich. 1993. Special features of phosphatidylcholine vesicles as seen in cryo-transmission electron microscopy. Eur. Biophys. J. 22: 329–340.

20 Guttmann, P., G.Schneider, J.Thieme, C.David, M.Diehl, R.Medenwaldt, B.Niemann, D.Rudolph and G.Schmahl. 1992. X-ray microscopy studies with the Göttingen X-ray microscopes. SPIE 1741:52–61.

21 Schmahl, G., D.Rudolph, B.Niemann, P.Guttmann, M.Robert-Nicoud, J.Thieme, G.Schneider, C.David, D.Diehl and T.Wilhein. 1992. Natural imaging of biological specimens with X-ray microscopes. *In* Synchrotron Radiation in Biosciences, vol.2. Cell Biology and Medicine. B.Chang et al., editors. Oxford Sciences, 1994. 539–562..

22 Rudolph, D., G.Schneider, P.Guttmann, G.Schmahl, B.Niemann and J.Thieme.1992. Investigations of wet biological specimens with the X-ray microscope at BESSY. In Springer Series in Optical Sciences, vol.67. X-ray microscopy III. A. Michette, G.Morrison and C..Buckley, editors. Springer Verlag Berlin. 404–407.

23 Thieme, J., C.David, N.Fay, B.Kaulich, R.Medenwaldt, M.Hettwer, P.Guttmann, U.Kögler, T.Maser, G.Schneider, D.Rudolph and G.Schmahl. 1993. Zone plates for high resolution X-ray microscopy. *In* X-Ray Microscopy IV. V.V.Aristov and A.I.Erko, editors. Bogorodskii Pechatnik Publ. Comp., Chernogolovka/Moscow, Russia. 487–493.

24 Niemann, B., G.Schneider, P.Guttmann, Rudolph, D., G.Schmahl. 1994. The new Göttingen X-ray microscope with object holder in air for wet specimen. *In* X-Ray Microscopy IV. V.V.Aristov and A.I.Erko, editors. Bogorodskii Pechatnik Publ. Comp., Chernogolovka/Moscow, Russia. 66–76.

25 Schmahl, G., D.Rudolph, P. Guttmann, G.Schneider, J.Thieme,. B.Niemann. 1995. Phase contrast studies of biological specimens with the X-ray microscope at BESSY. Rev.Sci.Instrum. **66:**1282–1286.

26 Schneider, G. and B.Niemann 1997. Cryo-X-ray microscopy experiments with the X-ray microscope at BESSY. This issue.

27 Guttmann, P. and B. Klösgen. 1994. In vitro imaging of arificial lipid membranes by use of X-ray microscopy. Proceedings of the 13th International Congress on Electron Microscopy. Les Editions de Physique, Les Ulis Cedex A, France. 795–796.

28 Guttmann, P. and B.Klösgen. 1994. X-ray microscopy studies of artificial lipid membranes. *In* X-Ray Microscopy IV. V.V.Aristov and A.I.Erko, editors. Bogorodskii Pechatnik Publ. Comp., Chernogolovka/Moscow, Russia. 217–239.

29 Schliebe,T., G. Schneider and H.Aschoff. 1997. High resolution zone plates in

nickel for X-ray microscopy. This issue.

30 Cazaux, J.1995. The role of the Auger mechanism in the radiation damage of insulators. Microc.Microanal.Microstruct.6: 345–362.

31 Klösgen, B. and W.Helfrich. 1996. Cryo-Transmission Electron microscopy of a superstructure of fluid dioleoylphosphatidylcholine (DOPC) membranes. Biophys. J. . submitted.

32 Schneider, G., B.Niemann, P.Guttmann, D.Rudolph, G.Schmahl. 1995. Cryo X-ray microscopy. Synch. Rad. News 8: 19–28.

33 Heide, H. and E.Zeitler. 1985. The physical behaviour of solid water at low temperatures and the embedding of electron microscopical specimens. Ultra-microscopy 16: 151–160.

34 Milner, S.T. and S.A. Safran. 1987. Dynamical fluctuations of droplet microemulsions and vesicles. Phys. Rev. A 36: 4371–4379.

35 Dunger, A., W. Helfrich, and B. Klösgen Observation of a disperse phase of lipids by freeze fracture electron microscopy. In preparation.

# X-Ray Microscopic Imaging
# of Magnetic Domains
# Using X-Ray Magnetic Circular Dichroism*

P. Fischer[1], T. Eimüller[1], G. Schütz[1], G. Schmahl[2], P. Guttmann[2],
D. Raasch[3]

[1] Univ. Augsburg, Exp. Phys. II, Memmingerstr. 6, D-86135 Augsburg, Germany
[2] Univ. Göttingen, Forschungseinrichtung Röntgenphysik, Geiststr. 11,
D-37073 Göttingen, Germany
[3] Philips Res. Labs, Weisshausstr. 2, D-52066 Aachen, Germany

**Abstract.** The combination of a transmission X-ray microscope based
on the zone plate technique providing a spatial resolution of $\approx 30 nm$
with the effect of X-ray magnetic circular dichroism, which gives a huge
contrast in the absorption mode is a new method to visualize in a quantitative and element-selective manner magnetic domains. We report on
the study of the variation of the shape and the magnetization of domains
in an applied magnetic field in a magneto-optical GdFe layered system
with a lateral resolution of $50 nm$.

## 1 Introduction

The magnetism of systems of reduced dimensionality (surfaces, thin films, multilayers, etc.) is attracting both scientific interest and technical importance. The
available techniques to prepare layers with defined structures and to charac-
terise their morphology allows new insights into the correlation of the complex
microstructure with macroscopic properties, as e.g. the occurence of giant mag-
netoresistance (GMR), oscillatory exchange effects and especially the magnetic
anisotropy perpendicular to the surface.

One important issue in understanding the micromagnetic performance is the
imaging of domains, wall structures and their dynamics driven by an applied
external field. Static domain properties can be visualized with very high spatial
resolution down to several $nm$ by powerful methods, as e.g. Scanning Electron
Microscopy with Polarization Analysis (SEMPA), Lorentz microscopy and Magnetic Force Microscopy (MFM) which, however, are inherent sensitive to the
surface or restricted to very thin layered magnetic structures. The dynamics
of the domains as a function of an external magnetic field can be studied by
Kerr-microscopy but with the resolution limit of optical microscopy.

Here we report on a development of a novel domain imaging technique using
the transmission X-ray microscope (TXM) at BESSY I in combination with a

* This work is funded by BMBF proj. no. 05 621 WAA and 05 644 MGA.

**X-Ray Microscopy and Spectromicroscopy**
Eds.: J. Thieme, G. Schmahl, D. Rudolph, E. Umbach
© Springer-Verlag Berlin Heidelberg 1998

contrast enhancement via X-ray magnetic circular dichroism (X-MCD). First results on a GdFe alloy are presented, where the hysteresis properties of the domain pattern with a spatial resolution of $50nm$ and their long term variation have been studied in real time transmission mode.

## 2   Principle of Dichroic Contrast

X-ray magnetic circular dichroism (X-MCD), i.e. the dependence of the core-level absorption coefficient of circular polarized photons on the
sample magnetization is a powerful X-ray spectroscopic method to study the magnetic characteristics of the electronic structure[1, 2].

Its physical origin is closely related to the polar magneto-optical Kerr effect. However, since in the X-MCD the initial state is a well defined core level state, X-MCD features an element-sensitivity and according to dipole selection rules also a symmetry-selectivity. Due to angular momentum conservation and spin-orbit interaction in the initial state the absorption of a circular polarised photon in an unpolarised initial state causes a photoelectron to exhibit an expectation value of both the spin and the orbital momentum projected onto the photon propagation direction. Hence X-MCD spectroscopy allows to determine with high relative accuracy ($< 10^{-2} \mu_B$) in appropriate cases as Fe,Co,Ni local magnetic moments separated into spin and orbital contributions invoking magneto-optical sum rules [3, 4].

The magnetic contrast can be described as the energy-dependent deviation $\Delta\mu(E)$ of the absorption coefficient relative to the polarization averaged absorption coefficient $\mu_{|i>}(E)$. It can be expressed by

$$\frac{\Delta\mu}{\mu_{|i>}}(E) = \frac{\sigma_c}{\sigma_{|i>}}(E)(\hat{m} \cdot \hat{e}_z)P_c \tag{1}$$

where $(\hat{m} \cdot \hat{e}_z)$ denotes the projection of the normalized magnetic moment $\hat{m} = \frac{\mathbf{m}}{|\mathbf{m}|}$ onto the propagation direction $(\hat{e}_z)$ of the photons with a degree of circular polarization $P_c$. The magnetic absorption cross section normalized to the polarization averaged atomic cross section $\frac{\sigma_c}{\sigma_{|i>}}$ reaches at the maximum of the Fe metal $L_3$ edge $\frac{\sigma_c}{\sigma_{|2p_{3/2}>}}(E = 706eV) \approx 23\%$ [5]. Following Eq.(1) the value of $\frac{\Delta\mu}{\mu_{|i>}}$ is a sensitive measure of $\hat{m} \cdot \hat{e}_z$, if $P_c$ is known. Taking into account the value of the absolute magnetic moment in the pure Fe-metal of $|\mathbf{m}| = 2.2\mu_B$ the observable experimental quantity reflects the absolute value of the spatial magnetization distribution.

Hence the dichroic effect in transmission can serve as a quantitative magnetic contrast suited even for multicomponent systems in spectroscopic x-ray imaging techniques which employ the absorption in the vicinity of an absorption edge.

Previous attempts reported in literature have succesfully combined the magnetic dichroism in photoemission [6, 7] to image magnetic domains with a resolution of some $\mu$m.

# 3   Experimental Details

The sample investigated was a binary $Gd_{27.7}Fe_{72.3}$ system with a Curie temperature of $T_c \sim 510K$. The amorphous film with thickness $h = 59 \pm 1nm$ was prepared by coevaporation from two electron-gun sources onto a $325nm$ polyimid foil as substrate and embedded between $10 - 15nm$ Al layers for chemical protection. The out-of-plane coercivity $H_c$ was determined both by Faraday effect, VSM and Kerr magnetometry measurements and in-situ with the X-ray microscope yielding $\sim 6(2)mT$ with a shift of the hysteresis loop of $\sim 8(2)mT$, which can be attributed to remaining magnetically hard regions.

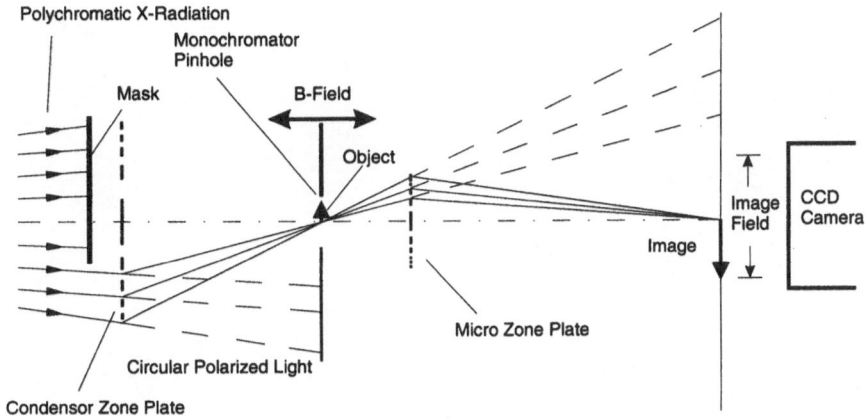

**Fig. 1.** The X-ray optical set-up of the TXM at BESSY I.

The x-ray optical set-up of the x-ray microscope (Fig. 1) which is based on the zone-plate technique is described in more detail in [8, 9]. It consists of a condensor zone-plate ($D = 9mm$) and a pinhole ($d = 20\mu m$), acting together as a linear monochromator with the monochromaticity given by $\lambda/\Delta\lambda = D/2d = 225$. Furthermore a microzone plate as a high resolution x-ray objective generates a magnified image of the object in the image field with a spatial resolution of about $50nm$, where it is recorded with a slow scan CCD camera with a thinned, backside illuminated CCD chip (DQE $\approx 70\%$) [10]. The total extinction of the radiation in the target $I/I_0$ for the GdFe system at a photon energy of $E = 706eV$ was determined to 90% with a contribution due to Fe $L_3$ absorption of 70%. The magnetic images were taken at the Fe $L_3$-edge ($\approx 706eV$) with exposure times for the x-ray images shown in Figs.2–6 of $1 - 3s$. The upper part of the condensor was masked so that the sample was illuminated by circular polarized X-rays with a degree of circular polarization of about 60%. Magnetic fields up to $80mT$ with the field direction pointing parallel/antiparallel to the photon beam propagation direction could be applied onto the sample.

## 4   Results and Discussion

While at photon energies below the Fe $L_3$ absorption edge the spatial intensity distribution is homogenous, significant structures appeared at $E = 706eV$ as demonstrated in Figs. 2 and 3. The applied external magnetic fields were $12mT$ and $25mT$ below $-H_c$, resp. The sign of the magnetic field corresponds to the magnetization consistent with the dichroic effect. Since the photons are left-handed the net magnetization of the darker/lighter shaded domains point into/out of the paper plane, i.e. antiparallel/parallel to the photon beam direction.

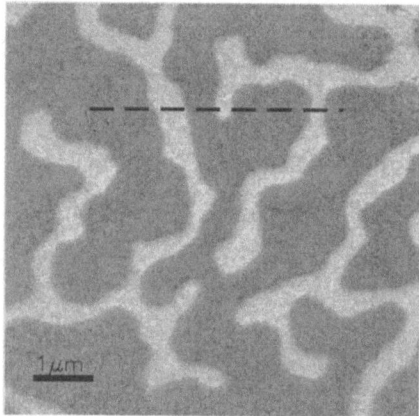
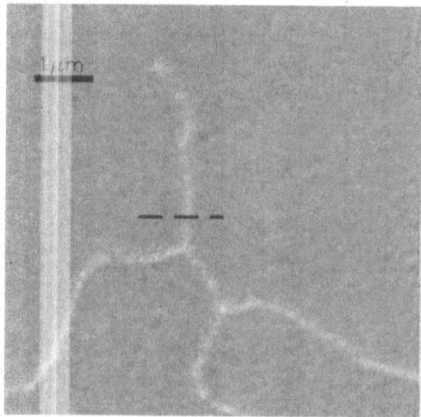

**Fig. 2.** Image of magnetic domains (field of view $7 \times 7\mu m^2$) at a magnetic field $12mT$ below $-H_c$.

**Fig. 3.** Image of magnetic domains (field of view $7 \times 7\mu m^2$) at a magnetic field close to the saturation.

The domains in Fig. 2 appear with irregular spotted patterns of average widths of the white (light) domains of $D \sim 0.5\mu m$ corresponding to a reduced domain size of $D/h \sim 10$. These serpentine-like structures tend to more worm-like ones with increasing field as seen in Fig. 3. The width of these worms of $100nm$ remain relatively stable by approaching the saturation field only a partly shortening of the tails with magnetic after-effects of some minutes are observed. Then they collapse suddenly at a field larger than $25mT$ below $-H_c$. Going back from the negative high field limit the state of the saturation (homogenous black (dark) picture) remains until the value of $-H_c$ is reached. Then suddenly within $0.2mT$ the irregular patterns are back again. A series of selected images within a complete hysteresis loop are shown in Fig.6. Figs. 2 and 3, which have been taken in different hysteresis loops, demonstrate that there is no evidence for domain pinning as the locations of the domains especially the worm like structures are completely irreproducable although the global characteristic of the pattern does not change.

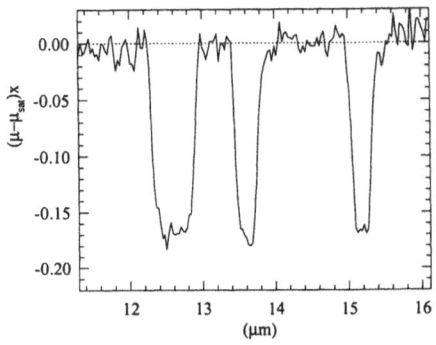

**Fig. 4.** Absorption scan across the region indicated in Fig.2.

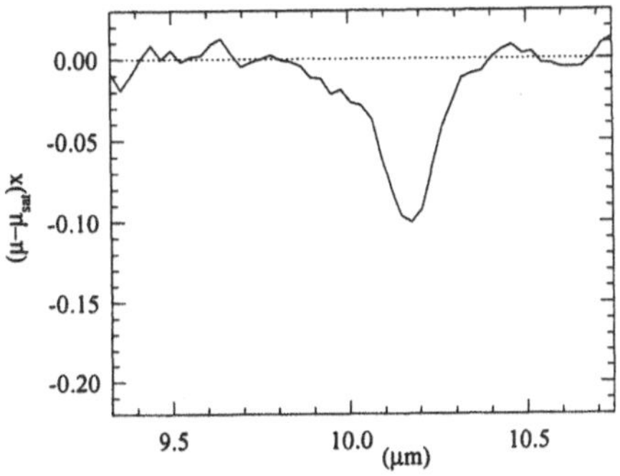

**Fig. 5.** Absorption scan across the region indicated in Fig.3.

In Fig. 4 the absorption profile $(\mu - \mu_{sat})$ along the line marked in Fig. 2 is shown. Hereby the intensity pattern of Fig. 2 is normalized to those for complete magnetization corresponding to fully dark image, where the maximum absorption $\mu_{sat}$ is achieved. An analysis of the wall profiles proves that their widths is significantly smaller than the experimental resolution of $50nm$ [11]. This is in excellent agreement with the estimation of the wall width $\delta_{dw} = \pi\sqrt{A/K_u} \sim 30nm$ depending on the exchange constant $A = 3.1 \pm 0.8 \cdot 10^{-12} J/m$, which has been determined by domain expanding and collapsing experiments [12], and the anisotropy constant $K_u \approx 0.33 \cdot 10^5 J/m^3$, which has been determined on reference glass substrates using a torque magnetometer. This demonstrates that the micromagnetic performance of the sample prepared on the polymere substrate does not differ much from the sample prepared on the glass substrate.

The difference of the absorption for reversed domain magnetization in Fig.4 yields values of $\sim 17\%$. Taking into account only the Fe $L_3$ absorption and the finite value of degree of circular polarization, the strengths of the dichroic effects corresponds to $\hat{m} \cdot \hat{e}_z = 1$ in eq.(1) proofing the complete alignment of the magnetic Fe moments onto the photon beam direction, i.e. perpendicular to the plane of the target.

## 5  Outlook and Conclusion

The combination of X-MCD with TXM allows to image magnetic domains with a resolution of $50nm$ (improvable to better than $20nm$ and to study the dynamics of the formation of magnetic domains in real-time and applied external field. The bulk-sensitivity and element-selectivity of this technique enables to investigate the concentration dependent magnetization distributions and the quantitative analysis of the magnetization distribution. The separation of spin and orbital moment distributions seems to be feasible at spin-orbit correlated absorption edges providing a unique method to learn more about the role of orbital polarization and spin-orbit effects on the micromagnetic performance especially for thin layers.

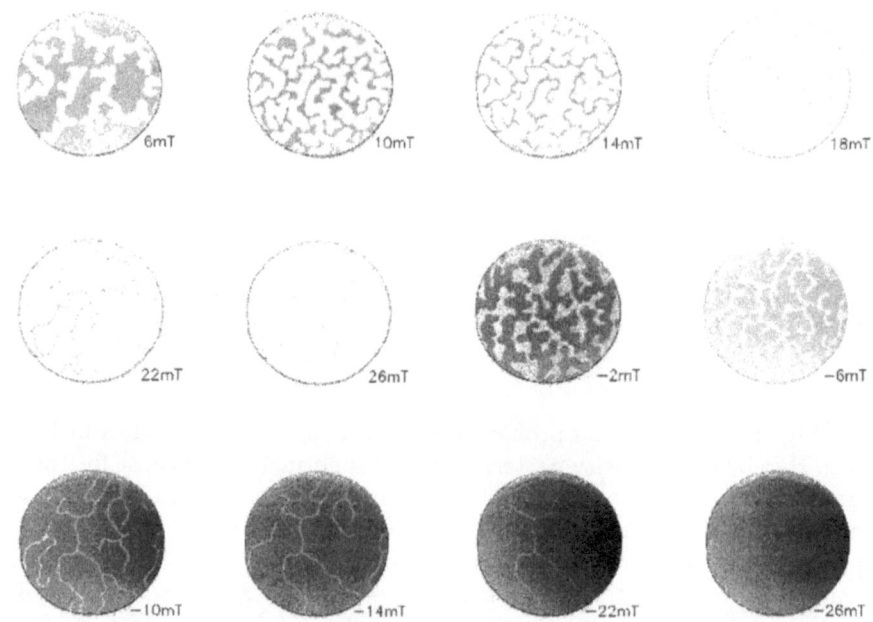

**Fig. 6.** Series of pictures (field of view diameter $16.9\mu m$) taken at different applied magnetic field strengths within a hysteresis loop. The dark and light regions correspond to different orientations of the local magnetization.

# References

1. G. Schütz et al, Phys. Rev. Lett. **58**, 737 (1987).
2. C.T. Chen et al., Phys. Rev. **B 42**, 7262 (1990).
3. B.T. Thole, P. Carra, F. Sette and G. van der Laan, Phys. Rev. Lett. **68(12)**, 1943 (1992) .
4. P. Carra, B.T. Thole, M. Altarelli and X. Wang, Phys. Rev. Lett. **70(5)**, 694 (1993).
5. G. Schütz et al., J. Appl. Phys. **76** (10), 6453 (1994).
6. C.M Schneider et al., Mat. Res. Soc. Symp. Proc. **313**,631 (1993).
7. J. Stoehr et al., Science **259**, 658 (1993)
8. B. Niemann et al., in *X-ray Microscopy IV*, V.V. Aristov and A.I. Erko eds., Begorodski Pechatnik Publishing Company, Moscow, 66 (1995),
9. G. Schmahl et al., Rev. Sci. Instruments **66(2)**, 1282 (1995).
10. T. Wilhein, Gedünnte CCDs: Charakterisierung und Anwendungen im Bereich weicher Röntgenstrahlung, Doktorarbeit, Verlag Shaker (1994).
11. P. Fischer et al., Z. f. Phys. B (1996) accepted.
12. D. Raasch and J. Reck, J. Appl. Phys. **74**, 1229 (1993); D. Raasch et al., J. Appl. Phys. **76**, 1145 (1994).

This article was processed using the LaTeX macro package with LLNCS style

# X-Ray Holography of Fast-Frozen Hydrated Biological Samples

S. Lindaas[1], B. Calef[1], K. Downing[1], M. Howells[1], C. Magowan[1], D. Pinkas[1], C. Jacobsen[2]

[1] Lawrence Berkeley National Laboratory, Cyclotron Rd, Berkeley CA 94720, USA
[2] Physics Department, State University of New York, Stony Brook, NY 11794, USA
E-mail: lindaas@afm1.lbl.gov

## 1 Introduction

Athough x-ray holography has a long history [11], it is only in recent times that short-wavelenth lasers and undulator x-ray sources have opened the possibility for competitive imaging performance. Experiments using the in-line (Gabor) geometry have been favored by most groups and have been carried out at Orsay [12], Brookhaven [9], Livermore [31], Tsukuba [34], Chilton [32], Osaka [27] and Grenoble [28] while the Fourier-transform geometry has been developed at Brookhaven [23]. Submicron resolution has been reported by the Orsay group [12] and by some of the present authors working at Brookhaven [10]. These holographic schemes all operated in or near the water window and were mostly conceived as technical developments toward imaging of natural biological material. Their motivations were mostly the practical simplicity of the hologram recording process, which is similar to contact imaging, and the fact that such recordings can be made to deliver both the amplitude and phase imgages. The chief disadvantages are that holography provides no excape from the large radiation dose that is required by all x-ray and electron imaging methods and that the in-line style of holography produces an inherent corruption to the desired image known as the twin image. We have discussed the question of how to exploit the potential of x-ray holography in light of these disadvantages in an earlier paper [18] and have concluded that part of the solution is to cool the sample to cryogenic temperatures. Under this condition, features as small as 5 nm remain intact at doses of up to $10^8$ Gray which thus allows a tilt series containing scores of holographic images to be taken.

In this paper we describe the design and operation of a new experimental apparatus with which we have recorded a number of holograms of natural wet biological samples at liquid nitrogen temperature. We also report how, after digitizing the holograms using an atomic-force microscope [18], we have addressed the problem of how to make a numerical reconstruction of the image. We discuss the nature of the twin image problem in the language of the theory of convex sets and show that standard reconstruction algorithms are effective in certain cases but inefficient in others.

**X-Ray Microscopy and Spectromicroscopy**
Eds.: J. Thieme, G. Schmahl, D. Rudolph, E. Umbach
© Springer-Verlag Berlin Heidelberg 1998

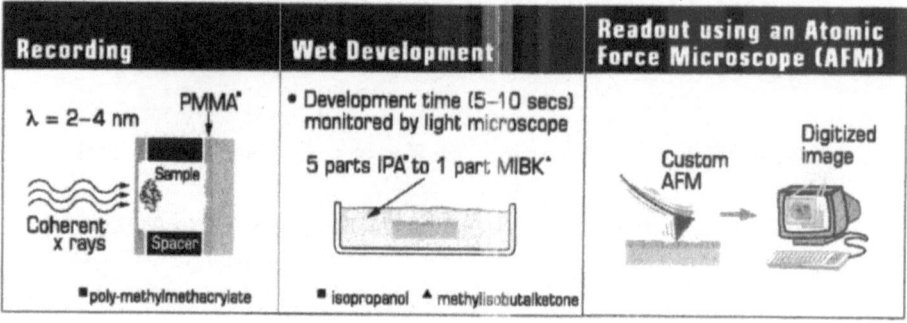

| Recording | Wet Development | Readout using an Atomic Force Microscope (AFM) |
|---|---|---|
| λ = 2–4 nm    PMMA* Coherent x rays | • Development time (5–10 secs) monitored by light microscope  5 parts IPA* to 1 part MIBK* | Custom AFM    Digitized image |
| *poly-methylmethacrylate | *isopropanol  ▲ methylisobutalketone | |

**Fig. 1.** The photoresist PMMA is used to record the hologram which consists of the interference pattern formed by the plane reference wave and the wave diffracted from the sample.The sample-photoresist spacing   (typically 400 μm) is determined in the recon- struction step (see Sec. VII). Some of the exposed polymer suffers mass losing during irradiation (self development) and more is later removed deliberately during wet develop- ment. Light development is necessary to avoid "washing out" low-contrast fringes [33]. The photoresist relief map (corresponding to incident X-ray intensity) is measured/digitized using our specially-made atomic-force microscope [18, 19]

## 2 In-Line Holography

The procedure that we have used for recording and readout of the holograms is out- lined in Fig. 1 and in our earllier publications [18].

## 3 Description of Experimental Apparatus

We have designed and built an experimental apparatus to enable us to record holograms at liquid-nitrogen temperature. Wherever possible we have taken advan- tage of commercially available systems developed for the electron microscopy community and have there- by obtained proven hard- ware for preparing the sample and loading it into the vacuum. In particular, we adapted a transmission electron        microscope (TEM) cryo-holder (Fig. 2) designed to hold a 3 mm diameter TEM grid so that it could accomodate a rela-

**Fig. 2.** Cryo-sample holder showing sample area

tively thick "package" consisting of the sample-grid, a spacer (≈ 300 μm) and a 500-μm-thick substrate covered with photoresist. The sample was cooled via a vold finger extending from the liquid-nitrogen dewar which allowed a temperature of - 160°C to be easily maintained at the sample as long as the dewar was partially filled. The cryo-holder was also equipped with a heater which was useful in deicing the tip between sample changes and could also be used as a means to freeze dry samples in vacuum. A thermocouple located next to the sample region allowed the temperature to be continuously monitored.

We also purchased the TEM airlock mechanism that mated with the cryo-sample holder and designed an interface for mounting it on an "xyzθ" stage [15] which allowed the sample to be accurately positioned at the center of our chamber. The whole system was capable of $10^{-9}$ torr although in the experiments reported here the

**Fig. 3.** Schematic of cryo-holgraphy chamber with cut-away of the airlock and ist interface

chamber pressure was about $10^{-5}$ torr. The cyro-sample holder and cryo work station plus the airlock were designed to enable samples that have been fast frozen to liquid-nitrogen temperature to be stored as long as needed at that temperature and then transferred into a vacuum chamber without warming above -160°C (at which tempereature damage due to ice crystal growth would begin to occur). A cut-

away diagram of the apparatus is shown in Fig. 3 and a photograph of it in place on beamline X1A at the National Syn-chrotron Light Source (NSLS) at Brookhaven National Laboratory is shown in Fig. 4.

The sample positio-ning system also inclu-ded a stereo micro-scope with a 7inch working distance. By this means one could look at the cryo-sample

**Fig. 4.** Cryo-holography chamber set up at X1A. The cryo-sample holder is inserted from the left and the stereo mi-croscope used to align the sample via the alignment mirror (see insert)

holder with the X-ray beam on via an X-ray transparent mirror (Al coated on a $Si_3N_4$ window) placed at 45 degrees to the beam. Phosphor was placed on the cryo-holder shutter (see Fig. 2) so that targeting of the beam on to a chosen sample area could be accomplished by using the microscope reticule.

## 4 Sample Preparation

The samples used in these experiments were red blood cells infected with the *Plasmodium falciparum* malaria parasite which were made by the Hematopoiesis/

Mammara Cell Biology Group at the Lawrence Berkeley National Laboratory (LBNL). The size of these cells made them well suited for x-ray holography and preparation techniques had already been worked out for imaging them in a transmission x-ray microscope [24].

Infected red blood cells were cultured *in vitro* as previously described [29]. Mature trophozoite stage infected erythrocytes were enriched to > 80 % by flotation in 0.5 % gelatin [25] and fixed with 1 % glutaraldehyde. All of the pathology and mortality associated with *Plasmodium falciparum* malaria is due to the blood stages of infection. The trophozoite is the developmental stage which is actively metabolizing hemoglobin and synthesizing new proteins that associate with the host erythrocyte membrane to alter ist morphology, antigenecity, and function [1, 13, 21, 30].

The cell culture war diluted till approximately one or two cells occupied each (40 μm) [2] grid area in test preparations. Using the fast-freezing workstation at Donner Laboratory (LBNL) we prepared frozen samples from this diluted cell culture. This workstation used immersion cooling [4] (via a spring-loaded plunger into liquid ethane) and was equipped to control the humidity around the sample to avoid sample dehydration, which can happen very quickly. The cryo-samples were placed under liquid nitrogen in modified storage trays, designed for electron-microscope grids, and then shipped to the Brookhaven using the "dry-shipper" type of cryogenic container.

## 5 Experimental Procedure

The fast frozen sample, the sample holder, the resist substrate and spacer were all introduced into the liquid-nitrogen-temperature environment of the cryo-work station and assembled there. Once in place the resist-spacer-sample package was covered with the shutter to minimize ice build-up during the transfer. The whole cryo-work station was then moved to the beam-line area and the cryo holder was inserted into the airlock.

The x-ray beam was then aligned over region of interest using the stereo microscope and the package was exposed. Using $\lambda = 2.4$ nm x-rays with $\lambda/\Delta\lambda \geq 150$ and a spatial coherence footprint of better than $(40 \ \mu m)^2$, we had $3 \times 10^8$ photons/$\mu m^2$ incident on the sample. For a sample with a 10 μm thick water/ice layer, this gives about $10^8$ photons/$\mu m^2$ on the resist and a sample dose of approximately $2 \times 10^6$ Grays. The exposure was monitored by photodiode measurements of the beam and optical-microscope examination of the self-developed resist. The magnified atomic-force-microscope image of the hologram is shown in Fig. 5.

Once the hologram was digitized, processing was carried out, using our own software package written in IDL, on a Pentium-Pro based computer with 256 Mbytes of RAM running Linux (version 2.0.0). The basic propagation operaiotn between the hologram and object plane took about one minute for a $2048^2$-pixel hologram and the corresponding speed of the twin-image algorithm was about 1.4 days per thousand iterations.

# 6 Results

We have recorded a number of holograms of natural, wet samples that were fast frozen in a manner consistent with producing amorphous ice and maintained at cryogenic temperature until after imaging. The fully processed version of the one we chose for our studies of twin-image suppression is shown in Fig. 5. By calculation of the power spectrum we have determined that there is no white-noise roll off so that information is apparently recorded in the photoresist to better than 30 nm. A reconstructed image obtained by the numerical equivalent of illumination of the hologram with the original reference wave is shown in Fig. 6 (the object-to-hologram distance was found by software to be 425 μm). The fringe artifacts due to the twin-image signal are readily apparent in the reconstructed image and we discuss this issue more fully in the next section.

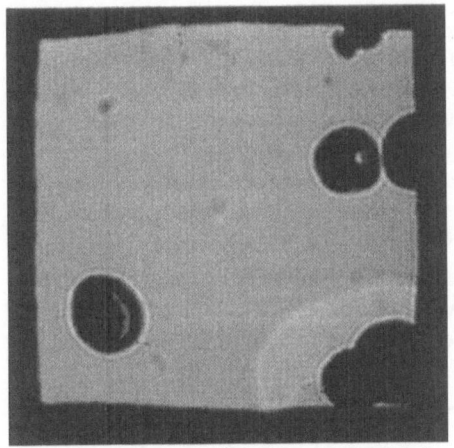

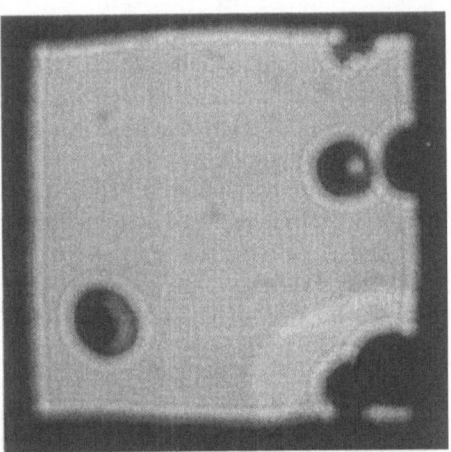

**Fig. 5.** Hologram of malaria infected red Blood cells. AFM readout: $(21 \text{ nm})^2$ pixels in a $(2048)^2$ array

**Fig. 6.** Numerically reconstructed image. The focus was found to be at $z = 425$ μm

# 7 The Twin-Image Signal and Its Suppression

A hologram is formed when a complex wavefield $a(\mathbf{x})$ scattered by the sample and a coherent reference wavefield $r(\mathbf{x})$ are made to interfere, registering an intensity $I(\mathbf{x}) = |r(\mathbf{x}) + a(\mathbf{x})|^2$ in the $\mathbf{x}$ plane. For an axial-plane-wave reference beam (Fig. 1), $r$ is a real constant which we take to be unity. Then

$$I(\mathbf{x}) = 1 + |a(\mathbf{x})|^2 + a(\mathbf{x}) + a^*(\mathbf{x}) \tag{1}$$

The four terms are known as the zero-order, indermodulation, virtual image and real image terms respectively. To calculate $a(\mathbf{x})$ suppose the object is represented by a two-dimensional complex transparency function $T(\mathbf{x}')$ which we prefer to write as $1 - A(\mathbf{x}')$ where $A(\mathbf{x}')$ is the complex *absorbency*. We can express the diffracted

wavefield at distanze $z$ downstream by means of a propagation operator $P_z$ with the following properties:
$P_z(1) = 1$, $P_z^* = P_{-z}$, $P_z P_{-z} = 1$, $P_{z1} P_{z2} = P_{z2} P_{z1} = P_{z1+z2}$. This operator represents the action of the diffraction integral on the initial wavefield and we have implemented it in software [20]. In terms of $P_z$ the field $1 + a(\mathbf{x})$ at the hologram is $P_z[1-A(\mathbf{x}')]$ and

$$I(\mathbf{x}) = 1 + |P_z[A(\mathbf{x}')]|^2 - P_z[A(\mathbf{x}')] - P_z^*[A^*(\mathbf{x}')]. \qquad (2)$$

Evidently the quantitiy $P_z(-A(\mathbf{x}'))$ is equivalent to the wavefield $a(\mathbf{x})$ used above.

The process of reconstructing the virtual image consists of illuminating the hologram $I(\mathbf{x})$ with a unit plane wave and backpropagating a distance z. In a numerical calculation, this leads to a complex amplitude distribution $A_1(\mathbf{x}')$

$$A_1(\mathbf{x}') = P_z^*[I(\mathbf{x})] = 1^* P_z^* |P_z[A(\mathbf{x}')]|^2 - A(\mathbf{x}') - P_{2}^*{}_z[A^*(\mathbf{x}')]. \qquad (3)$$

Therefore, with a reference wave of *exactly* unity, it is - $A(\mathbf{x}')$ that we reconstruct in the object plane along with unwanted signals derived from the intermodulation term and the out-of-focus real image (the twin-image). The so-called twin-image problem of in-line holography is to extract the object $A$ given only the hologram.

In some applications of in-line holography, the intermodulation term can be neglected. In such cases the cosines of the phases can be inferred from the hologram and particular approaches to the twin-image problem become appropriate [14, 22]. However, examination of equation (3) shows that neglect of the intermodulation term can only be justified when $|A(\mathbf{x}')|^2 \ll 1$. For this to be true the sample would have to be both a weak scatterer and a weak absorber and this is certainly not something we can normally assume in x-ray microscopy.

The twin-image problem is a member of a class of phase problems that arise commonly in optics. The class includes, image restoration from amplitude, for example in speckle interferometry [3], Michelson stellar interferometry [7], defocused pairs [2] or image-plane-diffraction-plane pairs [26], the design of computer-generated holograms [5] and many others. The general problem is to reconstruct a complex wave function given incomplete information about the function itself and the magnitude of ist transformation by a known linear operator.

In the above notation, the scenario of principal interest is that (i) we know that $A$ is different from zero only within a certain region of the object plane (ist support) and (ii) we know that the magnitude of $P_z(1-A)$ equals $\sqrt{I}$. There are two algorithms in widespread use for solving this type of problem: the error-reduction algorithm due to Gerchberg and Saxton [6] and the input-output algorithm of Fienup [5]. Their general structure is as follows: start in the hologram plane with the amplitude equal to $\sqrt{I}$ and guesses for the phases. Next transform (using $P_{-z}$) to the object plane and impose one or more constraints derived from a knowledge of the object or ist support (Gerchberg-Saxton) and from earlier iterations (Fienup). Finally transform back (using $P_z$) to the hologram plane, reimpose the hologram amplitudes and repeat. A detailed description of these algorithms together with practical advice for their application to optical problems is given by Dainty and Fienup [3].

A useful insight into the behavior of image recovery algorithms such as Gerchberg-Saxton is provided by a geometrical interpretation developed by Youla [36]. The images under consideration, with $M = NxN$ pixels, say, are regarded as points in an $M$-dimensional space. The effect of a constraint is then to limit the image to lie within a certain subset of the space. The support constraint ((i) above), for example, restricts the image to the set of points for which certain of the $M$ coordinates are prescribed to be zero. In the same language the hologram constraint requires the image to be a member of the set of all images the amplitude of whose transform (by $P_z$) is equal to the measured value $\sqrt{I}$. There can be any number of such constraint sets and the solution, if it exists, must be in the set whose members are common to all of the constraint sets, i. e. their intersection.

Suppose we want to find an image satisfying two constraints. This is represented pictorially in Fig. 7(a). If a unique solution exists, then there will be just one image that is a member of both

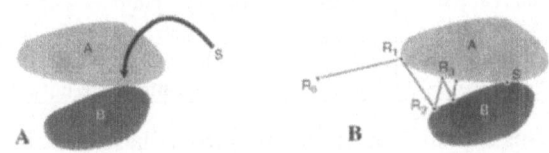

**Fig. 7.** Convex sets and possible unique solution S

constraint sets, namely the point S in the Fig.. The procedure for finding the point S is illustrated in Fig. 7(b). Let the starting image, which may be a guess, be represented by $R_0$. Let the projection of $R_0$ on the set A be $R_1$ (i. E. $R_1$ is the point in A closest to $R_0$). Let $R_2$ be the projection of $R_1$ on to B and so on. If the constraint sets A and B represent the support and hologram constraints respectively then the scheme in Fig. 7 is precisely Gerchber-Saxton.

One can see in a general way from Fig. 8 that if the shape of the constraint set is unfavorable, the alternating-projections procedure may

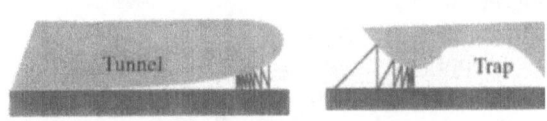

**Fig. 8.** Unfavorable constraint set shapes

converge to a local minimum rather than to the true solution at point S. The conditions needed to avoid this can be understood by means of the theory of convex sets [8, 17]. A set is said to be convex if for any points $X_1$ and $X_2$ in the set the point

$$Y = \lambda X_1 + (1 - \lambda)X_2 \quad (0 \leq \lambda \leq 1) \tag{4}$$

is also in the set, where $Y$ is called the convex combination of $X_1$ and $X_2$. In other words a set is convex if for any two points in the set, the line segment joining the two points is also in the set. One can see from Fig. 8 that it is plausible that strong convergence to the correct solution should occur in the event that all the constraint sets are convex and this has been shown rigorously by Youla [35]. It can also be shown [3] that the Gerchberg-Saxton procedure converges weakly, i. e. does not diverge, even if the sets are not convex. This means that for nonconvex sets, the procedure may converge or it may stagnate.

This leads us to consider whether or not the constraint sets we are interested in are convex and we consider the following four cases:

| Constraint Set=set of all Images with: | Convex ? | Argument |
|---|---|---|
| Certain prescribed pixels=0 (support) | Yes | If prescribed components of $X_1$ and $X_2$ are zero, then (by (4)) those of every Y must be also, so the Y's are in the set |
| Given phases | Yes | Complex numbers with given phase lie on a radius vector of the Argand plane, thence also their convex combinations |
| Given real parts | Yes | Complex numbers with given real part lie on an ordinate of the Argand plane , thence also their convex combinations |
| Given amplitudes | No | Complex numbers with given amplitude lie on a circle in the Argand plane but their convex combinations not |

This explains why restoration from phase is easier than from amplitude and why neglect of the intermodulation term is helpful (because then eq. (2) yields the real part of $P_z(A(\mathbf{x'}))$ ). However, since the hologram constraint set is nonconvex, the convergence of the algorithms we are using is not assured.

We have used the Gerchberg-Saxton and Fienup algorithms to seek solutions to the twin-image problem of in-line x-ray holography taking the hologram in Fig. 5 as an example. Our experience in applying the algorithms to both experimental and model data is as follows:

1. The algorithms work best if used in combination rather than separately. Best results have been obtained by using a fixed number of iterations (10–100) of each algorithm in turn.

2. Model objects that are sparse enough for in-line holography to work (> 80 % empty, say), reconstruct with

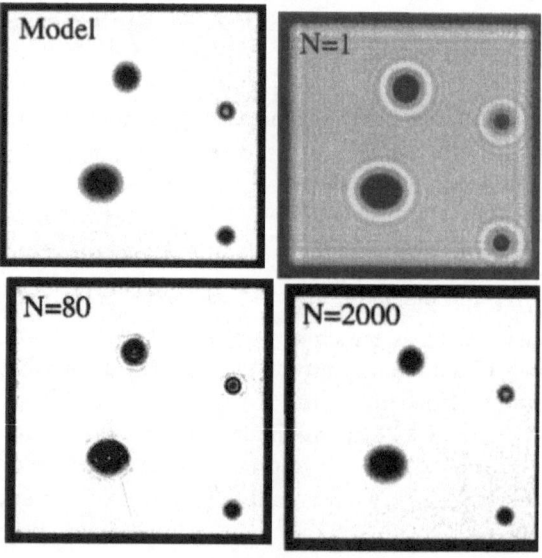

**Fig. 9.** The model objects shown, in a $(28\ \mu m)^2$ field, were used to test the twin-image suppression algorithm. The reconstructed hologram after N = 1 iteration shows substantial twin-image corruption. After N = 80 iterations the small objects has been recovered but the large (6 μm) object and the smaller donut-shaped object are not recovered until N = 2000 iterations

essentially no error (Fig. 9) and this remains true even when realistic amounts of random noise are added.

3. The number of iterations involved in may be quite large ($> 10^3$) and features in the centers of the largest objects require more iterations to reach a solution than those in smaller ones (see Fig. 9). It appears that this is due to the diminished effectiveness of the support constraint for features that scatter little energy into the region governed by the constraint due to their distance from the support boundary.

4. For experimental data (when only the hologram is given) there are difficulties in determining the best support boundary and in allowing for spatial variations in the incoming beam intensity. At the present time we refine the support boundary manually at least once during progress toward a solution and determine the beam intensity from a quadratic surface determined by fitting in the open areas.

5. For the reasons that were discussed in 3. above the algorithm tends to stagnate (that is it becomes trapped in a "tunnel") for thousands of iterations for the largest objects in our images which are blood cells about 7 - 8 μm wide. In spite of this we have obtained a reasonable image (Fig. 10) by brute force in the case at hand.

## 8 Discussion and Conclusion

The image shown in Fig. 10 has many of the characteristics of a successful x-ray-microscope image and in particular is almost free of the oscillations due to the twin-image effect that are so evident in the image obtained from just a single backpropagation (Fig. 6). Nonetheless we do not yet regard such images as ready for use in scientific applications of x-ray holography, although they certainly represent encouraging progress toward that. Our studies continue on how to improve the speed and accuracy of our reconstruction procedure.

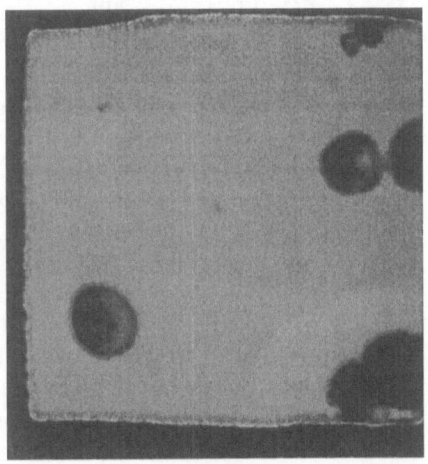

**Fig. 10.** Reconstructed hologram (see Fig. 5) with the twin-image suppression algorithm applied. The twin-image has been substantially reduced in the clear regions but the algorithm has not converged for the regions inside the objects

We believe that the use of cryo imaging essentially solves the radiation-damage problem and opens the way to multiview tomography. This raises the intriguing possibility that for a reasonably large number of views the twin-image signal may be *automatically* suppressed, as is found in certain types of electron holography [16].

## Acknowledgements

We thank J. Maser for working with us on parallel develeopment of the cryo-sample holder and interface. We thank S. Wirick for providing technical assistance and A. Osanna for providing assistance with the cryo setup at beamline X1A at the NSLS. We acknowledge useful discussions with R. Glaeser and J. Kirz. This research was supported in part by the Alexander Hollaender Distinguished Postdoctoral Fellowship Program (SL) sponsored by The US Dept. of Energy (DOE), Office of Health and Environmental Research, and administered by the Oak Ridge Institute for Science and Education, by the DOE under grant DE-FG02-89ER60858, Presidential Faculty Fellow award RCD 92-53618 (CJ), and Computational Science Graduate Fellowship Program (BC) sponsored by the DOE, and administered by Ames Laboratory. Holography exposures were carried out at the NSLS at Brookhaven National Laboratory, which is supported by the DOE.

## References

1    Aikawa, M., "Variations in structure and function during the life cycle of malarial parasites", *Bull. WHO*, **55**, 139 (1977)
2    Baltes, H.P., ed., *Inverse Source problems*, Vol. 9, Springer Series in Topics in current physics, Berlin, 1978.
3    Dainty, J.C., J.R. Fienup, "Phase retrieval and image reconstruction for astro-nomy", in *Image Recovery: Theory and Application*, Stark, H., (Ed.), Academic Press, Orlando, 1987.
4    Echlin, P., *Low-Temperature Microscopy and Analysis*, Plenum, New York, 1992.
5    Fienup, J.R., "Iterative Mehod Applied to Image Reconstruction and to Computer-generated Holograms", *Opt. Eng.*, **19**, 297-305 (1980).
6    Gerchberg, R.W., W.O. Saxton, "A practical algorithm for the determination of phases from image and diffraction plane pictures", *Optik*, **25**, 237-246 (1972).
7    Goodman, J.W., *Statistical Optics*, John Wiley, San Francisco, 1985.
8    Hadley, G., *Linear Algebra*, Addison-Wesley, Reading, USA, 1961.
9    Jacobsen, C., M.R. Howells, J. Kirz, K. McQuaid, S. Rothman, "X-ray Holographic Microscopy: Improved Image of Zymogen Granules", in *Short Wavelength Coherent Radiation: Generation and Applications*, Kirz, J., R. Falcone, (Ed.), Vol. **2**, Optical Society of America Conference Proceedings, Washington, 1988.
10   Jacobsen, C., M.R. Howells, J. Kirz, S. Rothman, "X-ray Holographic Microscopy Using Photoresists", *J. Opt. Soc. Am.*, **A7**, 1847-1861 (1990).
11   Jacobsen, C.J., *X-ray Holography: a History*, Japan Scientific Societies Press, Springer Verlag, Tokyo, 1990.
12   Joyeux, D., S. Lowenthal, F. Polack, A. Bernstein, "X-ray Microscopy by Holography at LURE", in *X-ray Microscopy II*, Sayre, D., M.R. Howells, J. Kirz, H. Rarback, (Ed.), Vol. **56**, Springer Series in Optical Sciences, Berlin, 1988.

13   Kilejian, A., A. Abati, W. Trager, "Plasmodium falciparum and Plasmodium coatneyi: immunogenicity of "knob-like protrusions" on infected erythrocyte membranes", *Exp. Parasitol.*, **42**, 157 (1977).

14   Koren, G., F. Polack, D. Joyeux, "Iterative Algorithms for Twin-Image elimination in inline holography using finite support constraints", *J. Opt. Soc. Am.*, **10**, 423-433 (1993).

15   Kurt Leskar, 1515, Worthington Ave, Clairton PA USA.

16   Len, P.M., S. Thevuthasan, C.S. Fadley, A.P. Kaduwela, M.A.V. Hove, "Atomic imaging by x-ray fluorescence holography and electron emission holography: a comparative theoretical study", *Phys. Rev. B.*, **50**, 275-278 (1994).

17   Levi, A., H. Stark, "Restorations from phase and magnitude by generalized projections", in *Image Recovery: Theory and Application*, Stark, H., (Ed.), Academic Press, Orlando, 1987.

18   Lindaas, S., M. Howells, C. Jacobsen, A. Kalinovsky, "X-ray holographic microscopy by means of photoresist recording and atomic-force microscope readout", *J. Opt. Soc. Am. A*, **13**, 1788-1800 (1996).

19   Lindaas, S., C. Jacobsen, M.R. Howells, "Development of a Linear Scanning Force Microscope for X-ray Gabor Hologram Readout", in *X-ray Microscopy*, Jacobsen, C., J. Trebes, (Ed), Proc SPIE, Vol. **1741**, Bellingham, 1992.

20   Lindaas, S. A., "X-ray Gabor Holography using a Scanning Force Microscope", Ph. D. Thesis, State University of New York at Stony Brook University, 1994.

21   Luse, S.A., L.H. Miller, "Plasmodium falciparum malarie. Ultrastructure of parasitized erythrocytes in cardiac vessels", *Am. J. Trop. Med. Hyg.*, **20**, 655 (1971).

22   Maleki, M.H., A.J. Devaney, "Noniterative reconstruction of complex-valued objects from two intensity measurements", *Obt. Eng.*, **33**, 3243-3253 (1994).

23   McNulty, I., J. Kirz, C. Jabobsen, E. Anderson, D. Kern, M. Howells, "High-resolution imaging by Fourier transform x-ray holography", *Science*, **256**, 1009-1012 (1992).

24   Meyer-Ilse, W., H. Medecki, J.T. Brown, J. Heck, E. Anderson, C. Magowan, A. Stead, T. Ford, R. Balhorn, C. Petersen, D.T. Attwood, "X-ray Microscopy at Berkeley", in *X-ray Microscopy and Spectromicroscopy*, Thieme, J., G. Schmahl, E. Umbach, D. Rudolph, (Ed), Springer Verlag, Heidelberg, 1996.

25   Pasvol, G., R.J.M. Wilson, M.E. Smalley, J. Brown, "Separation of viable schizont-infected red cells of Plasmodium falciparum from human blood", *Ann. Trop. Med. Parasitol.*, **72**, 87 (1978).

26   Saxton, W.O., *Computer Techniques for Image Processing in Electron Microscopy*, Academic Press, New York, 1978.

27   Schulz, M.S., H. Daido, K. Murai, Y. Kato, R. Kodama, G. Yuan, S. Nakai, K. Shinohara, I. Kodarma, T. Honda, H. Iwasaki, T. Yoshinobu, D. Neely, G. Slark, "Soft X-ray Holography using an X-ray Laser at 23.2/23.6 nm and 19.6 nm" in *Optical Society of America Topical Meeting; Shortwavelength V, San Diego*, Optical Society of America, Washington, 1993.

28  Snigirev, A., I. Snigireva, V. Kohn, S. Kuznetsov, I Schelokov, "On the possibilities of x-ray phase contrast microimaging by coherent high-energy synchrotron radiation", *Rev. Sci. Instrum.*, **66**, 5486-5492 (1995).

29  Trager, W., J. Jensen, "Human malaria parasites in continuous culture", *Science*, **193,** 673 (1976).

30  Trager, W., M.A. Rudzinska, P.C. Bradbury, "The fine structure of Plasmodium falciparum and its host erythrocytes in natural malaria infections in man", *Bull. WHO*, **35**, 883 (1966).

31  Trebes, J.E., S.B. Brown, E.M. Campbell, D.C. Mathews, D.G. Nilson, G.F. Stone, D.A. Whelan, "Demonstration of X-ray Holography with an X-ray Laser", *Science*, **238**, 517-519 (1987).

32  Turcu, I.C.E., I.N. Ross, M.S. Schulz, H. Daido, G.J. Tallents, J. Krishnan, L. Dwivedi, A. Hening, "Spatial coherence measurements and x-ray holographic imaging using laser-generated plasma x-ray source in the water window spectral region", *J. Appl. Phys.*, (1993).

33  Wang, S., C.J. Jacobsen, "A numerical study of resolution and contrast in soft x-ray contact microscopy", in *X-ray Microscopy and Spectromicroscopy*, Thieme, J., G. Schmahl, E. Umbach, D. Rudolph, (Ed), Springer Verlag, Heidelberg, 1996.

34  Watanabe, N., K. Sakurai, A. Takeuchi, S. Aoki, "Soft x-ray Gabor in-line holography using CCD camera", in *X-ray Microscopy and Spectromicroscopy*, Thieme, J., G. Schmahl, E. Umbach, D. Rudolph, (Ed), Springer Verlag, Heidelberg, 1996.

35  Youla, D.C., "Mathematical theory of image restoration by the method of projections on to convex sets", in *Image Recovery: Theory and Application*, Stark, H., (Ed), Academic Press, Orlando, 1987.

36  Youla, D.C., H. Webb, "Image reconstruction by method of convex projections: Part 1 - theory", *IEEE Trans. Medical Imaging*, **1**, 81-94 (1982).

# Applications of Laboratory Soft X-Ray Systems

A. G. Michette[1], C. J. Buckley[1], S. J. Pfauntsch[1], N. Khaleque[1], T. English[1],
M. Folkard[2], B. D. Michael[2], G. Schettino[1,2], I. C. E. Turcu[3], R. Allott[3], N. Lisi[3]

[1]Centre for X-Ray Science, Department of Physics, King's College London,
Strand, London WC2R 2LS, UK

[2]Cancer Research Trust Gray Laboratory, PO Box 100, Mount Vernon Hospital,
Northwood, Middlesex, HA6 2JR, UK

[3]Central Laser Facility, Rutherford Appleton Laboratory, Chilton,
Didcot, Oxon, OX11 0QX, UK

**Abstract.** Three systems which use laboratory scale X-ray sources and zone plate optics are under development: a dark field microscope, a scanning microscope and a microprobe for studies of cellular radiation response. Both microscopes are designed for use on a laser plasma source, but the dark field system can also be utilised at a storage ring. The microprobe uses either a microfocus source with a carbon target or a laser plasma source.

## 1 Introduction

Most applications of focused X-ray beams use synchrotron sources for microscopy of biological and non-biological material. However, microfocus and laser plasma sources, described briefly in section 2, have certain advantages, such as accessibility and cost, but for microscopy a compromise in terms of flux or brilliance has to be accepted [1, 2]. The development of modern non-synchrotron sources has also opened up possibilities for extending the range of applications of focused X-ray beams. In section 3 a dark field microscope designed for the laser plasma source is outlined; further details are given elsewhere in this volume [3]. A second microscope being developed for the laser plasma source, a scanned source system, is described in section 4 and a soft X-ray microprobe for studies of cellular radiation response, which uses either the laser plasma or the microfocus source, is discussed in section 5.

In experiments using laboratory X-ray sources it is important to have robust, well characterised and efficient zone plates. For the soft X-ray microprobe, in particular, ultrahigh resolution is not so important in the first instance. A new development in collaboration with the Institute of Microelectronics Technology, Chernogolovka, Russia, allows the manufacture of free standing nickel zone plates with outer zone widths of ~100 nm over diameters of ~100 μm, increasing to ~400nm for diameters of ~400 μm. It is intended to use these in the systems described below.

## 2 Laboratory Soft X-Ray Sources

### 2.1 The Microfocus Source

The microfocus source uses an electron beam, of up to 20 kV and 30 μA, focused by electron optics to a 1–2 μm diameter spot on a thin foil or solid target. This source was originally developed for the Wolter X-ray microscope at the National Physical

**X-Ray Microscopy and Spectromicroscopy**
Eds.: J. Thieme, G. Schmahl, D. Rudolph, E. Umbach

Laboratory. For the microprobe a solid carbon target is used, resulting in more flux than from a thin (1–2μm) film but a slightly increased source size. The electron energies used give a strong carbon K peak with spectrally well removed bremsstrahlung. The predicted and measured spectra for a 15 kV beam are plotted in figure 1, showing that the total bremsstrahlung contribution to the flux can be large. The predicted spectrum fits that measured (obtained over a short time during source tests) when the detector efficiency and response functions are taken into account.

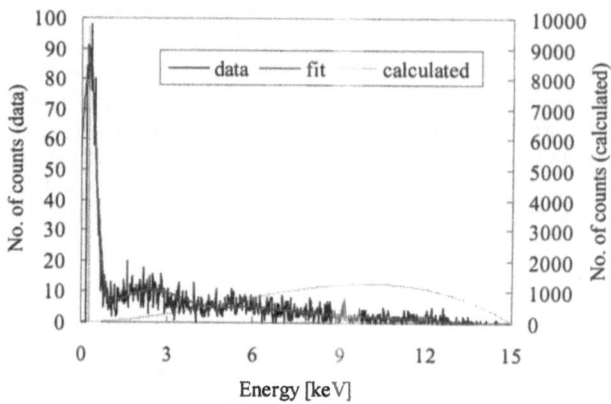

**Fig. 1.** Microfocus source spectrum for an accelerating voltage of 15 kV

## 2.2 The Laser Plasma Source

The laser plasma source used for developing the soft X-ray microscopes employs a small Nd:YAG laser at King's College. This gives insufficient flux for imaging, and so use is made of the laser plasma X-ray source at the Lasers for Science Facility (LSF) at the Rutherford Appleton Laboratory. This source, which has been described in detail elsewhere [1], delivers trains of eight 7ps sub-pulses, each with energy ~20mJ, in the 20ns pulse of a KrF excimer laser at a repetition rate of up to 100Hz (in the following, "pulse" will refer to the train of sub-pulses). The laser beam is focused to a ~15μm spot on a mylar tape target, a power density of ~$10^{20}$W m$^{-2}$. The resulting plasma is a strong X-ray line emitter from H- and He-like ions of carbon and oxygen, the major constituents of mylar; the C VI Lyman α line at λ = 3.37nm and the C V He α line at 4.02nm are the strongest. The source chamber contains He/N$_2$ at a pressure of ~20 torr, which acts as a buffer gas to prevent damage due to debris from the plasma, along with a helium jet blown across the tape surface. The nitrogen also acts as a filter to remove the oxygen lines, which are all efficiently absorbed being at wavelengths shorter than the N K edge, and to attenuate the 4.02nm line relative to that at 3.37nm, providing a quasi-monochromatic source with ~$10^{13}$ photons s$^{-1}$ mm$^{-2}$ mrad$^{-2}$ in 0.1% bandwidth at λ = 3.37nm, with only a small contamination of 4.02nm radiation which can readily be removed by using a small axial pinhole in conjunction with the zone plate.

By using a tape target consisting of a higher Z material, such as copper or iron [4], a quasi-continuum source can be obtained. This gives a much higher total X-ray output and offers the possibility of a tunable source, but any monochromating will result in a lower brilliance, for the same bandwidth, than with a low Z target.

## 3 The Dark Field X-Ray Microscope

The dark field microscope consists of a toroidal condenser, a post specimen zone plate which forms an image using X-rays scattered by the specimen, a central stop to prevent directly transmitted X-rays from reaching the detector, and a microchannel plate / phosphor / CCD combination detector. The design of the condenser – which may either be a mirror for use with a line source, or a grating for a continuum source – is described elsewhere in this volume [3]. The nickel coated mirror condenser has been made on a one inch square fused silica substrate with radii of 9.47cm and 11.015m for use at a grazing angle of 5.32°; the focal length of 1.021m means that the whole microscope is about 2m long. The laminar grating condenser, which operates at the same included angle of 169.36° as the mirror, will have 900 lines per millimetre, radii of 9.48cm and 12.113m and a groove depth of about 7nm.

Uses of the dark field microscope will include the location of a variety of deposits with elemental and, perhaps, chemical specificity [5]. The ability of a dark field system to detect features below the nominal resolution limit of the optics allows smaller deposits to be detected than with other forms of X-ray microscopy. At the LSF it should be possible to form a dark field image with one pulse of the source.

## 4 The Scanned Source X-Ray Microscope

Experiments with a prototype scanned source microscope have been reported previously [2]. The advantages of this arrangement, compared to the conventional one in which the specimen is scanned, include elimination of any acceleration induced specimen distortion and the coarser (and thus cheaper) scanning mechanism required, as the source movement is demagnified on the specimen by the zone plate. However, the system necessitates off-axis imaging which can cause an increase in aberrations (in practice always likely to be negligible) and it will be difficult to implement a source scanning system on a synchrotron.

Figure 2 shows a schematic diagram of the microscope. The arrangement requires seven movements: (i) the laser beam is focused onto the target by a lens which moves along the beam direction; (ii) the source is scanned by linked motion of the lens and target along and (iii) across the beam direction; (iv) the specimen is moved along the X-ray axis for focusing and (v) and (vi) across this axis to find areas of interest; and (vii) an optical alignment microscope is moved along the X-ray axis. Commercial linear drives are employed, with an eight axis controller linked to a computer. The computer is also used to fire the laser and to acquire and process images. One pulse of the LSF source will be sufficient to form one pixel of an image.

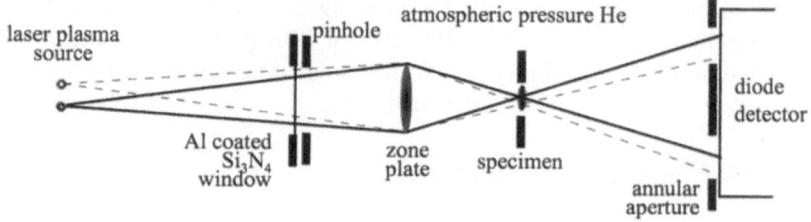

**Fig. 2.** Schematic diagram of the scanned source X-ray microscope

## 5 The X-Ray Microprobe

Studies of the response of biological cells to irradiation by X-rays have previously used unfocused beams [6]. The response is determined by a biological assay to measure the surviving fraction of cells, but no information is obtained on which parts of the cells are most susceptible to damage. Several advantages are obtained by irradiating the cells with a focused beam of X-rays. These include study of area specific sensitivity, the influence of spatial dose distribution, whether damaged cells can transmit signals to neighbouring, undamaged, cells and mechanisms of cell death other than by DNA damage. Using a pulsed source, e.g., a laser generated plasma, temporal effects can also be studied.

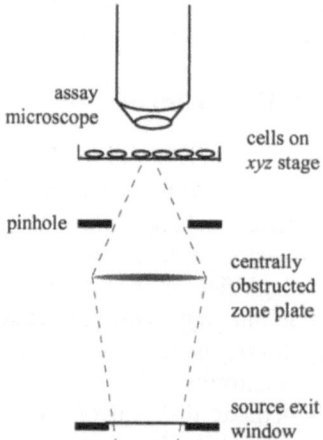

**Fig. 3.** Schematic diagram of the X-ray microprobe

A schematic arrangement of the microprobe is shown in figure 3. The assay microscope is also used to identify and locate cells, using UV radiation of sufficiently low level to not contribute to the radiation dose. The locations of the cells are stored so that each can be revisited, for irradiation, under computer control. Using X-rays with energies of a few hundred electron volts ($C_K$ or $Ly_\alpha$) doses of ~0.1–10Gy are required to give cell survival fractions of down to about 0.1% with unfocused X-rays. It is expected that similar doses will be needed with the focused probe, as damage takes place within a very short distance of the point where the X-rays are absorbed. In order to irradiate a sufficient number of cells in a reasonable time, these doses should be delivered on a timescale of about a second.

With the microfocus source the total bremsstrahlung flux is about 60 times higher than the carbon K flux. The bremsstrahlung is essentially unfocused by the zone plate and could, over time, contribute significantly to the dose. This can be reduced to effectively zero by reflection from a glass flat at 3.5°, which reduces the carbon K flux by only 35%. When the source is operated at 15μA and 20kV the dose rate is ~0.06Gy s$^{-1}$, so that a sample of 100 cells will take about an hour to irradiate. This dose rate will be improved by use of a larger and more efficient zone plate.

A pilot experiment with the LSF laser plasma source with a 5% efficient 92μm diameter zone plate has shown that a dose rate of ~0.3Gy per (four) pulse train can be achieved. Although this is sufficient to induce a measurable radiation response, higher doses are desirable. It is anticipated that improvements to both the X-ray and laser optics will yield more than an order of magnitiude improvement, such that biologically relevant doses can be delivered in a single 7ps pulse.

## 6 Conclusions

The discussions above indicate that laboratory soft X-ray sources can be used in a wide range of applications, including some which can be transferred from synchrotron sources and some novel uses. Although modern synchrotrons have the advantages of higher brilliance and greater tunability, with careful experimental design laboratory sources can be used to perform complementary research.

## Acknowledgments

The dark field X-ray microscope is supported by the Paul Instrument Fund of the Royal Society (ref 532003.G143) and by the King's College Research Strategy Fund. The scanned source microscope is funded by the Engineering and Physical Sciences Research Council (ref GR/K73381) and the microprobe by the Biotechnology and Biological Sciences Research Council (ref EO5297).

## References

1    A.G. Michette, I.C.E. Turcu, M.S. Schulz, G.R. Morrison, P. Fluck, C.J. Buckley and G.F. Foster, *Rev. Sci. Instrum.* **64**, 1478 (1993).
2    A.G. Michette, R. Fedosejevs, S.J. Pfauntsch and R. Bobkowski, *Meas. Sci. Technol.* **5**, 555 (1994).
3    S.J. Pfauntsch, A.G. Michette and C.J. Buckley, these proceedings.
4    I.C.E. Turcu, I.N. Ross and C. McCoard, *Ann. Rep. Central Laser Facility 1994–95*, 180 (Rutherford Appleton Laboratory, 1995).
5    C.J. Buckley, N. Khaleque, S.J. Bellamy and X. Zhang, these proceedings.
6    D.T. Goodhead, J. Thacker and R.Cox, *Int. J. Radiat. Biol.* **63**, 543 (1993).

# A Perspective on Biological X-Ray and Electron Microscopy

Chris Jacobsen[1], Robin Medenwaldt[2], Shawn Williams[3]

[1] Department of Physics, SUNY Stony Brook, Stony Brook NY 11794-3800, USA
[2] Institute for Storage Ring Facilities, University of Aarhus, Ny Munkegaade, DK-8000 Aarhus C, Denmark
[3] Boyer Center for Molecular Medicine, Yale University, New Haven CT 06510, USA

**Abstract.** We consider image contrast for electron microscopy of thick hydrated biological specimens, allowing for the use of phase contrast and energy filtering. This allows us to gain perspective on the relative roles of electron and soft X-ray microscopes. Radiation dose is found to depend strongly on ice thickness, with electrons offering lower dose if the ice thickness is less than about 500 nm, and x rays offering lower dose for thicker ice layers.

## 1 Introduction

Native biological structures are wet structures. The preferred tool for studying such structures often is visible light microscopy, provided the resolution is sufficient for the study at hand (near-field techniques [1] and deconvolution methods [2] can be used to study < 100 nm structures in favorable circumstances). However, both X-ray and electron microscopy offer higher spatial resolution as well as capabilities for elemental and chemical state mapping. These shorter wavelength probes come at a cost in loss of convenience and, more fundamentally, a cost in terms of damage caused by the ionizing radiation used if electrons or x rays are chosen. For that reason, cryofixation is often used for electron microscopy of hydrated structures, and it is beginning to find application in X-ray microscopy as well.

Which ionizing probe is better for high resolution studies: electrons or X-rays? On the one hand, arguments based on a high ratio of elastic to inelastic scattering and the low ($\sim 40$ eV as will be discussed) energy transfer of inelastic events point to electron microscopy as offering intrinsic advantages for atomic resolution microscopy [3, 4]. On the other hand, calculations by Sayre *et al.* [5] and others (see e.g., [6, 7]) suggest that X-ray probes offer lower radiation dose. While Monte Carlo models of image contrast for frozen hydrated specimens in electron microscopy have been considered by Schröder [8], we consider here an analytical model. The present work differs from that of Sayre *et al.* in that it assumes the availability of phase contrast in both electron and X-ray microscopy. Radiation dose is found to depend strongly on ice thickness, with electrons offering lower dose if the ice thickness is less than about 500 nm, and x rays offering lower dose for thicker ice layers.

**X-Ray Microscopy and Spectromicroscopy**
Eds.: J. Thieme, G. Schmahl, D. Rudolph, E. Umbach
© Springer-Verlag Berlin Heidelberg 1998

## 2    Electron Interactions

### 2.1    Cross Sections

When passing through a solid, electrons undergo elastic scattering with a wide range of angular deflections, and inelastic scattering. The cross sections for these interactions are well approximated by simple expressions as is described by Langmore and Smith [9], who extend earlier work by Langmore et al. [10] and Wall et al. [11]. For our purposes, we will speak of three cross sections: $\sigma_{el}$ for elastic scattering, $\sigma_{inel}$ for inelastic scattering, and $\eta$ for the fraction of elastic scattering events in which the scattered electron does *not* pass through the objective aperture in the microscope. The main case where these expressions are inaccurate is for inelastic scattering of hydrogen; we follow Langmore and Smith [9] in using their empirically determined value of $\sigma_{inel} = 8.8$ pm$^2$ at 80 keV with scaling to other voltages in our calculations. Finally, the angular distribution of inelastically scattered electrons is small enough to assume that all such electrons remain within the angular extent of the objective aperture.

These atomic scattering results can be extended to molecules and larger structures if one assumes the specimen to be amorphous over length scales smaller than the one of interest, such as is typically the case for dose-limited electron microscopy of thicker specimens. The net cross section $\sigma_m$ and objective aperture exclusion fraction $\eta_m$ for a collection of atoms (such as a molecule viewed at moderate resolution, where diffraction peaks from bonds are ignored) can be found by summation of effects on its constituent atoms:

$$\sigma_m \eta_m = \sum_{n=1}^{n=N_m} [\sigma(n)\eta(n)], \tag{1}$$

where $N_m$ is the number of atoms in the molecule. The probability for scattering is then

$$P = \sigma_m \eta_m \delta_m t_m, \tag{2}$$

where $\delta_m$ is the number density of molecules and $t_m$ is the thickness of the molecular layer. We can then speak of a probability $K$ per thickness for three interactions:

$$P_{inel} = K_{inel}t_m = \sigma_{inel}\delta_m t_m \tag{3}$$

$$P_{el,in} = K_{el,in}t_m = \sigma_{el}(1 - \eta_{el})\delta_m t_m \tag{4}$$

$$P_{el,out} = K_{el,out}t_m = \sigma_{el}\eta_{el}\delta_m t_m \tag{5}$$

$$P_{tot} = K_{tot}t_m = (K_{inel} + K_{el,in} + K_{el,out})t_m \tag{6}$$

where $P_{inel}$ refers to inelastic scattering, and $P_{el,in}$ and $P_{el,out}$ refer to elastic scattering *into* and *out of* the objective aperture, respectively. (All of these probabilities $P$ and probabilities per thickness $K$ are expressed in the limit of weak scattering per atom). Alternatively, we can also consider the mean free path for various interactions $\lambda = 1/K$; mean free paths calculated for electrons in vitreous ice ($\rho = 0.92$ g/cm$^3$) and for a "generic" protein ($H_{48.6}C_{32.9}N_{8.9}O_{8.9}S_{0.6}$,

$\rho = 1.35$ g/cm$^3$ as per [12]) are shown in Fig. 1. It should be noted that the inelastic mean free path used here differs significantly from that assumed in some calculations [8, 13], but agrees well with several experimental observations [14, 15].

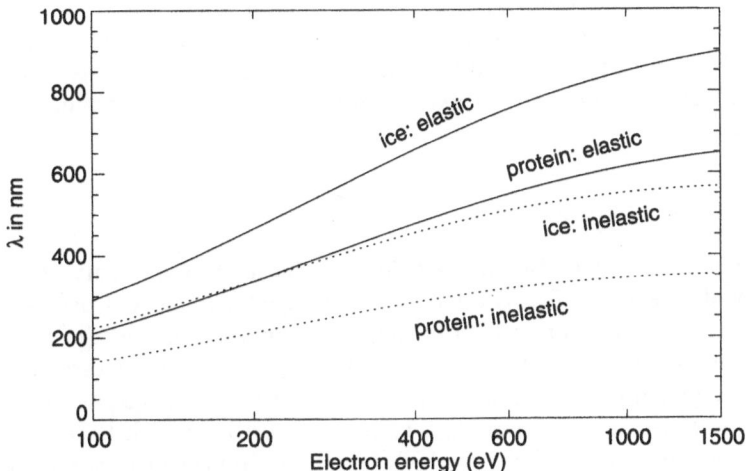

**Fig. 1.** Mean free paths $\lambda$ calculated for electrons in vitreous ice and in amorphous protein.

## 2.2   Energy Transfer

Inelastically scattered electrons cause ionizing radiation damage to the specimen. We must therefore consider the energy $\Delta E$ transferred in inelastic scattering events. For biological specimens, we will assume that inelastic scattering in ice dominates, since organic materials will be hydrated and distributed within ice. For radiation dose calculations, we wish to know what is the typical energy deposition $E$ for an inelastic scattering event. The low probability of high $\Delta E$ inelastic scattering events is partly offset by the large amount of energy they deposit. The typical energy deposition $E$ can be calculated from the probability distribution $s_1(E)$ for electron energy loss as

$$\frac{\int_0^E s_1(E)\,E\,dE}{s_1 E_{\text{tot}}} = \frac{1}{2} \text{ where } s_1 E_{\text{tot}} \equiv \int_0^\infty s_1(E)\,E\,dE. \tag{7}$$

Using EELS spectra of vitreous ice provided by R. Leapman of the National Institutes of Health (see also [15]), we calculate $E = 46$ eV for ice, whereas

Isaacson [16] gives $E = 37.5$ eV for dehydrated nucleic acid bases. The linear energy transfer

$$\frac{dE/dx}{\rho} = \frac{E}{\lambda_{\text{inel}}\rho} \tag{8}$$

calculated from these inelastic scattering cross sections and mean energy transfers is about a factor of two lower than that given in the commonly-used calculation of Berger and Seltzer [17]; this discrepancy is likely due to an overestimate of plasmon mode losses in Berger and Seltzer's calculations [18]. Using $E = 37.5$ eV and $\lambda_{\text{inel}} = 139$ nm at 100 keV to give $(dE/dx)/\rho = 2.0$ MeV $\cdot$ cm$^2$/g, one can estimate that an electron exposure of 1 $e^-$/nm$^2$ corresponds to a dose of $3.2 \times 10^4$ Gray.

## 2.3    Electron Categories

Using expressions given below, we calculate the relative intensity of electrons in a variety of categories: $I_{\text{noscat}}$ refers to electrons which have not been scattered at all; $I_{\text{1el}}$ refers to electrons which have undergone only one elastic scattering and which remain within the objective aperture; $I_{\text{multel}}$ refers to electrons which have undergone multiple elastic scatterings, no inelastic scatterings, and which remain within the objective aperture; $I_{\text{inel}}$ refers to electrons which have undergone at least one inelastic scattering yet which remain within the objective aperture; and $I_{\text{out}}$ refers to electrons which have been scattered outside the angular acceptance of the objective aperture. It can be shown that

$$I_{\text{noscat}} + I_{\text{1el}} + I_{\text{multel}} + I_{\text{out}} + I_{\text{inel}} = I_0, \tag{9}$$

so that we can consider all of the electrons to belong to one of these categories. While these expressions have limitations in describing some of the subtleties of plural scattering, they are useful as a first approximation. A plot of the fraction of electrons in each of these categories as a function of ice thickness is shown in Fig. 2.

## 3    Electron Microscopy Image Contrast and Dose

Image contrast of thin specimens in electron microscopy is well understood (see e.g., [19]). Amplitude contrast of specimens which are thick compared to a mean free path for elastic scattering (typically 100 nm) has been considered by several authors (see e.g., [5, 20]). However, the intrinsic contrast of native biological structures in vitreous ice tends to be rather low, so that in fact most work is done using defocus (to produce phase contrast) and energy filters which efficiently remove inelastically scattered electrons from the image plane of the microscope [21]. Schröder has made Monte Carlo calculations of a parameter related to image contrast of some specific model systems relevant to cryomicroscopy with energy filtering [8]. Our approach builds heavily upon the work of Sayre et al. [5] and Langmore and Smith [9].

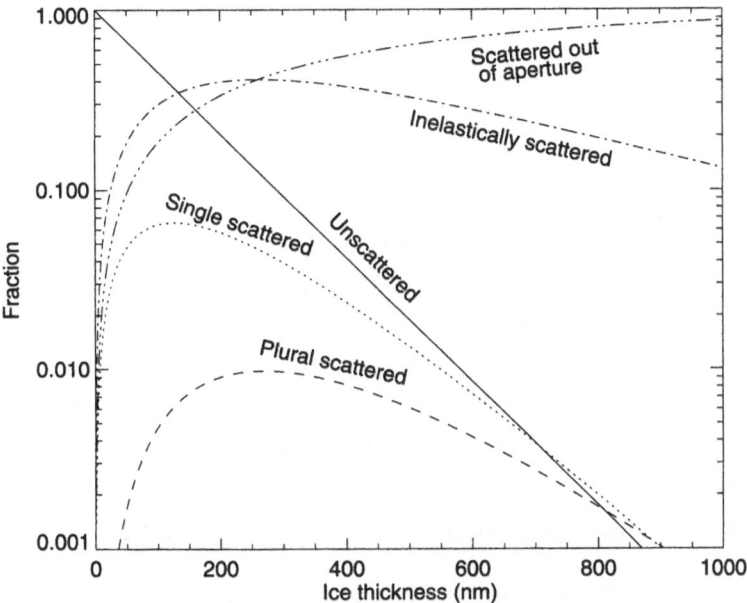

**Fig. 2.** Normalized intensity of 100 keV electrons in various categories as a function of thickness of vitreous ice. For phase contrast one desires interference between unscattered and single scattered electrons, but this signal declines steeply with increased ice thickness and furthermore the plural scattered signal rises to similar values. In thick ice layers it is excedingly important to remove inelastic scattered electrons using an imaging energy filter, for otherwise these electrons contribute an out-of-focus haze due to the fact that they are not well focused by electron optics which have intrinsic chromatic aberrations.

The above calculations allowed us to categorize electrons after passage through a thickness $t$ of an amorphous material. We now modify these calculations to consider passage through a thickness $t_f$ of a feature embedded within a thickness $t_b = t - t_f$ of a "background" material matrix. We then calculate the signal difference between the feature-present $I_f$ and the feature absent $I_b$ case and divide it by the square root of the sum of the two signals to account for electron statistics [22], giving a contrast $\Theta = (I_f - I_b)/\sqrt{I_f + I_b}$. The number of electrons $N$ required to see an object with a signal-to-noise ratio of 5:1 is then $N = 25/\Theta^2$. One can calculate $\Theta$ for a variety of imaging modes [5]; we restrict ourselves here to brightfield imaging with and without phase contrast, and with and without the use of energy filters. Using the expressions

$$I_{\text{noscat},b} = I_0 \exp[-K_{\text{tot},b}t] \tag{10}$$

$$I_{1\text{el},b} = K_{\text{el,in},b}\, t I_{\text{noscat},b} \tag{11}$$

$$I_{\text{in,noinel},b} = I_0 \exp[-(K_{\text{inel},b} + K_{\text{el,out},b})t] \tag{12}$$

$$I_{\text{multel},b} = I_{\text{in,noinel},b} - I_{\text{noscat},b} - I_{\text{1el},b} \tag{13}$$

$$I_{\text{in},b} = I_0 \exp[-K_{\text{el,out},b}t] \tag{14}$$

$$I_{\text{el,out},b} = I_0 - I_{\text{in},b} \tag{15}$$

$$I_{\text{inel},b} = I_{\text{in},b} - I_{\text{in,noinel},b} \tag{16}$$

(which were used to calculate Fig. 2) and

$$I_{\text{noscat},f} = I_0 \exp[-K_{\text{tot},b}t_b] \exp[-K_{\text{tot},f}t_f] \tag{17}$$

$$I_{\text{1el},f} = (K_{\text{el,in},b}\, t_b + K_{\text{el,in},f}t_f)I_{\text{noscat},f} \tag{18}$$

$$I_{\text{1el/f},f} = K_{\text{el,in},f}t_f I_{\text{noscat},f} \tag{19}$$

$$I_{\text{in,noinel},f} = I_0 \exp[-(K_{\text{inel},b} + K_{\text{el,out},b})t_b] \exp[-(K_{\text{inel},f} + K_{\text{el,out},f})t_f] \tag{20}$$

$$I_{\text{multel},f} = I_{\text{in,noinel},f} - I_{\text{noscat},f} - I_{\text{1el},f} \tag{21}$$

$$I_{\text{in},f} = I_0 \exp[-K_{\text{el,out},b}t_b] \exp[-K_{\text{el,out},f}t_f] \tag{22}$$

$$I_{\text{el,out},f} = I_0 - I_{\text{in},f} \tag{23}$$

$$I_{\text{inel},f} = I_{\text{in},f} - I_{\text{in,noinel},f} \tag{24}$$

we obtain results for imaging without phase contrast and with and without energy filtering of

$$\Theta_{\text{B}} = \frac{|I_{\text{in,noinel},f} - I_{\text{in,noinel},b}|}{\sqrt{I_{\text{in},f} + I_{\text{in},b}}} \tag{25}$$

$$\Theta_{\text{BF}} = \frac{|I_{\text{in,noinel},f} - I_{\text{in,noinel},b}|}{\sqrt{I_{\text{in,noinel},f} + I_{\text{in,noinel},b}}} \tag{26}$$

respectively. In the case of phase contrast with the optimum defocus for a spatial frequency of interest, we find that the contrast is given by

$$\Theta_{\text{B},\varphi} = \frac{|I_{\text{in,noinel},f} - I_{\text{in,noinel},b}| + 2\sqrt{I_{\text{noscat},f}I_{\text{1el/f},f}}}{\sqrt{I_{\text{in},f} + I_{\text{in},b}}} \tag{27}$$

$$\Theta_{\text{BF},\varphi} = \frac{|I_{\text{in,noinel},f} - I_{\text{in,noinel},b}| + 2\sqrt{I_{\text{noscat},f}I_{\text{1el/f},f}}}{\sqrt{I_{\text{in,noinel},f} + I_{\text{in,noinel},b}}} \tag{28}$$

for the cases of with and without energy filters, respectively. The additional terms for phase contrast assume that the two beams which interfere have a ±90° phase relationship between their amplitudes, which is the case for defocus phase contrast optimized for a particular spatial frequency.

Example calculations of dose using these expressions, along with calculations for phase contrast in X-ray microscopy, are shown in Fig. 3. For imaging 2 nm thick protein features in 60 nm ice at 100 keV using defocus phase contrast, these calculations predict an electron exposure requirement of 1200 $e^-/\text{nm}^2$, in reasonable agreement with experimental values of 600–700 $e^-/\text{nm}^2$ [9, 21].

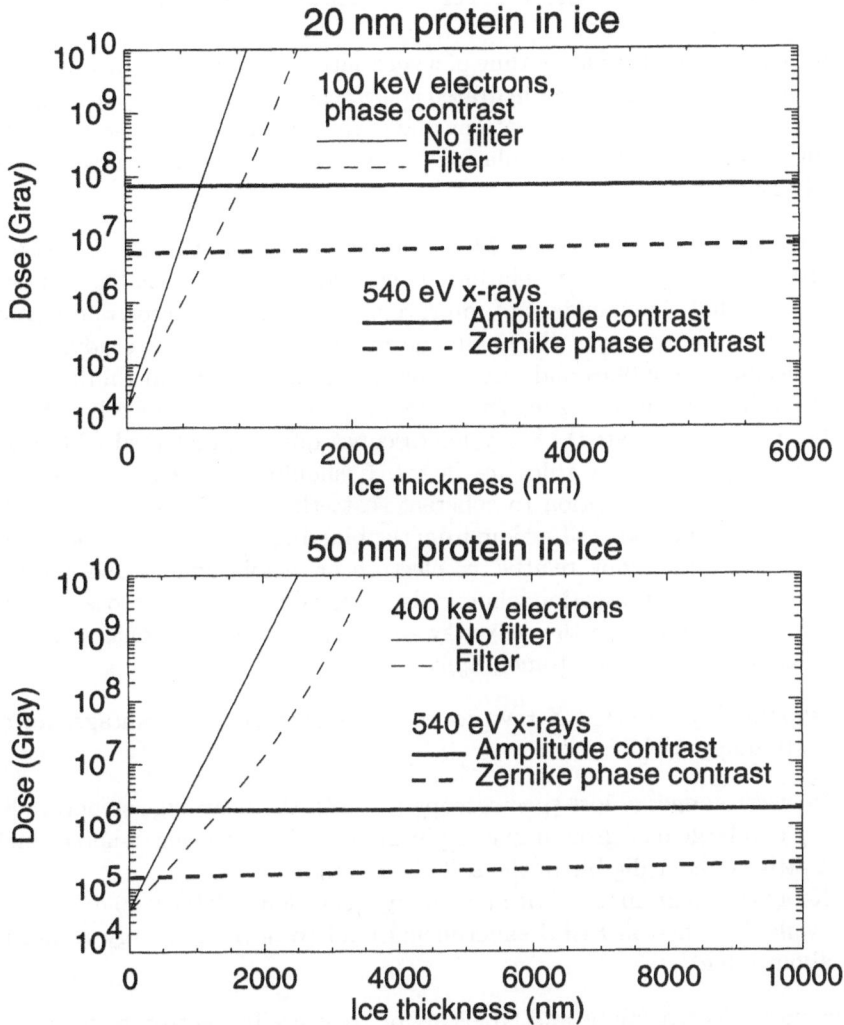

**Fig. 3.** Dose calculated for imaging protein features embedded in ice as a function of ice thickness. Note that these calculations assume that the feature is embedded exactly in the middle of the ice layer; furthermore, in the case of X-ray microscopy no allowance has been made for loss of high resolution signal due to microscope modulation transfer function. For the thinnest specimens (e.g., virus structures and macromolecular assemblies), electron microscopy provides higher resolution at low dose. For whole-cell-sized specimens, X-ray microscopy offers the ability to penetrate thick ice layers.

# 4    A Perspective on Electron and X-Ray Microscopy

We have presented here an outline of a calculation of image contrast and dose for electron microscopy of frozen hydrated specimens, and have compared the results both with experimental observations and with calculations for X-ray microscopy contrast and dose. These calculations suggest a dividing line between electron and X-ray microscopy:

- For specimens in ice layers of less than about 500 nm thickness, electron microscopy offers higher resolution at reduced radiation dose. One might then predict that, independent of improvements in X-ray microscope resolution, electron microscopy will remain the method of choice for studying macromolecular assemblies and virus structures in suspension in thin ice layers.
- For thicker ice layers (such as one might expect with whole cells which have been frozen hydrated), X-ray microscopes offer a great deal of freedom to handle exceptionally thick ice layers. It should be noted that the ratio of photoelectric absorption to coherent scattering for soft x rays is $\sim 10^4$, so that one would expect freedom from the blurring effects of multiple scattering in X-ray microscopy relative to electron or visible light probes. Note that while one can cryosection whole cells, up until now *serial* cryosectioning has not been possible so the only way to study three-dimensional structures in whole cells is through tomography.

One should also note that X-ray probes also offer intrinsic advantages for chemical state mapping [6]:

- In electron energy loss spectroscopy (EELS), the near-edge structure lies on top of a large background due to plural inelastic scattering, whereas inelastic scattering is negligible in the soft X-ray range.
- X-ray monochromators with an energy resolution of 0.05–0.10 eV are readily available, whereas EELS experiments tend to have an energy resolution of about 0.5 eV.

Of course, selective stains and labels are of great use in electron microscopy, and one can expect greater utility from further development of luminescence [23, 24] and gold sphere/dark field [25] labeling methods in X-ray microscopy as well.

Finally, it is worthwhile contemplating the ultimate limits of resolution in X-ray microscopy of frozen hydrated biological specimens. In electron microscopy, frozen hydrated specimens will "bubble" at exposures of $10^3$–$10^4$ $e^-/nm^2$, or at doses of between $3 \times 10^7$ and $3 \times 10^8$ Gray. The bubbles are comprised of hydrogen gas [26], and one may therefore expect bubbling to depend on the *rate* of dose deposition if hydrogen gas can diffuse through vitreous ice. Indeed, preliminary reports of X-ray cryomicroscopy by Schneider *et al.* indicate no observable radiation damage at the 50 nm level to doses of about $10^{10}$ Gray [27] delivered over 2 hours. If bubbling is not a limit, will cryogenic specimens prove to be indestructable? The experience of electron microscopy suggests otherwise; for example, loss of $\sim 10$ nm resolution structural detail is observed to take place

[28] at exposures of only 30 $e^-/\text{nm}^2$ or $10^6$ Gray (see also e.g., [29, 30]). Cryo methods may stop secondary radiolytical reactions, but they are powerless to prevent initial bond breakage.

# Acknowledgements

We wish to thank Bob Glaeser, Michael Isaacson, Janos Kirz, and Richard Leapman for helpful discussions. This work was carried out under support from the National Science Foundation (ECS division) under Presidential Faculty Fellow award RCD 92-53618 (CJ).

# References

1. E. Betzig, J. K. Trautman, T. D. Harris, J. S. Weiner, and R. L. Kostelak. Breaking the diffraction barrier: optical microscopy on a nanometric scale. *Science*, 251:1468–1470, 1991.
2. W. A. Carrington, R. M. Lynch, E. D. W. Moore, G. Isenberg, K. E. Fogarty, and F. S. Fay. Superresolution three-dimensional images of fluorescence in cells with minimal light exposure. *Science*, 268:1483–1487, 1995.
3. J. R. Breedlove, Jr. and G. T. Trammel. Molecular microscopy: fundamental limitations. *Science*, 170:1310–1313, 1970.
4. R. Henderson. The potential and limitations of neutrons, electrons and X-rays for atomic resolution microscopy of unstained biological molecules. *Quarterly Reviews of Biophysics*, 28(2):171–193, 1995.
5. D. Sayre, J. Kirz, R. Feder, D. M. Kim, and E. Spiller. Transmission microscopy of unmodified biological materials: Comparative radiation dosages with electrons and ultrasoft x-ray photons. *Ultramicroscopy*, 2:337–341, 1977.
6. M. Isaacson and M. Utlat. A comparison of electron and photon beams for determining micro-chemical environment. *Optik*, 50:213–234, 1978.
7. P. Gölz. Calculations on radiation dosages of biological materials in phase contrast and amplitude contrast x-ray microscopy. In A. G. Michette, G. R. Morrison, and C. J. Buckley, editors, *X-ray Microscopy III*, volume 67 of *Springer Series in Optical Sciences*, pages 313–315, Berlin, 1992. Springer-Verlag.
8. R. R. Schröder. Zero-loss energy-filtered imaging of frozen-hydrated proteins: model calculations and implications for future developments. *Journal of Microscopy*, 166:389–400, 1992.
9. J. P. Langmore and M. F. Smith. Quantitatitve energy-filtered electron microscopy of biological molecules in ice. *Ultramicroscopy*, 46:349–373, 1992.
10. J. P. Langmore, J. Wall, and M.S. Isaacson. The collection of scattered electrons in dark field electron microscopy: I. elastic scattering. *Optik*, 38:335–350, 1973.
11. J. Wall, M. Isaacson, and J.P. Langmore. The collection of scattered electrons in dark field electron microscopy: II. inelastic scattering. *Optik*, 39:359–374, 1974. of Emin should be 2.
12. R. A. London, M. D. Rosen, and J. E. Trebes. Wavelength choice for soft x-ray laser holography of biological samples. *Applied Optics*, 28:3397–3404, 1989.
13. C. Dinges, A. Berger, and H. Rose. Simulation of TEM images considering phonon and electronic excitations. *Ultramicroscopy*, 60:49–70, 1995.

14. S. Q. Sun, S. L. Shi, and R. D. Leapman. Water distribution of hydrated biological specimens by valence electron energy loss spectroscopy. *Ultramicroscopy*, 50:127–139, 1993.

15. R. Grimm, D. Typke, M. Bärmann, and W. Baumeister. Determination of the inelastic mean free path in ice by examination of tilted vesicles and automated most probable loss imaging. *Ultramicroscopy*, 63:169–179, 1996.

16. M. Isaacson. Inelastic scattering and beam damage of biological molecules. In B. M. Siegel and D. R. Beaman, editors, *Physical aspects of electron microscopy and microbeam analysis*, pages 247–258, New York, 1975. Wiley.

17. M. J. Berger and S. M. Seltzer. Tables of energy-losses and ranges of electrons and positrons. Technical Report Publication 1133, Committee on Nuclear Science, National Research Council, National Academy of Sciences, Washington, D.C., 1964. Chapter 10, pp. 205–268, Library of Congress catalogue number 64-60027.

18. M. Isaacson, 1994. Personal communication.

19. L. Reimer. *Transmission electron microscopy: physics of image formation and microanalysis*. Springer-Verlag, Berlin, third edition, 1993. Springer Series in Optical Sciences **36**.

20. A. V. Crewe and T. Groves. Thick specimens in the CEM and STEM. I: Contrast. *Journal of Applied Physics*, 45:3662–3672, 1974.

21. R. R. Schröder, W. Hofmann, and J.-F. Ménétret. Zero-loss energy filtering as improved imaging mode in cryoelectronmicroscopy of frozen-hydrated specimens. *Journal of Structural Biology*, 105:28–34, 1990.

22. R. M. Glaeser. Limitations to significant information in biological electron microscopy as a result of radiation damage. *Journal of Ultrastructure Research*, 36:466–482, 1971.

23. C. Jacobsen, S. Lindaas, S. Williams, and X. Zhang. Scanning luminescence x-ray microscopy: imaging fluorescence dyes at suboptical resolution. *J. Microscopy*, 172:121–129, 1993.

24. M.M. Moronne, C. Larabell, P.R. Selvin, and A. Irtel von Brenndorff. Development of fluroescent probes for x-ray microscopy. In G. W. Bailey and A. J. Garratt-Reed, editors, *Proceedings of the 52ⁿᵈ Annual Meeting of the Microscopy Society of America*, pages 48–49, San Francisco, 1994. San Francisco Press.

25. H. N. Chapman, J. Fu, C. Jacobsen, and S. Williams. Dark-field x-ray microscopy of immunogold-labeled cells. *Journal of the Microscopy Society of America*, 2(2):53–62, 1996.

26. R. D. Leapman and S. Sun. Cryo-electron energy loss spectroscopy: observations on vitrified hydrated specimens and radiation damage. *Ultramicroscopy*, 59:71–79, 1995.

27. G. Schneider, B. Niemann, P. Guttmann, D. Rudolph, and G. Schmahl. Cryo x-ray microscopy. *Synchrotron Radiation News*, 8(3):19–28, 1995.

28. J. F. Conway, B. L. Trus, F. P. Booy, W. W. Newcomb, J. C. Brown, and A. C. Steven. The effects of radiation damage on the structure of frozen hydrated HSV-1 capsids. *Journal of Structural Biology*, 111:222–233, 1993.

29. M. K. Lamvik. Radiation damage in dry and frozen hydrated organic material. *Journal of Microscopy*, 161:171–181, 1991.

30. M. F. Schmid, J. Jakana, P. Matsudaira, and W. Chu. Effects of radiation damage with 400-kV electrons on frozen, hydrated actin bundles. *Journal of Structural Biology*, 108:62–68, 1992.

This article was processed using the LATEX macro package with LLNCS style

# Carbon Index Measurement Near K Edge, by Interferometry with Optoelectronic Detection

Denis Joyeux[1] and François Polack[2]

[1]Institut d'Optique Théorique et Appliquée (IOTA),
BP 147 - 91403 Orsay cedex, France
E-mail: denis.joyeux@iota.u-psud.fr

[2]Laboratoire pour l'Utilisation du Rayonnement Synchrotron (LURE),
Campus d'Orsay, bat 209d, 91405 Orsay cedex, France
E-mail: polack@lure.u-psud.fr

**Abstract.** After discussing the design principles of soft X-ray interferometers, we present in some details the implementation and recent results of an experiment currently developed at Orsay, to measure the dispersion of the carbon refractive index, near the K-edge. A particular stress will be given to the moiré-based detection system, which provides a quick, nearly "real-time" measurement of the sample optical thickness.
More generally, as interferometric techniques provide access to the optical phase of X-ray wavefronts, they should bring a powerful new toolbox for doing physics with soft X-rays. A short overview on some interferometric experiments, in preparation or already attempted, will be given.

## 1 Introduction

As the *luminance* (flux per unit of beam étendue) of recent synchrotron radiation (SR) sources is improved by orders of magnitude with respect to older X-ray sources, the corresponding increase of the coherent flux available opens the way to the realization of true interferometric experiments, i. e. experiments that give direct access to the phase of propagating wavefronts [1]. As a matter of fact, spatially coherent experiments started in the past 10–15 years with the focusing of a diffraction limited scanning spot (for scanning microscopy), and the recording of Gabor holograms [2]. Since then, the most recent progress has made feasible truly interferometric experiments, i. e. involving interferometers, in the usual sense.

In our opinion, interferometry should provide new tools for soft X-ray physicists. This claim is well illustrated by the present situation of refractive index determination. The refractive index (or its microscopic equivalent, the $f_1$ form factor) is a basic parameter of the photon-matter interaction, and also an important one for the design of some X-ray optics. In the soft X-ray spectral range, most determinations are based on absorption measurements, and the use of the Kramers-Kronig relations. The result is far from being reliable [3] in the resonance regions, most interesting for condensed matter studies. Other methods are based on angular reflectivity data. However, the processing must take into account the roughness of the surface and it is impaired by the strong influence of absorption [4]. In summary, the refractive part is not reliably known at least in some important spectral regions. That is why, after an early experi-

**X-Ray Microscopy and Spectromicroscopy**
Eds.: J. Thieme, G. Schmahl, D. Rudolph, E. Umbach
© Springer-Verlag Berlin Heidelberg 1998

ment performed in difficult experimental conditions by Aoki in 1986 [5], we proposed a possible way to measure the index interferometrically [6, 7]. The principle is the usual one, namely to produce some simple fringe pattern by means of a two-arms interferometer, and to measure the fringe shift when a sample is introduced in one arm. We report here in some details the present state of this experiment, with a particular emphasis to the detection system.

However, as the soft X-rays place specific constraints on practical realizations, we shall first discuss the design and implementation of interferometers for this spectral range.

## 2 Designing the Soft X-Ray Interferometer

### 2.1 Basic Design Principles [8]

Three particular issues must be considered: a) the necessary optical quality of wavefronts, b) the large absorption of materials; and c) the necessary mutual coherence of wavefronts, separating as usual the temporal and the spatial terms.

Within the soft X-ray spectral range, only *plane mirrors,* preferably under grazing incidence, can deliver interferometric quality wavefronts (this is far from being the case with any focusing component). Single metallic coatings can be considered for improved reflection efficiency, if necessary. Simple considerations show that $\lambda/10$ plane wavefronts ($\lambda \approx$ a few nm or larger) can be obtained on a limited surface (a few mm$^2$), using today's top level optical polishing. Accordingly, the surface roughness can be reduced to 0.2–0.3 nm, or even less, down to 0.05 nm.

Absorption makes the use of true reflecting/transmitting beam splitters very difficult. Therefore we chose to eliminate the beam splitting problem by considering only true *wavefront division interferometers.* This choice implies some minimal spatial coherence.

Besides, flux is usually a limiting factor. It is therefore advisable to use as large a spectral bandwidth as possible, restricting the path difference (i. e. the temporal coherence) to what is just necessary. As a matter of fact, the ultimate monochromaticity that is practically available from SR sources does not allow large path differences, e. g not more (and usually much less) than 20 μm at $\lambda$=4 nm with $\lambda/\delta\lambda$=5000 (an excellent value in this spectral range). From the technical point of view of the design and realization, this is a *near zero path difference interferometer.*

The required spatial coherence can be obtained by tuning the source divergence and size, and/or the source to experiment distance. It must be reminded that, if properly made, this operation does not change the available coherent flux, which is determined by the source luminance and the wavelength, not by the intrinsic coherence of the source [1].

This analysis is based on simple considerations, with the view of implementing a system that does not require sophisticated optical and mechanical components. Other choices are no doubt possible [9, 10] at the cost of increased complexity. In addition, we would like to remind the almost perfect solution proposed and realized by Bonse et al [11]. As it consists of a Mach-Zehnder interferometer, manufactured from a single perfect monocrystal of silicon, the reflection and beam splitting are obtained by the Bragg effect and the wavefront quality relies on the lattice regularity. Unfortunately, this is not applicable to wavelengths much longer than $\lambda \approx 0.1$ nm.

## 2.2 Practical Implementation for Optical Thickness Measurements

According to the previous discussion, we choose the Fresnel mirrors interferometer (or similar configurations) as a good basic design for interferometric experiments. In practice, we used the same implementation as described in ref. 6.

In order to obtain about 50 fringes with a reasonable spacing (in view of the pattern analysis), the tilt angle was set to 2.25 arcmin and the observation plane was placed 120 mm downstream from the mirror's common edge. This sets the required spatial coherence to $\approx 160$ $\mu$m. As the source is practically at infinity (10 m), the resulting 3-D interference pattern consists of parallel planes, with 3.36 $\mu$m spacing at $\lambda = 4.4$ nm, and modulated by Fresnel diffraction fringes from the mirror's common edge. The recorded field is approximately 8 mm in the direction of the fringes.

The source was the exit slit of a TGM monochromator on the SU7 undulator beamline at SuperACO (Orsay). It was adjusted so as to give sufficient temporal coherence (in fact $\Delta\lambda/\lambda \approx 1/400$) as well as spatial coherence ($\approx 0.65$ in the fringe pattern, from the Zernike theorem).

The bimirror was realized as a single optical element, with a stable tilt angle. In order to minimize the optical workshop difficulties, two flats with parallel faces (10x15 mm$^2$) were first polished with special care to the zone near the common edge (the active region in fact), and to the roughness. These flats were then bonded together on a metallic base, providing the correct tilt angle. The gluing technique was such that the angle defined by the metallic parts is not changed during and after assemblage (fig. 1), neither by the glue layer nor by any stress. To this end, we used an optical UV bond, putting drops into spaces that were specially left between the glass and the metal. Note however that it is not critical to get an accurate predetermined value to better than a few percents.

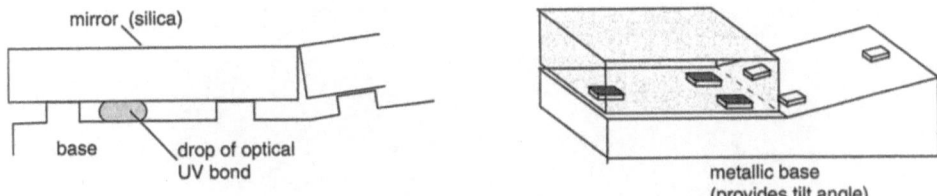

**Fig. 1.** Realization of the bimirror as a single optical part.

Exactly as in "visible" interferometry, a transmitting sample (under normal incidence) suitably introduced in front of one of the two mirrors shifts the fringe pattern, according to the relation: (shift/ fringe spacing)=(n-1)×thickness/$\lambda$. The next step is to measure the fringe shift.

## 3 Experiments with Photographic Detection

As a first try, we used high resolution photographic plates (Kodak HR 1A) to record the interferogram. This implies simultaneous recording of a reference pattern and the shifted pattern. To this end, the sample introduction was realized so as to "shadow" half the width of one mirror, as depicted on Fig. 2.

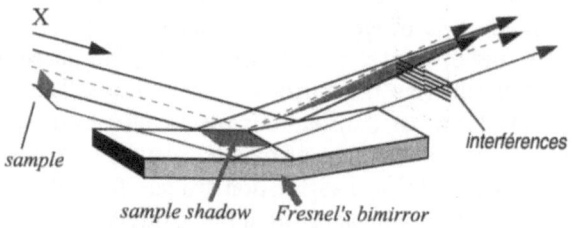

**Fig. 2.** Fresnel's bimirror arrangement, for simultaneous production of sample and reference patterns.

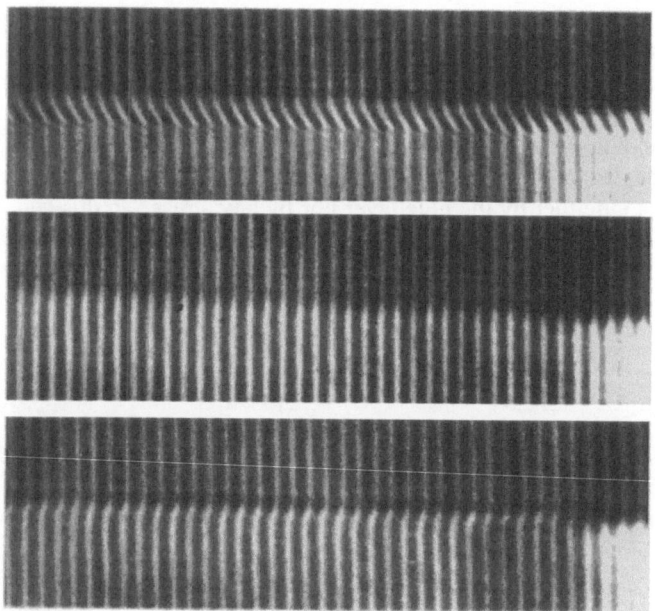

**Fig. 3.** Three interferograms, recorded at successive wavelengths, with $\Delta\lambda$=0.02 nm ($\lambda\approx$4.4 nm). The reference patterns are at the top of each image, and the sample patterns at the bottom. The sample is a 3 μm thick layer of photoresist.

The photographic plates were  tilted to receive X-rays at a 6° grazing incidence in order to increase the apparent spacing of the recorded fringes up to ≈30 μm. A few tens of such records have been obtained at different wavelengths in the K-edge region. Figure 3 shows a partial view of three interferograms, obtained with a sample made of a layer of carbon polymer (a photoresist). They show clearly the sign inversion of δ=1-n in the K edge region.

Later, we recorded a series of interferograms using a free standing foil of evaporated carbon, with a mass surface of 103.6 µg/cm² (provided by the Goodfellows Co, Great Britain). These were processed by Fourier analysis according to ref. 7. Although we were able to draw a partial dispersion curve (from ≈15 interferograms), we concluded that photographic recording could in no case be a routine detection method. We therefore designed an optoelectronic detection system, which allows a direct and quick determination of the fringe shift produced by the sample.

## 4 Experiments with Direct Detection of the Fringe Shift

### 4.1 Principle and Realization

The system is based on the production and detection of a moiré pattern between the fringes and a binary grid of equal spacing and orientation. This moiré consists of a uniform pattern (zero spatial frequency), and the total flux transmitted by the grid is sinusoidally modulated, when the relative position of the grid and the fringes varies linearly. This allows to track and measure with an excellent sensitivity any change in

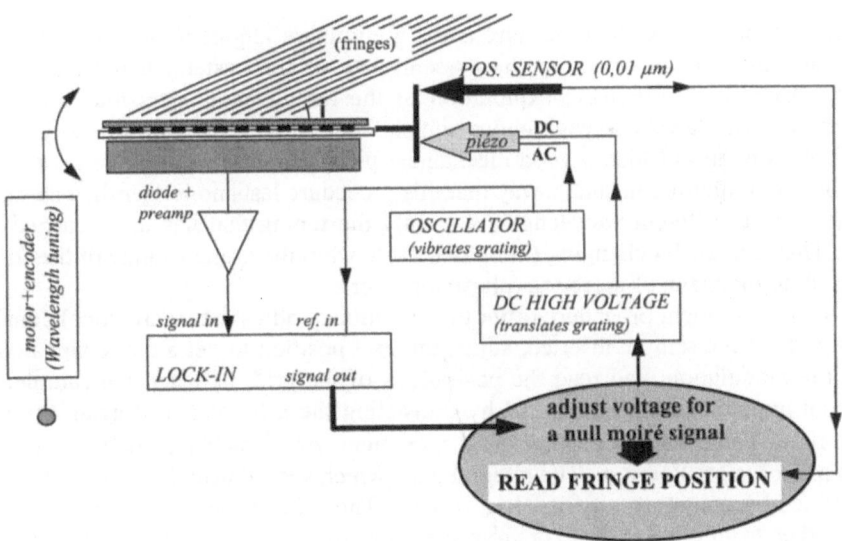

**Fig. 4**. Principle of the moiré detector, for direct determination of the fringe shift.

the pattern/grid relative position (provided that the position of the grid can be monitored), e. g. by keeping the moiré signal maximum.

In practice, we used the system depicted in Fig. 4. The X-ray fringe pattern is first converted into a visible fringe pattern by a thin layer of very fine phosphor. This layer is deposited onto a mask consisting of a binary grating etched by microlithographic techniques (for accuracy) into a metallic layer coated on a glass plate. As with the photographic detection, the plate is tilted so as to make the X-ray beam grazing at an angle of about 9 deg; thus the visible fringe pattern has a spacing of about 20 μm. Finally, the grid orientation relative to the fringe direction, and the fringe grazing incidence onto the grid plane can be adjusted in order to tune exactly the grating and fringe spatial frequencies.

The phosphor-grid plate can be translated and vibrated parallel to its own plane and perpendicular to the grating lines, and the grid average position is monitored by a high sensitivity position sensor. A single photodiode integrates the flux transmitted by the grating through the glass plate, giving the moiré signal, averaged on the whole fringe pattern.

With this configuration, the key signal is the time modulation of the moiré signal which is synchronous with the grid vibration. For best sensitivity, the vibration amplitude is about 1 pitch peak-to-peak, and the moiré signal is detected by means of a lock-in amplifier with a narrow bandwidth (a few seconds time constant). The monitoring of the fringe/grid position is obtained by tracking the zeros of the synchronous signal (i.e. extrema of the moiré signal), through the translation of the grid plate. The resulting sensitivity (experimental noise limit) on the fringe position is presently about $\lambda/600$ for high contrast fringes, i. e. when absorption is small.

### 4.2 Operating Procedure

The most tricky part of the procedure is the preliminary alignment and tuning of the fringe and the grid, because, in the present state of the system, it must be made blindly. Therefore, systematic exploration of the tuning parameters must be done, until the lock-in detects a synchronous signal. Then, this signal can be easily maximized. As a matter of fact, a good mechanical prealignment is enough to restrict the domain of exploration, in such a way that this procedure lasts no longer than about 15 minutes. For a different wavelength, it is only the tuning that has to be carried out again. This is made by changing (in a predictable way) the grazing angle of the fringe pattern onto the plane of detection (phosphor layer).

The measurement procedure (after the preliminary adjustments are done) consists of: 1) without the sample inserted, adjust the grid position to get a moiré signal with null time modulation, and read the position $x_1$ of the grid; 2) insert the sample, retrieve an unmodulated moiré signal by translating the grid, and read again the new position $x_2$. Then, the optical thickness increment $\Delta t$ is such that $\Delta t/\lambda$ is the fractional part of $(x_2-x_1)$/period . The integral part (which should usually be 0 or 1) can be determined by continuity or from known data. The sign of the increment can be derived either from a priori data, or from the sense of the shift and the analysis of the interferometer geometry.

This principle was first tested in visible light on an optical bench [12], with the same geometrical parameters as in the X-ray experiment, i.e. 50 fringes with 30 μm spacing, but without the phosphor layer. We found that a sensitivity of about $\lambda/800$ was achievable in these conditions. Finally, it should be noted that a full automatization (PC based) of the measurement procedure is not difficult to implement.

## 4.3 Results and Discussion

The moiré detection system was completed and tested recently with a free standing carbon foil (evaporated carbon, mass surface: 103.6 μg/cm²), producing the dispersion data shown in figure 4 with black diamonds. It is noteworthy that this curve is in excellent agreement with the former determination from photographic records (crosses) obtained with the same carbon foil. A plot of the imaginary part β of the complex refractive index, obtained from absorption data of the sample, is also shown. Care has been taken to get rid of the unavoidable scale shifts occurring on SR beam lines, keeping the wavelength scales consistent for all three plots. This was made by monitoring the position of the spectral dip arising from absorption by the carbon pollution on optical surfaces in the beam line. This provides a recalibration accuracy better than .002 nm.

Data from a standard table (available from CXRO, on internet) are given for comparison (white diamonds). As our wavelength scale was not absolutely calibrated, we shifted the table data so as to align the index peak with the absorption edge (middle of the β plot, from experimental absorption data), which is consistent with the data provided by tables.

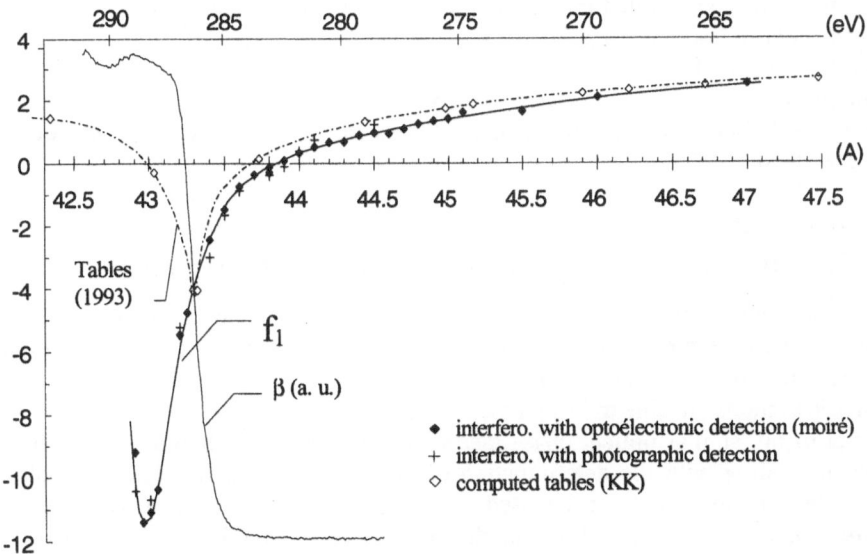

**Fig. 5.** The $f_1$ form factor of an evaporated carbon foil, near K edge vs. Wavelength. Two sets of data are shown: black diamond are obtained by optoelectronic measurement of the fringe pattern shift, and crosses are obtained by processing photographic records of interferograms. $\delta = 1-n$ is proportional to $f_1 \lambda^2$. β is derived from measurements of the sample absorption.

Figure 5 shows two major differences between our experimental data and the theoretical ones: the peak depth, and a small but significant shift between the peak position and the absorption edge position (less than 1% of the edge position). Although the first difference is not surprising, the second one is, as a shift was never reported or predicted. We have therefore to be cautious, not excluding the possibility of an experimental artifact. First of all, the consistency of such a shift with the Kramers-Kronig relations must be checked theoretically: is such a shift forbidden by the mathematical properties of the KK relations ? Although it is forbidden when a simple model of one single edge is used, it is not clear whether the result holds for actual spectra or not. Besides, carbon is known to be quite complicated, and our sample was not well characterized (amorphous, graphite or a mixture of different species).

With this in mind, the following comments can be made:

1. The good agreement between determinations with the moiré system and from photographic records excludes errors due to some parasitic electronic signal, (e. g. taking precedence over the true signal when absorption becomes large). As a matter of fact, photographic records show a fringe shift of about 1 fringe at the wavelength of the peak.

2. It is noteworthy that theoretical and experimental curves seem to join in the long wavelength branch. Conversely, they start separating when the absorption is still small, thus excluding artifact due to large absorption.

3. Multiple reflection in the carbon foil is completely excluded, owing to the large absorption (on the small wavelength side) and small reflectivity.

4. Spectral pollution can be invoked, as it does exist. Although we have not as yet found a mechanism that may produce a spurious index peak, this point is worth of further investigations.

In conclusion, further measurements with an improved system, and a thinner sample (to decrease absorption) have to be performed, and the influence of a spectral pollution of the large absorption side by diffusion from the small absorption side has to be evaluated.

# 5 Conclusion:
# A Quick Overview on Further Interferometric Experiments

Various improvements are planned in the experimental detection system in order to assess and increase the reliability of the index determination. However, and despite the possibility of an artifact in the data presented, there is no doubt that interferometric systems are able to bring new and valuable data.

First of all, similar optical phase measurement techniques should benefit all experiments dealing with the determination of optical constants. In addition to direct index determination, they can be used, as presented here, for the measurement of *complex reflection factors*. To this end, the measured reflecting layer can be deposited on one of the two mirrors, provided its thickness is known or compensated for. This additional measurement can even be unnecessary, if the phase variation is related to some external influence, as is the case for studying the field related properties of magnetic layers.

However, it is most likely that we have so far only a faint idea of the new experimental capabilities that will be brought by interferometric techniques. In our opinion, they might be an important future development of the physics made with soft X-rays. The following experiments are good early examples: some are presently planned, and some have already been attempted.

In the field of microscopic imaging, new developments were presented in this conference. A *phase sensitive scanning imaging* system, working in a differential interference mode, has been proposed [13] to extend the field of soft X-ray scanning microscopy to phase structures. Preliminary results have been presented [14]. Phase imaging in hard X-rays, using Bonse' interferometer [15], has also been presented.

*Interferometric plasma tests,* again based on a Mach-Zehnder configuration, and using the X-ray laser, have already been reported [10]. Similar investigations, with a Fresnel mirrors interferometer, are presently under development at the intense laser facility LULI (Palaiseau, France), in collaboration with LSAI (Orsay). Here the source will be the X-ray laser itself at 21.2 nm.

Finally, the feasibility of high resolution *Fourier transform spectroscopy* in the XUV-soft X-ray range has been analyzed, and a project based on a Mach-Zehnder configuration is presently being developed for a resolution of several 100,000 [9]. Our group is presently considering the realization of a different configuration [8], according to the discussion of section 2.

In our opinion, coherent techniques might be a key tool for the future development of the physics made with soft X-rays.

## Acknowledgments

This work was partly supported by EEC grant, contract n° CHRX CT94 0600.

It was made with the collaboration of Jan Svatos during his PhD thesis. The interferometer was designed by Daniel Phalippou (research engineer, Institut d'Optique), and manufactured by the Institut d'Optique workshops. We are indebted to these persons, whose skill and competence made these results possible.

## References

1    D. Joyeux, P. Jaeglé, and A. L'Huillier, in *Trends in Optics*, vol 3 (Academic Press, 1996).

2    See for instance: *X-Ray microscopy II,* eds D. Sayre, M. Howells, J. Kirz and H. Rarback, (Springer, Berlin, 1988).

3    C. T. Chantler: Radiat. Phys. Chem., 41, 759 (1993).

4    E. Spiller, Appl. Opt., 29, 19 (1990).

5    S. Aoki, S. Kikuta, AIP conf. proc. 147, 49 (1986).

6    F. Polack and D. Joyeux, in *X-ray Microscopy III*, A. G. Michette, G. R. Morrison, and C. J. Buckley eds. (Springer series in optical science vol. 67, Springer, 1992).

7    J. Svatos, D. Joyeux, D. Phalippou, F. Polack, Opt. Lett., 18, 1367 (1993).

8    F. Polack, D. Joyeux: Rev. Sci. Instrum., 66, 2 (1995).

9    M. R. Howells, K. Frank, Z. Hussain, E. J. Moler, T. Reich, D. Möller, D. A. Shirley: Nucl. Instrum. Methods, A347, 182, (1994).

10   B. DaSilva et al: Phys. Rev. Letters, 74, 3991 (1995).

11   U. Bonse, H. Lotsche, and A. Henning, J. X-ray Sci. Technol., 1, 107 (1989).

12   J. Svatos, Ph. D. thesis, (Univ. Paris XI, Orsay, France, 1994).

13  F. Polack, D. Joyeux, *X-ray Microscopy IV*, edited by A. I. Erko and V. V. Aristov, (Bogorodski Pechatnik, Chernogolovka, Moscow, 1994).

14  F. Polack and D. Joyeux, "Phase contrast experiments on the NSLS-X1A scanning microscope", this conference.

15  U. Bonse, F. Beckmann, F. Busch, and O. Günnewig "X-ray microtomography using interferometric phase contrast", this conference.

# Microspectroscopy and Spectromicroscopy

# Atmospheric Inputs and Species Interactions

# NEXAFS and X-Ray Linear Dichroism Microscopy and Applications to Polymer Science

H. Ade

Dept. of Physics, North Carolina State University, Raleigh, NC 27695, USA

**Abstract**. We review the development of transmission Near Edge X-ray Absorption Fine Structure (NEXAFS) microscopy and linear dichroism microscopy over the last few years utilizing the X1-Scanning Transmission X-ray Microscope (X1-STXM) at the National Synchrotron Light Source and present some of its applications. NEXAFS provides excellent specificity to various functional groups and moieties in organic molecules and polymeric materials. This specificity can be utilized to map the distribution of various compounds in a material, or to micro-chemically analyze small sample areas. Examples of applications include the study of various phase-separated polymers, multicomponent polymer blends, and polymer laminates. Linear dichroism microscopy furthermore explores the polarization dependence of NEXAFS in (partially) oriented materials, and can determine the orientation of specific functional groups. Applications of linear dichroism microscopy have focused so far on determining the relative degree of radial orientation in Kevlar® fibers on a semi-quantitative basis.

## 1 Introduction

Scanning and transmission x-ray microscopes have originally been developed over the last two decades primarily with the goal to image biological specimen in the wet/hydrated state. This is also the case for the Stony Brook Scanning Transmission X-ray Microscope (X1-STXM) located at the National Synchrotron Light Source (NSLS). More recently, the X1-STXM has also been utilized to investigate other materials, which most frequently were synthetic polymers. This is primarily due to the realization that compositional sensitivity can be achieved via Near Edge X-ray Absorption Fine Structure (NEXAFS) spectroscopy [1] while orientational sensitivity if provided via X-ray linear dichroism microscopy [2].

We illustrate the chemical finger printing capability of NEXAFS spectroscopy by showing NEXAFS spectra of poly(ethylene terephthalate) (PET), polycarbonate (PC), polyarylate (PAR), and poly (p-phenylene terephthalamide) (PPTA) (Kevlar®) in Fig. 1 and acquired with the X1-STXM at an energy resolution of about 0.3 eV. Although these polymers have similar chemical functional groups (carbonyl, aromatics, etc.) their spectra are quite different and unique. Even the transitions near 285 eV which are due to the aromatic groups in the these polymers have pronounced different lineshapes. This reflects the fact that the molecular orbitals probed by NEXAFS spectroscopy are complex and can extend across many atoms. Particularly, conjugated $\pi$ orbitals can be delocalized across many atoms, as is the case for PET,

**X-Ray Microscopy and Spectromicroscopy**
Eds.: J. Thieme, G. Schmahl, D. Rudolph, E. Umbach
© Springer-Verlag Berlin Heidelberg 1998

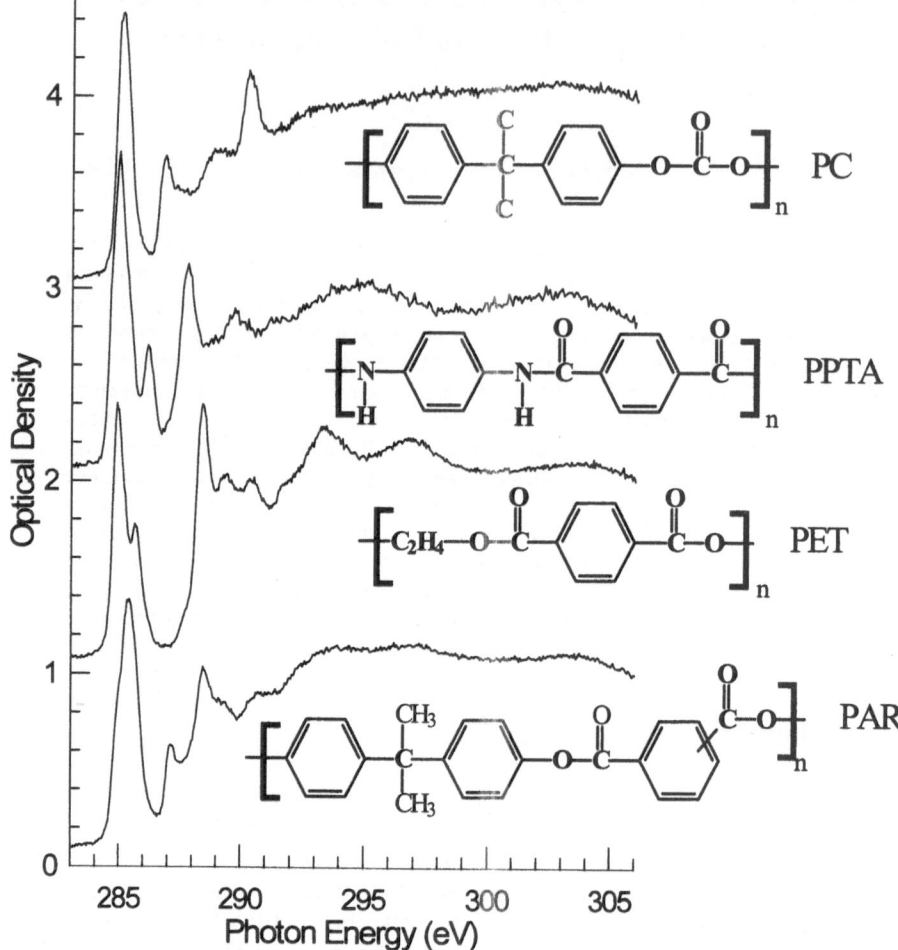

**Fig. 1.** Carbon NEXAFS spectra of a variety of polymers exhibiting a unique spectral signature, despite the common presence of certain functional groups. All spectra were acquired with the Stony Brook X1-STXM with an energy resolution of about 0.3 eV.

PC, PAR and PPTA, with a significant influence on the NEXAFS spectrum of the respective material. NEXAFS spectra of related polymers can be found in Refs. [3] and [4].

Given the excellent characterization capabilities of NEXAFS spectroscopy, it is not difficult to see that NEXAFS microscopy would make an excellent analytical tool. The first demonstration of NEXAFS microscopy relied primarily on the spectral differences of saturated versus unsaturated bonding by imaging the morphological characteristics of a binary blend composed of polypropylene (PP) and poly(styrene-*r*-

crylonitrile) (SAN) [1]. A subsequent, illustrative example of the capabilities of NEXAFS microscopy was the imaging of the morphology of PET-PC blends without staining [5]. High contrast and contrast reversal was achieved with a change in photon energy of only 350 meV in an energy range near 285 eV that was sensitive to the substitutional groups of the phenylene in either polymer (see Fig. 1). The image contrast and contrast reversal was thus based on the presence or absence of carbonyl or ester groups next to the phenylene, respectively, and the interaction between these groups. We will focus our subsequent discussion on the X1-STXM and additional applications with it, as it is this instrument that has been utilized first for transmission NEXAFS microscopy and is presently the X-ray microscope used most extensively for polymer research.

## 2 Experimental

The Stony Brook X1- STXM uses as its X-ray source a high brightness undulator at the NSLS that is demagnified by a zone plate to a small micro-probe. The spot size achieved with the zone plate determines the spatial resolution of the microscope. Features as small as 35-40 nm have been observed [6,7]. The transmitted photon flux of a thin sample is detected with a gas flow counter. The optical density of the sample is thus measured and its two dimensional (X,Y) variation is utilized to provide contrast in a raster scanned image. The sample is located in a He purged, atmospheric pressure enclosure and is investigated at room temperatures.

In addition to imaging, the focused beam can remain on the same sample spot while the photon energy is adjusted to acquire an energy scan. Simultaneously the sample zone plate distance is adjusted to remain in focus. In order to get absorption spectra, an energy scan (I) from the sample is recorded, and subsequently or just prior to it another energy scan ($I_0$) is recorded without a sample or through an open area of the sample. The negative log ratio of these energy scans ($-\ln(I/I_0)$) is an optical density spectrum in units of absorption lengths. It takes a few minutes to acquire a chemically sensitive image, and about the same time to record several energy scans from small sample areas and normalization scans from open areas. Spectra and images are acquired with an energy resolution of about 0.3-0.4 eV. Energy calibration is provided *in-situ* by leaking $CO_2$ into the He enclosure of the microscope while the sample is in place. The $CO_2$ procedure utilized also easily reveals problems with the instrument [8].

Typically, sections 100-200 nm in thickness are utilized for carbon K-edge NEXAFS. This thickness provides the best compromise between good signal to noise and distortions of spectral features due to excessive thickness and the presence of some background signal. However, samples ranging in thickness from 30 to 500 nm have been investigated successfully so far. If required, spectra can be normalized for thickness and density variations between different sample locations by utilizing the vacuum continuum cross-section above the edge which is devoid of chemical sensitivity (above about 320 eV for carbon NEXAFS). Similarly, density and thickness variations in images can be detected and corrected for by acquiring an image above 320 eV, or by isolating the chemical composition information via ratio images. Additional details about the BNL STXM and its performance is provided in articles by Jacobsen *et al.* [6] and Zhang *et al.* [9].

In X-ray linear dichroism microscopy, none of the hardware of the X1-STXM has to be changed. One simply takes explicit advantage of the linear polarization of the X-ray source if one is interested in the (average) orientation and degree of orientation of certain molecules or bonds in the sample [2].

## 3 NEXAFS Microscopy Applications

Several studies have been undertaken since the first demonstration of NEXAFS microscopy and linear dichroism microscopy. Additional work includes the study of the morphology of polymer blends [10-12] and phase-separated polymers, such as precipitates in polyurethanes [5] and liquid crystalline polyesters [13], the study of layered polymers or polymer laminates [14], diffusion at interfaces [15], orientation in Kevlar fibers [16], heat treated polyacrylonitrile fibers [17], development of exposure strategies for poly(methyl methacrylate) resists [18], and studies of biological [19,20] and organic geochemical [21-23] samples. We will review some of these in some detail below.

At times, the spatial resolution afforded by the X-ray microscopes is also a great help in acquiring NEXAFS spectra from samples that are difficult to prepare as a uniform bulk material. An example might be the acquisition of NEXAFS spectra from a soft segment polyurethane model polymer, which is a viscous liquid and can be supported on a holely carbon grid owing to surface tension, or small flaky materials such as certain polyurea model polymers [10,24].

### 3.1 Polymer Blends

Since NEXAFS microscopy provides direct compositional information at relatively high spatial resolution, it might be an invaluable tool to characterize multi-component systems, such as ternary or quaternary polymer blends. Traditionally, conventional microscopies, particular electron microscopy in conjunction with staining methods, are used for the characterization of these materials. Many systems of interest contain, however, components that have very similar absorption rates for staining agents. It is then impossible to delineate these components. Researchers have thus started to investigate a variety of multi-component elastic polymer blends with the X1-STXM. As an example, we show the investigation of the morphology of a ternary blend of poly(ethylene terephthalate), PET, low density polyethylene, LDPE, and Maleated Kraton (a modified styrene-butadiene-styrene block copolymer) [10]. Of particular interest is the distribution of the Kraton, a rubbery component, and more specifically whether the Kraton is also located at the PET-LDPE interface. Careful inspection of the micrographs in Fig. 2 clearly suggests that indeed the Kraton has a preference to also be distributed at the PET-LDPE interface, rather than just inside the LDPE domains. The features in the lower right hand corner of this figure make this particularly obvious: dark domains are touching each other (Kraton around the LPDE domain) in Fig. 2a, while by comparison the LDPE domains are sharply delineated and separated in Fig. 2b. In contrast, the interpretation of electron micrographs of stained samples of this materials yielded ambiguous results concerning the distribution of Kraton.

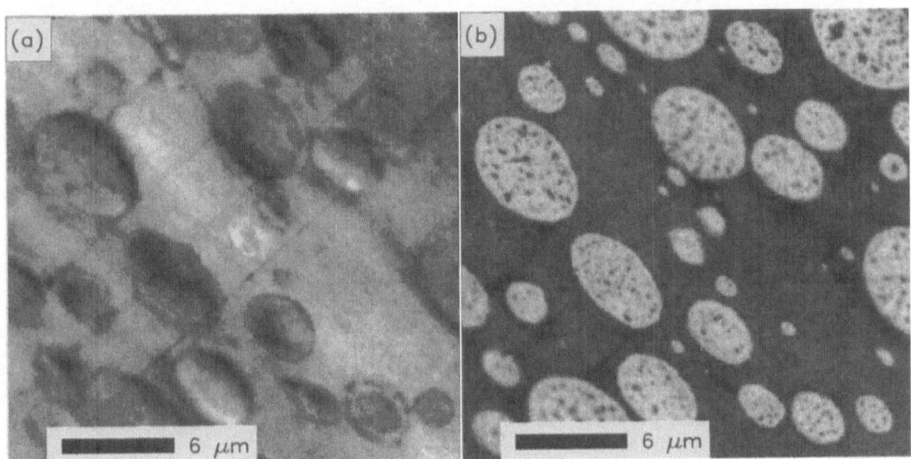

**Fig. 2. (a)** Micrograph of LDPE, PET and Kraton ternary blend acquired at a photon energy of 299 eV. Both LDPE and Kraton appear relatively dark due to their high concentration of single C-H bonds. **(b)** Micrograph acquired near 285 eV. LDPE is very transparent and appears bright while PET and Kraton are dark.

Another relatively complex polymer system shown as an illustration here are polycarbonate-ABS blends [10]. These blends are complex mixtures consisting of three polymeric components, polycarbonate (PC), styrene-acrylonitrile copolymer intermediate gray, while the PB is very bright. Fig. 3b (at higher energy) shows the SAN as the darkest phase. The SAN is thus shown to accumulate at the interface between the continuous PC phase and the dispersed SAN-g-PB particles. There is also evidence of free SAN in the PC matrix. The titanium oxide can be located and emphasized in images acquired below the carbon edge in energy (not shown here) and is typically associated with the SAN-(SAN-g-PB) agglomerates. We are presently ascertaining whether NEXAFS microscopy has enough sensitivity to differentiate SANs with different nitrile percentages so that the free SAN and the grafted SAN can be mapped independently if their nitrile content is different.

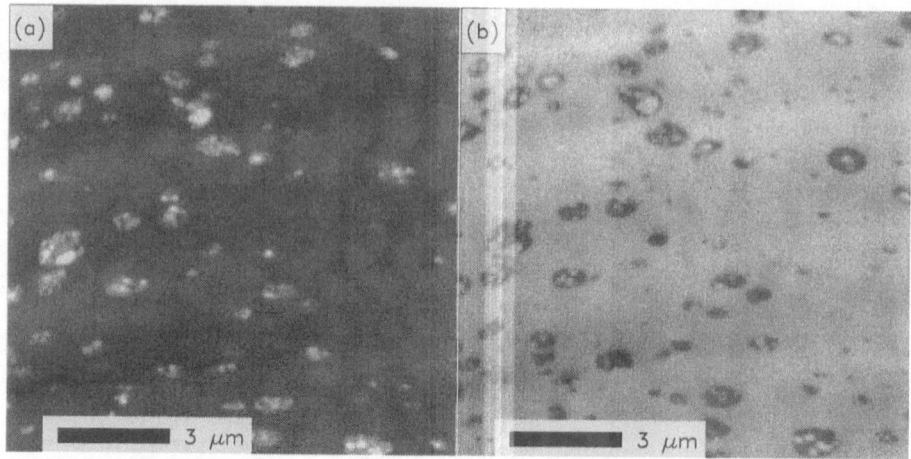

**Fig. 3.** NEXAFS images of polycarbonate-ABS blend acquired at **(a)** 285.43 eV and **(b)** 286.55 eV. The brightest features in (a) are predominantly PB, while the darker areas in (b) are regions with high nitrile concentrations (SAN). In (b) SAN appears as the darkest phase and is predominantly distributed between the PC matrix and the SAN-(SAN-g-PB) agglomerates. (Note a small amount of drift between these two images.)

## 4 X-Linear Dichroism Microscopy of Kevlar® Fibers

X-ray linear dichroism microscopy has been used to obtain a semi-quantitative determination of the relative lateral orientational order of various poly(p-phenylene terephthalamide) (PPTA) Kevlar® fiber grades [16]. Micrographs of thin sections (45° with respect to the fiber axis) of these technologically important, high crystallinity fibers exhibit a certain pattern when imaged at photon energies specific to certain chemical functionalities. This pattern has alternating higher and lower absorbing, rotated quadrants, and for scaled extrema the theoretical optical density follows a $\cos^2$ law with azimuthal angle. It is reminiscent of butterfly wings and we refer to it as 'butterfly' pattern (see Fig. 4). It reflects the average lateral orientation of functional groups and shows, for example, that the average aromatic ring planes and carbonyl groups are pointing radially outwards. This observation is qualitatively consistent with the idealized radially symmetric sheet-like structure for these fibers [25]. The observed contrast of the 'butterfly' pattern reflects the relative degree of orientational order, i.e. the partial orientational order, between the different fiber grades and the rank order observed correlates with the relative crystallinity of these fibers (Kevlar 149® is largest and Kevlar® 29 is smallest [25]) even though these parameters are not directly related.

(a)    (b)    (c)

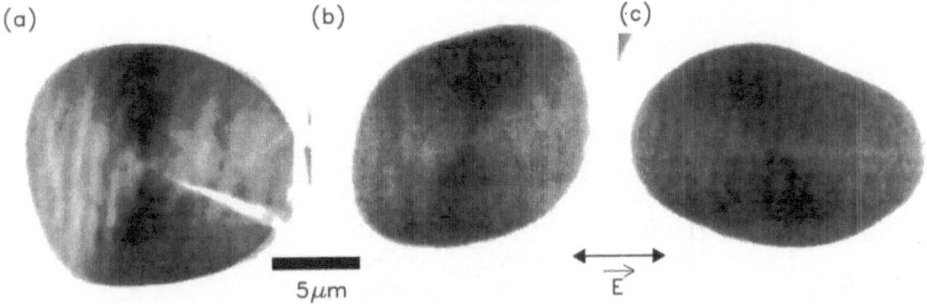

5μm    E

**Fig. 4.** Micrographs of thin films (200 nm thick, sectioned at 45° relative to fiber axis) of **(a)** Kevlar® 149, **(b)** Kevlar® 49, and **(c)** Kevlar® 29, imaged at a photon energy of 285.1 eV with the direction of the electric field vector as indicated. This energy is characteristic of the aromatic groups of the fiber polymer, and the 'butterfly' patterns observed in all three grades of Kevlar® fibers are due to the radial symmetry and partial orientational order of these fibers. Images are displayed with the same nominal contrast and the relative difference in apparent contrast observed reflects the difference in degree of radial orientational order between fiber grades. Kevlar® 149 is the most ordered and Kevlar® 29 the least ordered.

In order to semi-quantify the lateral orientational order, carbon K-edge absorption spectra were acquired from locations within the fiber with the polarization direction parallel and perpendicular to the radial position vector (see inset Fig. 5). The differences in these spectra were extracted by least squares fitting the peak intensities. A 'molecular orientation parameter' was defined by Smith and Ade as the difference of the OD in the perpendicular and the parallel locations divided by their sum [16]. Using this measure, they have found average values of 0.20 for Kevlar® 149, 0.12 for Kevlar® 49, and 0.09 for Kevlar® 29 for the spectral peak dominated by the carbonyl functionality (287 eV). Considerable variations within the same fiber grade have been observed. The orientation parameter reflects the degree of radial orientational order although presently these numbers do not express the absolute degree of radial orientational order. Nevertheless, the higher this orientation parameter the larger the radial order. In addition, the relative orientational order between fiber grades can be estimated by computing ratios of the orientation parameter. Utilizing the first three spectral features of the Kevlar® fibers, Smith and Ade estimate that Kevlar® 149 is about 1.6 and 2.3 times as radially oriented as Kevlar® 49 and Kevlar® 29, respectively. Nitrogen and oxygen NEXAFS [26] will most likely allow to determine the absolute degree of orientational order in these fibers.

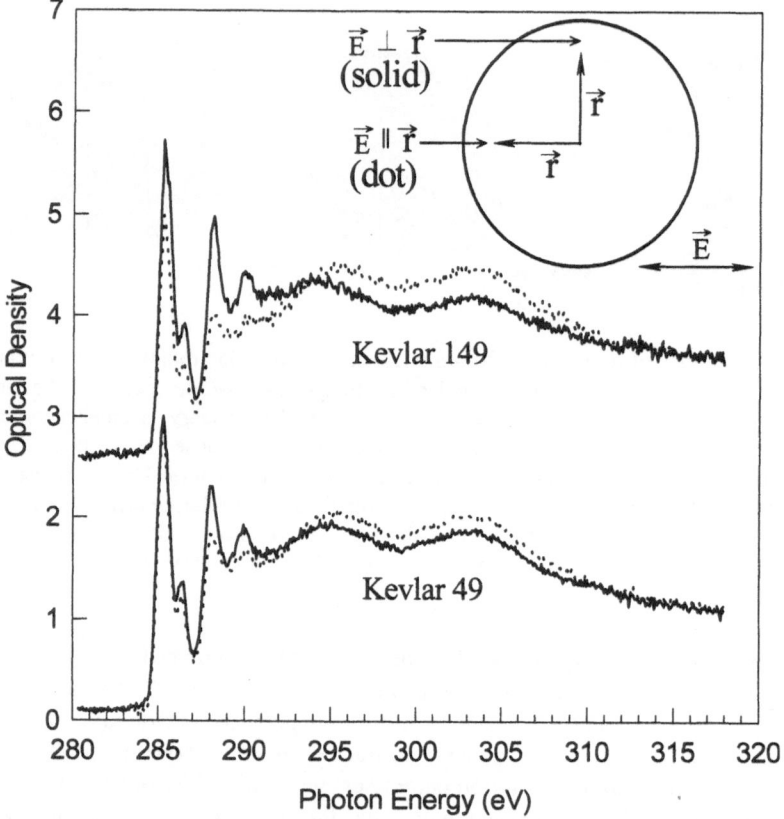

**Fig. 5.** Spectra of Kevlar® 149 and Kevlar® 49 fibers obtained from locations within the fiber as indicated. The differences in the peak intensities are due to differences in the degree of radial orientational order in these fiber.

## 5 Comparison of NEXAFS Microscopy to EELS Microscopy

The same core excitation information provided by NEXAFS spectroscopy can be obtained, if care is exercised to select dipole transition conditions, with electron energy loss spectroscopy (EELS). EELS, which is extensively utilized in the gas phase to characterize the electronic structure of small molecules, can also be performed in an electron microscope with an energy filter, resulting in a general material analysis tool with high spatial resolution capabilities [27,28]. A comparison between NEXAFS and EELS microscopy and the relative damage associated with each technique has been performed recently by Rightor *et al.* [29] utilizing the damage threshold of PET. In this study the EELS spectra where recorded in a scanning transmission electron microscope (TEM) (Vacuum Generator model HB 501) equipped with a field emission source and a parallel detection electron spectrometer (Gatan model 666). The X-ray data was recorded with the scanning transmission X-ray microscopes at both the NSLS and ALS. EELS data was acquired at 100 K to reduce radiation damage, while the NEXAFS data was acquired at room temperatures. Generally, the data acquired with NEXAFS microscopy provide more spectral details due to better energy resolution and require a lower radiation dose. Much of the difference regarding radiation damage can be readily understood in that in transmission NEXAFS most core excitations result in a useable signal, while in EELS many excitations, particularly excitations of valence electrons, occur that increase the administered dose to the sample but do not provide a signal. The theoretical resolution of the TEM-EELS can be very high, i.e. sub-nanometer, but can in practice not be utilized in polymeric studies due to radiation damage. Overall, the rule of thumb was derived that NEXAFS microscopy can spectrally analyze areas about 500 times smaller then TEM-EELS given the same radiation damage. Even though the NEXAFS microscope has a higher damage threshold, radiation damage is always a possibility. At the present spatial resolution we have, however, generally not encountered serious problems with radiation damage. Additional advantages of NEXAFS microscopy include the easy energy calibration *in situ* with $CO_2$ and the fact that it does not have to be performed in a high vacuum or at cryogenic temperatures.

# 6 Future of NEXAFS and Linear Dichroism Microscopy

As a result of the relatively low damage and the high quality of the core-excitation spectroscopic data generated with NEXAFS microscopes it seems very likely that there will be a myriad of additional applications in materials science in general and in the field of polymer science in particular. It is interesting to note that the far field diffraction limit near the carbon edge corresponds to a spatial resolution of about 2.2 nm. As zone plate technology advances the spatial resolution of X-ray microscopes will thus be improved and it seems only a matter of time until X-ray microscopes will achieve a spatial resolution of less than 10 nm, a threshold length-scale to study problems associated with polymer interfaces and lamellae.

NEXAFS microscopy can also, in principle, be made surface sensitive analogous to NEXAFS experiments without spatial resolution [3]. This would allow a whole new class of samples (i.e. surfaces) to be studied with chemical sensitivity at high spatial resolution that are not easily possible in an EELS microscope. An alternative approach to surface microscopy with X-rays are XPS microscopes with the full complementary information made possible with XPS. Several microscopes combining XPS and NEXAFS operating modes are presently being commissioned worldwide and it is only a matter of time until 'simultaneous' XPS and NEXAFS microscopy become routinely available.

One of the major developments influencing the impact of X-ray microscopy is not only further instrument development, but also the growth in the number of advanced synchrotron radiation facilities worldwide. This in turn will increase the number of available X-ray microscopes, a growth already evident particularly in the number of efforts aimed at XPS microscopy. We are at this point in time at the early stages of X-ray microscopy applications to polymeric systems and an exciting and productive future seems to be laying ahead of us.

## Acknowledgments

We thank J. Kirz and C. Jacobsen from SUNY@Stony Brook and their groups for the development and maintenance of the X1-STXM. The zone plates utilized in the X1-STXM have been provided through an IBM-LBL collaboration between E. Anderson, D. Attwood, and D. Kern. The author would also like to thank B. Hsiao, S. Subramoney, B. Wood and I. Plotzker from DuPont, B. Young, W. Lidy, and E. Rightor from Dow Chemical, C. Sloop from IBM, D.-J. Liu, S. C. Liu, J. Chung, J. Marti, and A. Monisera from AlliedSignal, and A. P. Smith, G. R. Zhuang, R. Spontak, R. Fornes, and R. Gilbert from North Carolina State University for providing samples and assisting in numerous ways. We also gratefully acknowledge our collaboration with A. Hitchcock and S. Urquhart who provide valuable NEXAFS spectroscopy insight and advice. This work is supported by a National Science Foundation Young Investigator Award (DMR-9458060), a DuPont Young Professor Grant and a grant from Dow Chemical. The NSLS and ALS is supported by the Department of Energy, Office of Basic Energy Sciences.

## References

1  H. Ade, X. Zhang, S. Cameron, C. Costello, J. Kirz, and S. Williams, Science **258**, 972 (1992).
2  H. Ade and B. Hsiao, Science **262**, 1427 (1993).
3  J. Stöhr, *NEXAFS Spectroscopy* (Springer-Verlag, Berlin, 1992).
4  J. Kikuma and B. P. Tonner, J. Elec. Spectros. Relat. Phenom., in press (1996).
5  H. Ade, A. Smith, S. Cameron, R. Cieslinski, C. Costello, B. Hsiao, G. Mitchell, and E. Rightor, Polymer **36**, 1843-1848 (1995).
6  C. Jacobsen, S. Williams, E. Anderson *et al.*, Opt. Commun. **86**, 351 (1991).
7  S. Spector, C. Jacobsen, and D. Tennant, in *X-ray Microscopy and Spectromicroscopy*, J. Thieme, G. Schmahl, E. Umbach, and D. Rudolph, Eds., (Springer-Verlag, Berlin, 1997).
8  A. P. Smith, T. Coffey, and H. Ade, in *X-ray Microscopy and Spectromicroscopy*, J. Thieme, G. Schmahl, E. Umbach, and D. Rudolph, Eds., (Springer Verlag, Berlin, 1997).
9  X. Zhang, C. Jacobsen, and S. Williams, in *Soft X-ray Microscopy, SPIE Proc. 1741*, C. Jacobsen and J. Trebes, Eds., (1992), pp. 251.
10 H. Ade, A. P. Smith, G. R. Zhuang *et al.*, in *Mater. Res. Soc. Symp. Proc.*, L. Terminello, S. Mini, H. Ade, and D. Perry, Eds., (in press 1996).

11 D.-J. Liu, S.-C. Lui, V. Zhuang, H. Ade, A. Monisera, J. Marty, and J. Chung, 1995 National Synchrotron Light Source Activity Report (1996).

12 A. P. Smith, J. H. Laurer, H. W. Ade, S. D. Smith, A. Ashraft, and R. Spontak, Macromolecules, in press (1996).

13 H. Ade, B. Wood, and I. Plotzker, 1994 NSLS Activity Report (1995).

14 G. Mitchell, M. Cheatham, Y. Chonde, J. Marshall, H. Ade, and V. Zhuang, 1995 National Synchrotron Light Source Activity Report (1996).

15 C. Zimba, A. P. Smith, and H. Ade, 1996 NSLS Activity Report (1997).

16 A. P. Smith and H. Ade, Appl. Phys. Letters (in press) (1996).

17 B. P. Tonner, D. Dunham, T. Droubay, J. Kikuma, J. Denlinger, and E. Rotenberg, J. Electron Spectrosc. Relat. Phenom. **75**, 309 (1995).

18 X. Zhang, C. Jacobsen, S. Lindaas, and S. Williams, J. Vac. Sci. Technol. **B 13**, 1477-1483 (1995).

19 X. Zhang, R. Balhorn, J. Mazrimas, and J. Kirz, J. Struc. Biol. **116**, 335-344 (1996).

20 C. J. Buckley, N. Khaleque, S. J. Bellamy, M. Robbins, and X. Zhang, in *X-ray Microscopy and Spectromicroscopy*, J. Thieme, G. Schmahl, E. Umbach, and D. Rudolph, Eds., (Springer-Verlag, Berlin, 1997).

21 R. E. Botto, G. D. Cody, J. Kirz, H. Ade, S. Behal, and M. Disko, Energy & Fuels **8**, 151-154 (1994).

22 G. D. Cody, R. E. Botto, H. Ade, S. Behal, M. Disko, and S. Wirick, Energy & Fuels **9**, 525-533 (1995).

23 G. D. Cody, R. E. Botto, H. Ade, S. Behal, M. Disko, and S. Wirick, Energy & Fuels **9**, 153 (1995).

24 A.P. Smith, H. Ade, and E. Rightor, 1995 NSLS Activity Report (1996).

25 H. H. Yang, *Aromatic High Strength Fibers* (Wiley-Interscience, New York, 1989).

26 T. Warwick, H. Ade, A. P. Hitchcock, H. Padmore, B. Tonner, and E. Rightor, submitted to J. Elect. Spectros. Relat. Phenom. .

27 R.F. Egerton, *Electron Energy Loss Spectroscopy in the Electron Microscope* (Plenum Press, New York, 1986).

28 M. M. Disko, C. C. Ahn, and B. Fultz, Eds., *Transmission Electron Energy Loss Spectrometry in Materials Science* (Minerals, Metals and Materials Society, Warrendale, 1992)

29 E. G. Rightor, A. P. Hitchcock, H. Ade *et al.*, J. Phys. Chem., in press (1996).

# An Improved Microprobe
# Using Direct Undulator Radiation

M. R. Weiss[1], V. Wüstenhagen[2], C. Heske[1], R. Fink[1], E. Umbach[1,2]

[1] Universität Würzburg, Experimentelle Physik 2,
Am Hubland, D-97074 Würzburg, Germany
[2] Universität Stuttgart, 4. Physikalisches Institut,
Pfaffenwaldring 57, D-70550 Stuttgart, Germany

**Abstract.** The concept of a microspectrocope using direct undulator radiation is described. This instrument utilizes the quasi monochromatic beam from the first undulator at BESSY without monochromatization. Thus X-ray induced Auger spectroscopy as well as photoemission using the 2 eV wide first harmonic of the undulator are possible. Spatial resolution is achieved by an adjustable aperture in front of the sample yielding a diffraction limited resolution of 3–4 $\mu$m. An improved concept using an ellipsoidal mirror is described resulting in a spatial resolution in the submicrometer range. As an example a microanalysis of $Cu(InGa)Se_2$, a material used for thin film solar cells, is briefly discussed.

## 1 Introduction

Microscopes which provide spectroscopic information are presently under development because of their benefits in materials research. Especially photoelectron spectroscopy has been implemented since one can analyze not only the composition but also the chemical state of a sample surface. The spatial resolution is obtained either by imaging the emitted photoelectrons or by scanning focused photons over the surface. Both methods, scanning or imaging, have specific advantages and disadvantages. The imaging method is very well suited for surveys and spatially evolving patterns since it gives an image of the surface for a certain electron energy in short time. Experimental setups based on this concept have reached the highest spatial resolution today [1, 2, 3, 4]. This is mostly due to the fact that in the past many efforts were concentrated on the imaging technique in electron microscopy.

The scanning method, on the contrary, allows to direct all photons onto the spot in question, thus providing faster spectroscopic analysis of this spot with better signal-to-noise ratio. It is, in principle, as fast as imaging provided that the time for the scanning motion is short. In the scanning mode one can also use additional analysis techniques, for example mass spectroscopy of photo-desorbed atoms [5].

**X-Ray Microscopy and Spectromicroscopy**
Eds.: J. Thieme, G. Schmahl, D. Rudolph, E. Umbach
© Springer-Verlag Berlin Heidelberg 1998

The focusing of a UV- or X-ray spot can be achieved by Fresnel micro zone-plates [6], by normal incidence X-ray mirrors, e.g. in Schwarzschild geometry with multilayer coatings [7, 8], or by grazing incidence, (often) non-spherical X-ray mirrors [9, 10].

The photon sources for these microscopes are dipole magnets or insertion devices at synchrotron radiation facilities. Present laboratory sources do not provide the flux and brightness necessary for such low numerical aperture optics. For high spectral resolution X-ray monochromators are used which reduce the flux considerably. To overcome the shortcomings of other concepts, namely low flux, no adjustable photon energy and no availability of other methods of analysis, we have developed a microspectroscope of the scanning type which is described in the following.

For very high flux and hence short measurement times we use direct undulator light. This source is quasi-monochromatic by interference effects of its periodic magnet structure. Since the linewidth is rather broad (here: $\Delta E \approx 2eV$), this microspectroscope is best suited for X-ray induced Auger electron spectroscopy (XAES) for which monochromaticity is not required. Still, X-ray photoemission (XPS) analysis is possible, with reduced energy resolution. The specific advantages of this concept which we named Photon Induced Scanning Auger Microscope (PISAM) are discussed in more detail in section 2.

For first tests, focusing was achieved by a simple aperture which demonstrated the possibilities of the concept and yielded first results. This apparatus and one example of an application are also described in section 2. For improved spatial resolution a focusing optics based on two grazing incidence mirrors was designed (see Fig. 1).

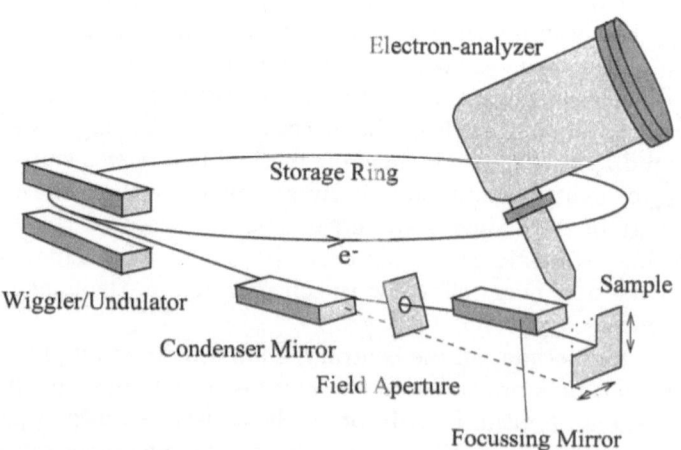

**Fig. 1.** Scheme of the new PISAM II concept. The condenser mirror increases the flux at the field aperture and the ellipsoidal focusing mirror provides the high spatial resolution.

Although the highest possible spatial resolution has not yet been demonstrated for a grazing incidence ellipsoidal mirror it was chosen because it allows to focus a broad range of photon energies. Thus the proper conditions can be selected by the experimentalist, e.g. to gain higher count rates or additional information by resonant spectroscopy. The design considerations and first results of the improved setup can be found in section 3.

## 2    The First Apparatus: PISAM-1

### 2.1    Description of the Setup and Its Performance

The first experiment using this concept was conducted with a set of apertures which reduced the direct undulator beam down to a width of approximately 4 $\mu m$. The sample was mounted on a scanning stage motorized by steppermotors (1.25 $\mu m$ per step). The excited photoelectrons were detected by a VG CLAM-2 spherical sector analyzer. A preparation chamber with standard equipment (sputter gun, mass spectrometer, etc.) and a sample transfer system completed the apparatus [11, 12, 13].

We performed several experiments to characterize the potential of this simple setup [14, 15, 16]. Measurements of microstructured samples demonstrated a photon energy-dependent spatial resolution of 15 $\mu m$ at 45 eV and 3-4 $\mu m$ at 280 eV. The accessible photon energies ranged from 40 eV to 550 eV covering the absorption edges of sulfur, carbon, nitrogen, and oxygen.

The surprisingly good resolution regarding the simple setup and the diffraction limit is partly due to the proximity of the sample to the aperture (12 mm). The resulting diffraction is of the Fresnel type which leads to a smaller spot than expected by Fraunhofer diffraction calculations [13, 14]. One might object that the spatial resolution of conventional electron-induced Auger microscopes (SAM) is well beyond this resolution (typically 50-200 nm, even 5 nm reported [17]), but the comparison of spatial resolutions leaves out other important aspects. SAM resolutions of this kind can only be reached by highly accelerated electrons (10 keV up to several 100 keV). Unfortunately, these electrons contribute to a large background of inelastically scattered electrons which reduce the signal-to-background ratio (S/B) considerably. For good spectroscopic analysis longer sampling times are hence required. Additionally, sensitive adsorbates and samples cannot withstand the required high fluxes and energies in SAM instruments. Photons as excitation source have the advantage that much less damaging secondary electrons are produced, and hence that less background and a better S/B ratio result. This can even be improved by tuning the photon energy to a resonance or threshold which also leads to additional spectroscopic information [14].

Furthermore, we performed experiments to utilize the microanalytical potential of this setup (for an example see the next section). Additionally, we tested microfabricational applications, for example writing of microwires either by depositing molybdenum from an organometallic precursor [15] or by photolytic polymerization of condensed monothiophene[14, 16].

## 2.2    Analysis of Cu(In,Ga)Se$_2$ Thin Films for Solar Cells

Thin CuInGaSe$_2$ (CIGS) films are very promising absorber materials for low cost and high efficiency solar cells. A typical cell consists of a soda lime glass substrate, a 1$\mu m$ thick Mo back-electrode, a 1-2$\mu m$ CIGS absorber, a CdS buffer layer, and a ZnO top-electrode. The efficiency of such a solar cell is found to correlate with the XPS-derived sodium content segregating from the glass substrate through the Mo back-electrode into the CIGS absorber film [18]. A laser-cutting technique is applied on the Mo electrode for microstructuring and connecting the single absorber cells in a module. Because this laser-cut drastically influences crystal growth and segregation chemical microanalysis of the CIGS film on top of this microstructure and its correlation with the electrical performance of a solar cell in operation is of particular interest.

An microscopic image of the vicinity of a laser-cut in the Mo back-electrode (width ≈30 μm) written prior to CIGS deposition and formation is shown in Fig. 2 a). The film has a changed morphology near the cut indicated by differ-

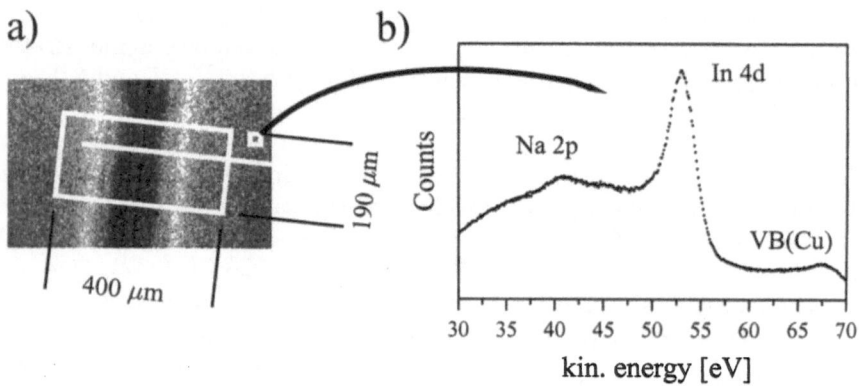

**Fig. 2.** (a) Optical microscopic image of the vicinity of a laser-cut. Brighter areas show locations of different morphology. The analyzed position, line, and area are indicated. (b) Valence spectrum using the PISAM microprobe set at the indicated small area on the sample. The photon energy of the first harmonic was 70 eV.

ent grey scales. Measurements of the optical beam induced current (OBIC) [19] yielded areas of increased efficiency at the position of the brighter strips. The microspectroscopical analysis of a spot (Fig. 2 b) displays the occurrence of sodium, indium, and copper in the surface region indicated by the photoemission structures from Na 2p, In 4d and valence band, the latter being predominantly due to copper, for h$\nu$=70 eV.

With the electron analyzer set at the corresponding energies we scanned across the laser-cut as indicated by the white box in the left frame of Fig. 2.

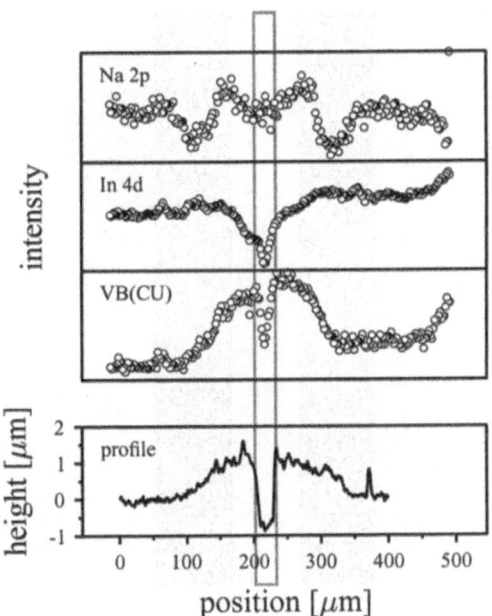

**Fig. 3.** Line scans across the laser-cut with the analyzer set at the Na 2p, In 4d, and valence (i.e. Cu VB) energies indicating the relative concentration of the elements. The narrow rectangular box in the middle (position: 200-230 μm) indicates the actual size of the cut, the grey filled areas show regions of high OBIC signal.

The resulting line scans are displayed in Fig 3. A close examination of these and similar data (not shown) leads to the following conclusion. At the laser-cut all intensities appear to drop strongly. This is probably an artefact of the measurement which is due to due to charging effects since the absorber material has no back-electrode there. An area of ≈100 μm width left and right from the cut shows reduced indium and enhanced copper concentration indicating a new Cu-rich phase. In the areas of increased OBIC signal (grey boxes in Fig. 3), the sodium concentration shows a strong variation. This variation cannot be explained by topographical effects alone, such as height variation of the film (as indicated by the height profile in Fig. 3, bottom). We rather believe that sodium influences the size of the crystallites [18], which are larger in this area than elsewhere. Sodium is also expected to segregate predominantly to the grain boundaries [18]. The distribution of elements displayed in Fig. 3 can be found all along the cut as seen in area scans. Fig. 4 shows 2D grey-scale intensity images for sodium and indium displaying the 2D-distribution of these elements along the laser-cut.

Obviously, the patterning step of the Mo-back contact leads to compositional variations on a significantly larger length scale than that of the actual laser-cut.

a)

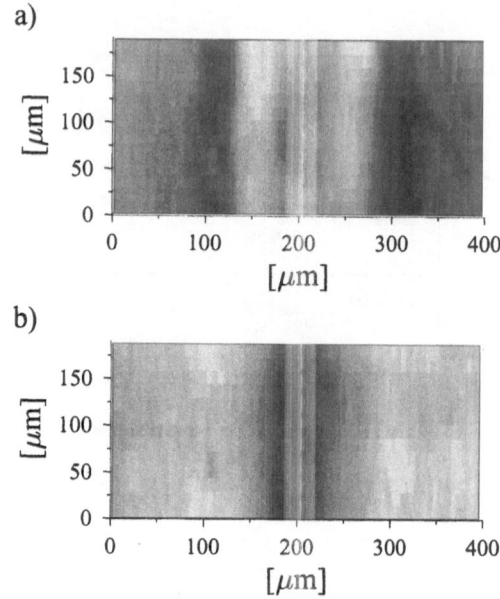

b)

**Fig. 4.** 2D-intensity maps for sodium (**a**) and indium (**b**) in the vicinity of a laser-cut in the Mo back-electrode. The scanned area is about $400 \times 190 \mu m^2$.

An improvement of the complete solar cell therefore must be combined with improved patterning methods. Moreover, new structuring techniques or substrate modifications might be necessary in order to optimize the spatial distribution of the sodium content in the absorber material.

## 3    Improved Design: PISAM-2

After successful testing of the performance of PISAM-1 we thought of ways to improve the spatial resolution. An obvious solution is to image the aperture demagnified onto the sample. Several alternatives are available to focus photons in the XUV region from 10 eV to 600 eV, as mentioned above. Microzoneplates have reached spatial resolutions down to 100 nm [6]. Unfortunately, their efficiency is very low, and they focus only for a fixed preselected photon energy. Bragg-Fresnel lenses provide higher efficiency but are still in the exploration phase and hence still limited in their spatial resolution. Normal incidence mirrors are coated with a multilayer to improve their reflectivity at a certain photon energy, e.g. 100 eV. These yielded the highest spatial resolution achieved at present with mirrors: 100 nm [7]. This is mostly due to the fact that normal incidence mirrors are rotational symmetric and spherical (or only little aspheric), and hence can be manufactured with very high precision.

We decided to use grazing incidence optics. Although still difficult to manufacture they are the only focusing elements with energy-independent focusing. Therefore, one needs not to sacrifice the tunability of the photon energy which can be an important experimental parameter in photoemission and allows to perform X-ray absorption experiments. Thus, we implemented an ellipsoidal mirror for demagnification and also added a condenser mirror between undulator and aperture to even increase the photon flux (see Fig. 1). To find optimized parameters for the system we calculated the expected focus for each possible configuration, characterized by demagnification factor and grazing angle of incidence and limited by the manufacturing errors: tangential error and microroughness [20, 21]. Based on this consideration an optimized mirror was selected with a grazing angle of 6° and a demagnification factor of 20. Larger angles would lead to lower transmission at high photon energies, smaller angles to increased curvature and therefore to even more problems to obtain the required surface quality. Higher demagnification also leads to unsurmountable problems in producing the mirror. The described evaluation process resulted in a design with the following properties:

- focus width: $0.2 - 0.3 \, \mu$m
- focus height: $0.9 - 1.1 \, \mu$m
- photon flux: $>2 \cdot 10^{12}$phot/s 1%BW in the smallest possible focus $(0.2 \times 0.9 \, \mu\text{m}^2)$.

From the ray tracings a focal intensity-distribution as shown in Fig. 5 was derived. Also, the sensitivity to maladjustments was determined by ray tracing. It

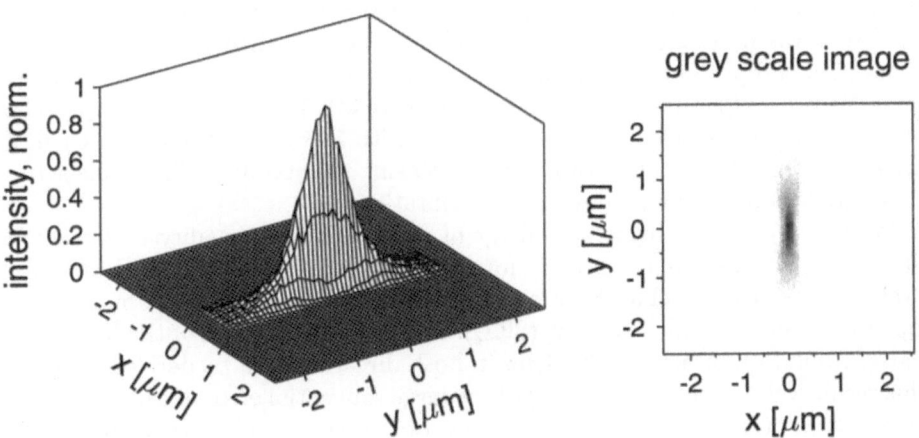

**Fig. 5.** Calculated intensity distribution of the PISAM-2 in the image plane including all broadening effects for a photon energy of $600 \, eV$.

resulted in stringent requirements for the mirror mount and fine adjustment: the mirror must be positioned with an accuracy of $1 \, \mu m$ and oriented within $1 \mu rad$ for optimum conditions.

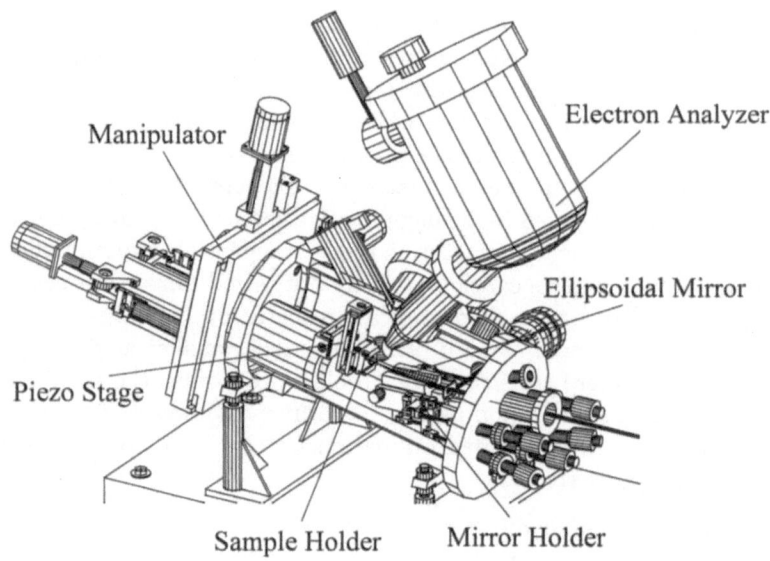

**Fig. 6.** Cut open view of the central experimental PISAM-2 chamber showing details of the sample stage and mirror holder.

The recently completed PISAM II apparatus (see Fig. 6) appears to meet these requirements and is briefly described in the following. To reach the goal of high accuracy and stability, a complete reconstruction of the original PISAM-1 apparatus was neccessary: The sample positioning was divided in coarse movements with stepper motors and fine movements by piezo-stacks in flexure hinge frames (see Fig. 6), thus giving a full scan range of 25 mm in 50 nm steps. The ellipsoidal mirror is held by a mirror mount with 5 degrees of freedom, each of which acts directly on the mirror, thus increasing the stability. The whole setup is mounted on a marble plate to reduce vibration.

With a prototype ellipsoidal mirror of reduced surface quality first tests of the new apparatus were recently performed on an integrated circuit. We scanned across wire-edges on a chip. High count rates indicate the conductor material as opposed to the adjacent insulator ($SiO_2$). From the steep edges of the linescan a spatial resolution of better than $1\mu m$ in both directions can be derived (Fig. 7). This result was unexpected because of the large tangential error of this prototype mirror. It is probably due to the fact that only a very small part of the mirror was used which has a much higher surface quality than the rest of the mirror.

## 4   Conclusion

With this project we have further developed the idea of spatially resolved spectroscopy. The result is an apparatus that has reached submicron resolution utilizing the direct undulator radiation. The microspectroscope can be operated in

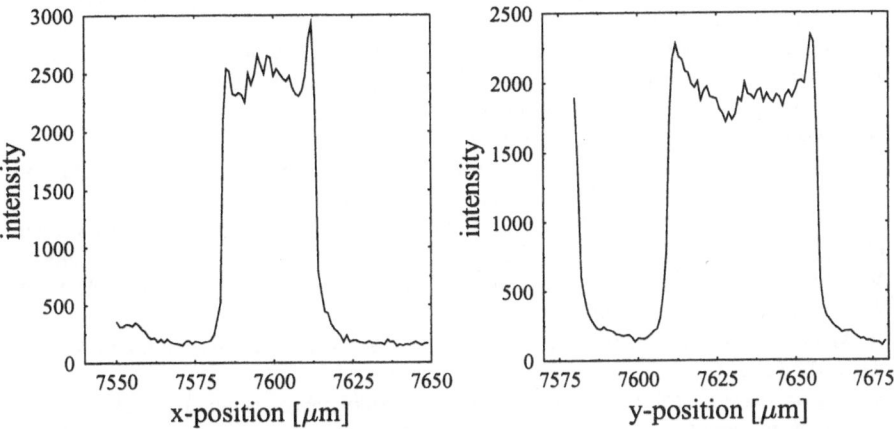

**Fig. 7.** Line-scans across a wire of an integrated circuit. The steep edges indicate a resolution of better than $1\mu m$ in both directions.

the Auger or photoemission mode with variable photon energies and provides the highest photon density ever achieved in the sampling spot of such an instrument. Thus PISAM-2 is an interesting microprobe that may be useful for several applications one of which, a structured film of a $Cu(In,Ga)Se_2$ solar cell, has been sketched as example.

## Acknowledgements

We like to thank Dr. W. Riedl, Siemens Corp., for the solarcell samples. We also like to thank the BESSY-crew and especially Prof. W. Peatman for support regarding all stages of development. Financial support by the BMBF (projects 05 644WWA and 05 5WWAXB) is gratefully acknowledged.

## References

1. E. Bauer, Rep. Prog. Phys. **57**, 895 (1994).
2. M. Keenlyside and P. Pianetta, J. Electron Spectrosc. Relat. Phenom. **66**, 189 (1993).
3. B. Tonner, G. Harp, G. Koranda, and J. Zhang, Rev. Sci. Instrum. **63**, 564 (1992).
4. R. Fink, M.R. Weiss, E. Umbach, D. Preikszas, H. Rose, R. Spehr, P. Hartel, W. Engel, R. Degenhardt, R. Wichtendahl, H. Kuhlenbeck, W. Erlebach, K. Ihmann, R. Schlögl, H.-J. Freund, A. M. Bradshaw, G. Lilienkamp, T. Schmidt, E. Bauer, and G. Benner (to be published in this issue).
5. J. Voss, M. Fornefett, J. Friedrichs, C. Kunz, M. Ptrorius, A. Ranck, K. Berens v. Rautenfeld, M. Schroeder, H. Sievers, and V. Wedemeier (to be published in this issue).

6. H. Ade, J. Kirz, S. Hulbert, E. Johnson, E. Anderson, and D. Kern, Appl. Phys. Lett. **56**, 1841 (1990).
7. C. Capasso, W. Ng, A. Ray-Chaudhuri, S. Liang, R. Cole, Z. Guo, J. Wallace, F. Cerrina, J. Underwood, R. Perera, J. Kortright, G. De Stasio, and G. Margaritondo, Surf. Sci. **287/288**, 1046 (1992).
8. K. Ninomya and M. Hasegawa, Surf. Sci. **287/288**, 1051 (1993).
9. J. Voss, H. Dadras, C. Kunz, A. Moewes, G. Roy, H. Sievers, I. Storjohann, and H. Wongel, J. X-Ray Sci. Technol. **3**, 85 (1992).
10. R. Nyholm, M. Eriksson, K. Hansen, O.-P. Sairanen, S. Werin, A. Flodström, C. Törnevik, T. Meinander, and M. Sarakontu, Rev. Sci. Intr. **60**, 2168 (1989).
11. V. Wüstenhagen, J. Taborski, E. Umbach, I. Storjohann, and J. Voß, HASYLAB Technical Report, 555 (1989).
12. V. Wüstenhagen, M. Schneider, J. Taborski, W. Weiss, and E. Umbach, Vacuum **41**, 1577 (1990).
13. V. Wüstenhagen, *Photoemissions- und photoinduzierte Auger-Mikroskopie unter Verwendung von Undulatorstrahlung*, Dissertation, Universität Stuttgart (1992).
14. M. Weiss, V. Wüstenhagen, R. Fink, and E. Umbach, J. Electr. Spectrosc. Relat. Phenom. **84**, 9 (1997).
15. P. Väterlein, V. Wüstenhagen, and E. Umbach, Appl. Phys. Lett. **66**, 2200 (1994).
16. P. Väterlein, M. Weiß V. Wüstenhagen, and E. Umbach, Appl. Surf. Sci. **70/71**, 278 (1993).
17. G. G. Hembree, J. S. Drucker, F. C. H. Luo, M. Krishnamurthy, and J. A. Venables, Appl. Phys. Lett. **58**, 1890 (1991).
18. C. Heske, R. Fink, and E. Umbach, Appl. Phys. Lett. **68**, 3431 (1996).
19. W. Riedl, private communication.
20. M. Weiß, *Hochortsauflösende Elektronenspektroskopie mit Undulatorstrahlung*, Diplomarbeit, Universität Stuttgart (1993).
21. M. Weiss, V. Wüstenhagen, and E. Umbach, in: *Proc. SPIE*, vol. 2279, 2 (1994).

This article was processed using the LaTeX macro package with LLNCS style

# Cathode Lens Spectromicroscopy
# with a Low-Energy Electron Microscope

G. Lilienkamp, C. Koziol, T. Schmidt, E. Bauer

Physikalisches Institut der Technischen Universität Clausthal, Leibnizstr.4,
D-38678 Clausthal-Zellerfeld, Germany
E-mail: Gerhard.Lilienkamp@TU-Clausthal.de

**Abstract.** An imaging band pass filter added to the Clausthal low energy electron microscope (LEEM) makes cathode lens spectromicroscopy with high spatial resolution possible. The performance as well as the limitations of the spectromicroscope are discussed and compared with the standard imaging modes of the instrument.

## 1 Introduction

The low energy electron microscope has demonstrated repeatedly its unique performance in surface microscopy [1,2]. Its outstanding properties are: speed (video frequency), resolution (down to 10 nm), and the variety of contrast mechanisms. Present fields of applications are: in situ studies of epitaxy and growth [3], chemical reactions [4], defect structures [5], phase transitions[6], and magnetic imaging with the help of a spin polarized gun [7]. But spectroscopic contrast was still missing. There are promising approaches to submicron spectroscopy like micron-scale x-ray absorption near edge structures (micro-XANES) in a photoemission electron microscope (PEEM)[8] or scanning x-ray microscopy. The first suffers from the chromatic aberration which results in a limited resolution while the other is slow. The designs for sub 100 nm resolution are only suitable for a narrow range of photons [9]. Moreover the spectroscopic LEEM is capable of other contrast mechanisms and small area electron diffraction that enables the operator to apply many electron beam methods nearly simultaneously to the same small sample area. Further information is to be found in the article by L. Veneklasen [10].

## 2 Experimental

The LEEM microscope is a direct imaging instrument. It is capable of imaging electrons irrespective of their origin, starting angle, and energy up to a few keV. To reduce the electron optical aberrations the range of starting angles and the energy spread however has to be limited. Figure 1 shows a schematic of the spectroscopic LEEM. The basic parts of the system are: cathode lens objective, beam splitter, transfer optics, energy filter, two different projectives with image converter, and the electron illumination optics.

For the simplest mode of operation, the PEEM-mode e.g. with mercury short arc lamp illumination, only the objective, the transfer optics, and one of the projectives are needed. The photoelectrons starting with about 0...2 eV are accelerated to about 20 keV within the objective. The image produced by it is transferred to the projective which magnifies it further onto a chevron channelplate array. The intensified image on a phosphorous screen can be recorded by a video camera. The energy filter is

**X-Ray Microscopy and Spectromicroscopy**
Eds.: J. Thieme, G. Schmahl, D. Rudolph, E. Umbach
© Springer-Verlag Berlin Heidelberg 1998

turned off, so that the electrons are not deflected. The range of accepted angles can be limited by introducing a contrast aperture into an intermediate diffraction pattern inside the transfer optics of the instrument. Whith high melting point materials thermionic emission microscopy is also possible.

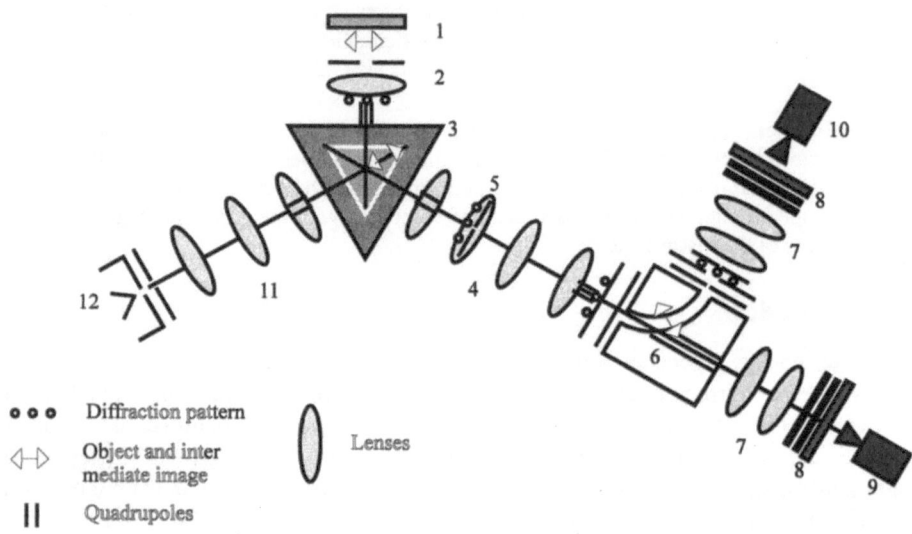

**Fig. 1.** Schematic of the spectroscopic LEEM. (1): Sample, (2) Objective Lens, (3) Beam Splitter, (4) Transfer Optics, (5) Contrast Aperture, (6) Energy Filter, (7) Projective, (8) Image Converter, (9) Video Camera, (10) Slow Scan CCD camera, (11) Illumination Optics, (12) Electron Gun with $LaB_6$ Cathode.

In the XPEEM (x-ray photoelectron emission microscopy) mode illumination by x-rays e.g. from a synchrotron creates a wide energy distribution of the emitted electrons. The energy window of the energy filter in our instrument is set to somewhere between a few eV and about 0.5 eV to enable imaging with characteristic photoelectrons, Auger-electrons, energy loss electrons, or energy selected secondaries. The main demand on the design of the filter was simplicity so that the instrument could be handled easily. The optical elements are a deceleration lens, a dispersive element (90° spherical capacitor), and an acceleration lens. Only three additional voltages are needed for the operation of the filter. Adjustment of the electron energy, which is used for imaging, requires only a change of the sample potential and a slight correction of the focal power of the objective. The settings of the transfer and projective optics and of the filter remain constant. The light intensity in spectroscopic imaging is usually not high enough to record the image at video rates. This and the need of a wider dynamic range needed for spectroscopy made it necessary to use a cooled slow scan CCD camera (here: Princeton Instruments, 16 bit, Peltier cooled). The analogue in lateral integrating surface science is X-ray photoelectron spectroscopy (XPS) and ultraviolett photoelectron spectroscopy (UPS). In principle also angular resolved UPS (ARUPS) imaging is possible by selecting the angle with the contrast aperture in reciprocal space, but a reliable determination of the angle is not realized in our instrument.

Electron illumination of the sample requires a beam splitter to separate the incoming and the reflected or emitted electrons. In our case the beam splitter is designed as a multiple prism array consisting of an outer triangular polepiece and an independantly excitable inner part. The fringing fields in the gap between the two pole pieces determine the out of plane focal power of the beam splitter. By choosing a proper ratio of outer and inner excitation the beam splitter can be made nearly stigmatic. The basic LEEM operation mode uses the beam splitter for imaging with elastically scattered electrons. The contrast mechanisms are: bright field imaging with the zero order diffraction peak, dark field imaging, and additional interference effects like step contrast and quantum size contrast. Here the incoming electrons are decelerated to about 0...100 eV, an energy range where the reflection coefficient of the electrons is very sensitive to adsorbed material. By readjusting the transfer optics the (LEED) diffraction pattern can be observed on the phosphorous screen. By limiting the beam diameter or selecting a small sample area by an aperture in an intermediate image plane diffraction patterns from small regions (< 1 μm) can be obtained. For sample potentials below 0 V electrons are deflected before they touch the surface and the contrast is determined by the sample potential. This regime is called mirror microscopy. It is applicable to all kinds of solid and flat samples (inclusive ceramics and other insulators), the images are bright and the microscope is very easy to adjust in this mode, but because of distortion due to strong potential variations specially at small charged particles the image is not always easy to interpret. The corresponding methods in laterally averaging surface science are measurements of the reflection coefficient for slow electrons and the diode method for determining work function changes. In mirror microscopy local work function changes can be determined.

Additional tuning coils allow different electron energies for illumination and imaging. This is important for Auger microscopy as well as for imaging with secondaries and inelastically reflected electrons. The excitation energy can be chosen between 0 eV and 3 keV. Hence also Auger spectroscopy and electron energy loss spectroscopy (EELS) can be performed laterally resolved. The electron gun for this inelastic regime is the same as for the elastic modus: a triode gun with a $LaB_6$ emitter. $LaB_6$ is a good compromise between high coherence for elastic imaging and the LEED mode and the high current needed for Auger imaging.

Other features of the spectroscopic LEEM are: fast entry sample load lock, and sample cleaning by ion sputtering in a separate preparation chamber. The base pressure in the main chamber is $10^{-10}$ mbar.

The XPEEM images were taken at the TGM 5 undulator beam line at BESSY in Berlin. The energy selection slit of the beam line is five times demagnified onto the sample by a toroidal mirror, thus giving a reasonable flux density at the specimen for valence band and core level studies. The upper limit of the photon energy at the TGM5 which is suitable for imaging is about 100 eV.

# 3 Results

Figure 2 gives an impression of what is possible and what the limitations of the instrument are. It shows a W(110) surface covered with a layer of lead in the bright field LEEM mode. This image has been taken with the energy filter switched off. Clearly monoatomic steps are visible. One of them originates at a screw dislocation. The shortest distance at which two steps still can be separated is a good measure for the line resolution. We find here less than 8 nm. Of course the energy filter as a nonideal optical element introduces aberrations but we still could achieve a line resolution in the LEEM mode of about 15 nm with the energy analyzer turned on.

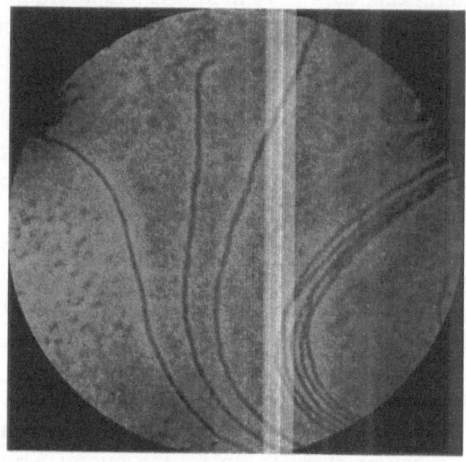

**Fig. 2.** Pb/Mo(110). LEEM image. The field of view was 700 nm x 700 nm, the electron energy at the sample 7 eV.

Can we expect the same for emission microscopy? With respect to the electron optics there should be no big difference, provided that the starting conditions like angular acceptance, intermediate magnifications and final energy spread are the same. But there are of course other limitations and whether the same starting conditions could be fulfilled is a question of intensity. At BESSY there is not enough intensity to work with the same aperture angle as in LEEM. Figure 3 shows an example. The four images show the same sample (Pb and Ag epitaxy on Mo(110)) in different imaging modes. Additional features in Fig. 3b..d compared to Fig. 3a are due to the continuing Pb deposition. Image 3a is taken in the PEEM mode, 3b is an image of secondary electrons (SEEM) and 3c and 3d are taken with characteristic photoelectrons from the Pb 5d level and the Ag 4d band. What do we learn from these images? First the PEEM and SEEM images show nearly equally well at comparable exposure times of the order of 10s the topography of the sample. 3c and 3d allow us to distinguish Pb and Ag islands, the exposure time is, however, 120s. The 15/85 edge resolution in 3c and 3d is 70...80 nm. A better resolution could be obtained only for a smaller field of view (3.6 µm compared to 7.5 in Fig.3): below 40 nm in SEEM and about 40 nm with characteristic photoelectrons. The reason lies in the energy analyzer, which is limiting the number of transmitted pixels to about 200 x 200. A reduction of the field of view

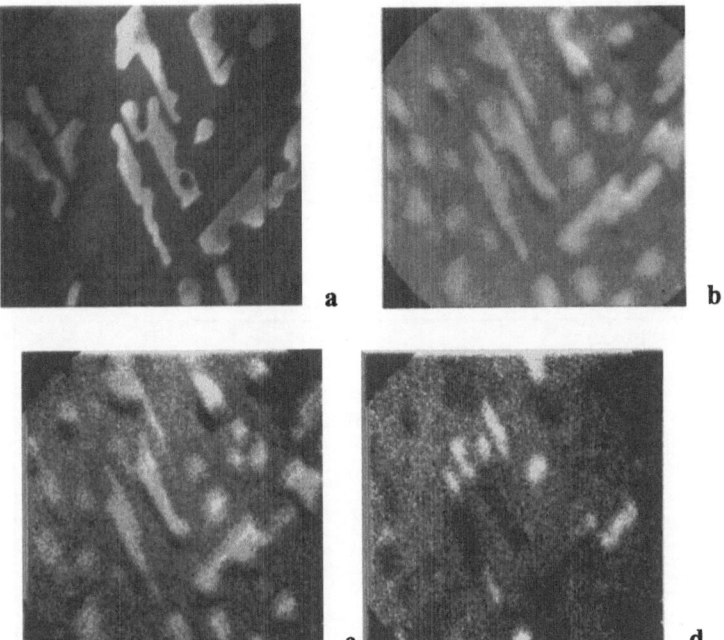

**Fig. 3.** Codeposition of Ag and Pb on Mo(110). a) PEEM, b) Imaging with 6.4 eV secondary electrons, c) with 36.0 eV Pb 5d electrons, d) with 49.5 eV electrons from the Ag 4d band. The field of view was 7.5 μm x 7.5 μm, the exposure time 10 s for a) and b) and 120 s for c) and d).

below 3 μm does not improve the situation because of lack of intensity and too high noise.

What is the energy resolution of the spectroscopic LEEM? Again Pb is a good test candidate because of the narrow 5d levels. Fig.4 shows a part of a series of images of Pb islands on W(110) taken at different kinetic energies. The spectrum below is obtained by integrating the intensity of the islands (points) over an area of 0.4 μm², the crosses represent the spectrum for the surrounding Pb monolayer. We use the full width at half maximum (FWHM) as a measure of the energy resolution. It is determined by fitting a broadened Lorentz function, resulting in less than 500 meV FWHM. Taking into account the natural line width and the energy spread of the photons we estimate the energy resolution of our filter to 400 meV. The unfamiliar slope of the spectrum compared to laterally averaging experiments is due to the decreasing transmission of the objective lens at higher energies.

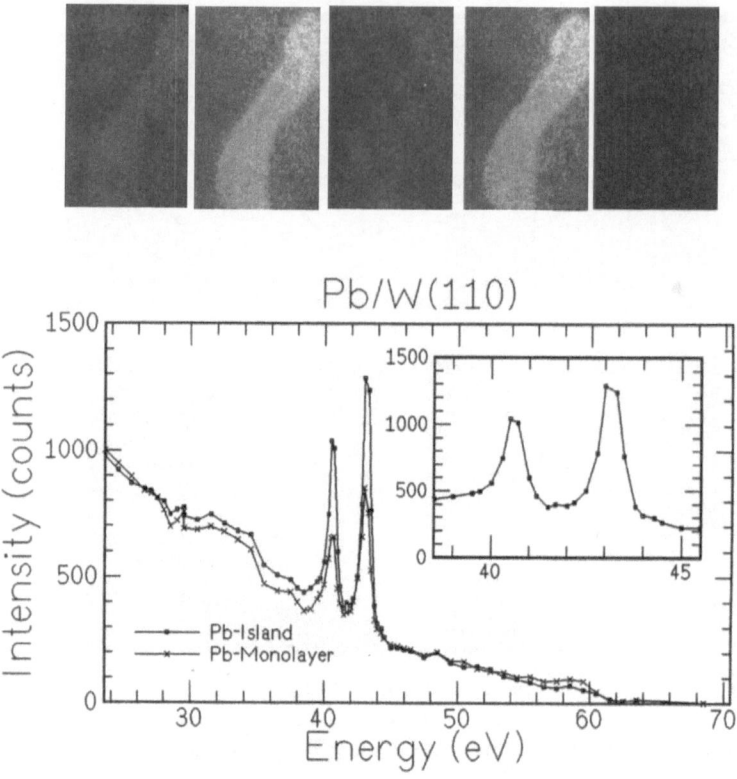

**Fig. 4.** Pb islands on W(110) surrounded by a monolayer of Pb. Top: A sequence of exposures with different kinetic energies. From left to right:39 eV, 40.5 eV, 42 eV, 43 eV, 45 eV. Photon energy: 60.8 eV. Field of view: 2.3 μm x 3.7 μm. Bottom: The spectra were obtained by integration of the intensity of a small sample area (0.4 μm² on the island and 0.75 μm² on the surrounding monolayer).

Auger spectroscopy with electron excitation does not need synchrotron radiation and gives sometimes additional information. Fig. 5 shows an example of the Auger performance of the spectroscopic LEEM. The sample is a Si(111) substrate with Ag islands on it. The spectrum is obtained similar as in Fig. 4. Signal to background ratio and half width of the peaks are similar to laterally averaging but angular resolved measurements [11]. The spatial resolution for this images is only 200 nm, probably because of the above mentioned pixel problem. Fast Auger image recording at video rates was possible at fields of view of about 50 μm. An example of what kind of new details can be observed is the Pb island on Si(111) in Fig. 6 which was hexagonal after preparation. Upon annealing to about the melting point of Pb the island forms a spherical droplet, but a hexagonal fingerprint of the island remains. The Auger spectrum in this fingerprint region shows still lead, but the peak shape is completely different from the Pb monolayer and bulk Pb. This might be due to surface alloying. Zhao, Jia, and Yang [12] reported such intermixed Pb/Si(111) and Pb/Si(001) interfacial phases, in contrast to the previous findings by LeLay, Hricovini, and Bonnet [13].

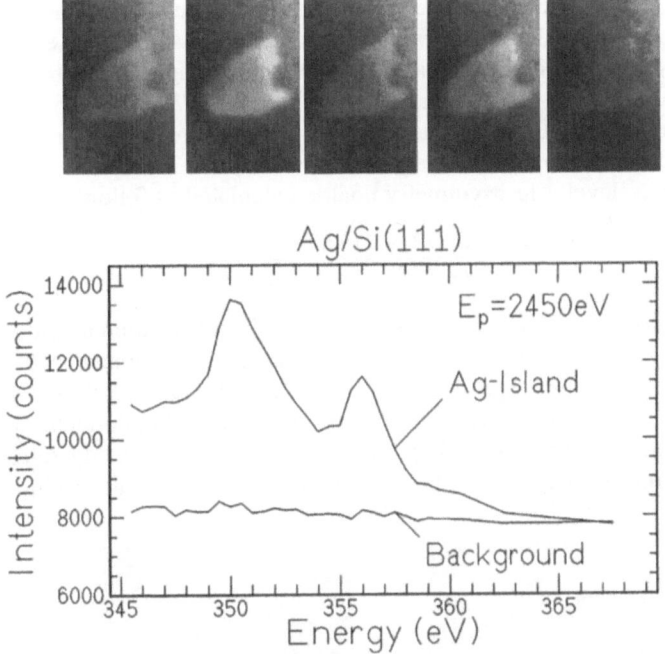

Fig. 5. Auger images of an Ag island on Si (111).The images were taken at 346 eV, 350 eV, 354 eV, 356 eV, and 360 eV. The primary energy was 2450 eV and the field of view 4.5 μm  x 7 μm. The spectra were obtained by integration of the intensity of a small sample area.

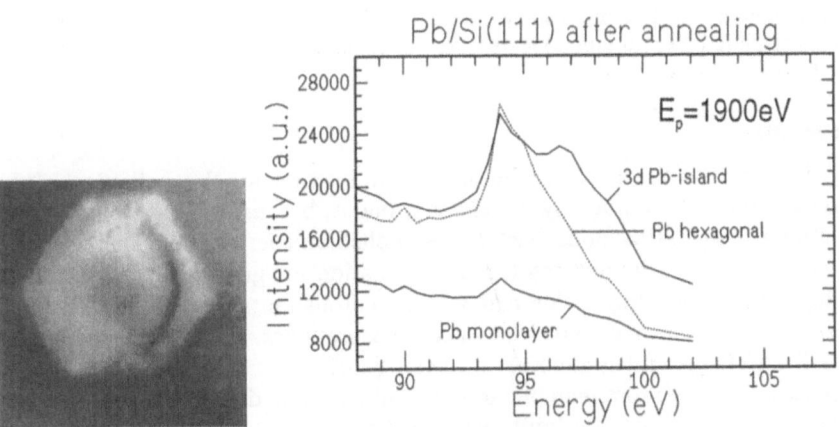

Fig. 6. Auger image from a Pb island on Si(111), which has become spherical upon annealing. The electron energy was 94 eV. The field of view was 6.4 μm x 6.4 μm. The spectra have been taken from the Pb droplet, the surrounding hexagon, aund the Pb monolayer.

That XANES-microscopy with cathode lenses is a very good method to study magnetic domains was already shown by Stöhr et al. [14]. In that experiment the magnetic x-ray circular dichroism allowed to obtain magnetic contrast, when the sample was illuminated with circular polarized light. Fig. 7 is another example taken with our spectroscopic LEEM by illumination with left circular polarized light at the BESSY SX700III beamline. Two images were taken at the absorption edges of the Fe $2p_{1/2}$ and the $2p_{3/2}$ level. The asymmetry images calculated as follows are shown:

$$\frac{I_{\sigma+}(2p_{3/2})-I_{\sigma+}(2p_{1/2})}{I_{\sigma+}(2p_{3/2})+I_{\sigma+}(2p_{1/2})} \tag{3.1}$$

The sample is a piece of magnetic stainless steel, polished, sputtered, and annealed up to 600°C (Fig. 7a).  Fig. 7b was taken after flashing the sample to 650°C which caused a breakup of the 'large domain into several domains. Selected area LEED shows that the domain in Fig. 7a is confined to a grain in (111) orientation [15]. The next grain on the left has also (111) orientation but is azimuthally rotated by 77°. Hence LEED makes a correlation between magnetic and crystalographic orientation possible.

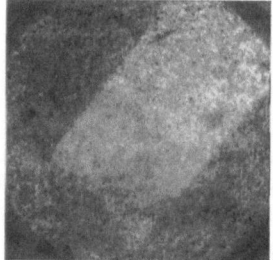 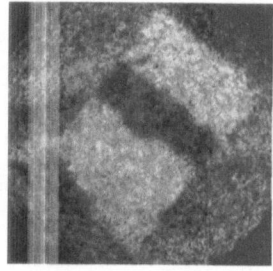

Fig. 7. The magnetic structure of a stainless steel sample. The field of view was 25 μm x 25 μm and the exposure times 10 min. a) shows the sample after cleaning and b) after a flash to about 650°C.

## 4 Discussion

We will first discuss the final resolution of our instrument. We believe that the present resolution is not limited by the electron optics, because we have obtained 15 nm resolution in the LEEM mode even with switched on energy filter. There are two possible reasons for the limited resolution in emission imaging modes: low photon flux at the sample and stray fields and vibration from the surroundings. The first reason can not simply be overcome by longer exposure times, because of unavoidable sample drift and longer integration of  disturbing influences from the surroundings. With exposure times of more than one minute and a sample drift in the range of 20 - 30 nm we are surely close to a limit, even for temperature stabilized samples. The second reason as well has very likely a strong influence. The magnetic AC stray fields at the BESSY TGM5 has been measured to be more than 100 mG peak to peak. This is about a factor of 100 more than what is recommended by electron microscope manufacturers, and the level of vibration and noise is for sure far beyond the usual levels for electron microscopes. That there is an influence is confirmed by the fact that even for the fast LEEM mode where sample drift does not play a role the resolution was only slightly better than 30nm at the BESSY beamline. What are the benefits of a third generation synchrotron? Will we be able to improve the resolution?

Bauer [16] estimated that microspectroscopy (detection of a Ag island with monolayer height on a Si (111) substrate) with characteristic photoelectrons on a time scale of 1s and a resolution of about 20 nm should be possible at a third generation synchrotron assuming an ideal energy filter and detector. The transmission of the energy filter strongly depends on the aberrations of the filter and the settings of the energy selection slits. A reasonable estimate should be 0.5. Detector systems such as our chevron channelplate system with phosphorous screen and cooled CCD camera have been discussed by Hermann and Krahl [17]. The DQE (detective quantum efficiency) of the channelplates is low (about 0.3), our light optics has a transmission of about 0.2 and a cooled CCD camera should not be too far away from DQE = 1. Smaller losses are due to reflection at the interfaces. A total factor of about 100 should be added to Bauers result, if not a different kind of detector (e.g. phosphorous screen + fibre optics + cooled CCD camera) is used. With this factor we are close to our exposure times at BESSY TGM5 for Pb islands on Si, the resolution of Bauers assumption still being a factor of about 2 to 3 better than what we have achieved in our images. This means we should increase our exposure time by a factor of about 10 to have the same signal to noise ratio or use a brighter illumination. Different cross sections for Pb (our experiment) and Ag (Bauers estimation) and different background should also contribute to not more than a factor 10. This means, that if the third generation synchrotron delivers a factor of 100 higher flux, which can be expected, we should be able to do spectromicroscopy with a resolution of 20 nm on a time scale of 100s on the model system Ag/Si(111), assuming that stray fields and vibration at the new light sources could be kept lower by one order of magnitude.

Further improvements can only be achieved if the exposure times could be decreased. One way is to use an aberration corrected instrument which increases the transmission of such a microscope by one to two orders of magnitude [18]. But besides the gain of resolution and access to the smallest particles in nanostructured materials and quantum size effects in small structures one will be able to perform dynamic studies in spectroscopy which may be just as important.

Quantitative analysis of course is in principle possible. One has to master the instrumental parameters like energy-dependent transmission but this is straight-forward. Difficulties arise here as well as for laterally averaging experiments because of the angular dependence of the emission [16] and one should keep in mind that the microscope is an angular resolving instrument with an angle of acceptance down to a few mrad, depending on electron energy and resolution. The second problem is that one should know something about the elemental distribution in the sample: is it homogeneous? is it only a top layer? or a buried layer?or....before applying e.g. an exponential decay function which describes the intensity decrease of the signal from the deeper layers. This is sometimes an unsolvable problem, but to have as many complementary probes as possible investigating the same sample area is a big step forward. Our realization of such an instrument is the spectroscopic LEEM.

# 5 Summary

Experience from laterally averaging experiments tells us that reliable statements are seldom possible with only one probe applied to the sample. Our approach to Surface Science on a nanometer scale is the spectroscopic LEEM where now Auger spectroscopy, XPS and UPS have been realized, in addition to the other fast operating modes like LEEM, LEED, mirror microscopy, PEEM, and SEEM. Future work at ELETTRA in Trieste will give us the chance to extend the resolution to its instrumental limits and open up the full necessary photon energy window to step from test experiments to exciting applications.

## Acknowledgments

We thank the Volkswagen Foundation and the BMBF for continuous support.

## References

1    E. Bauer, Rep. Prog. Phys. 57, 895 (1994).
2    E. Bauer, in Handbook of Microscopy Vol. 1, edit. by S. Amelinckx, D. Van Dyck, J. Van Landuyt, and G. Van Tendeloo (Verlag Chemie; Weinheim, 1996).
3    E. Bauer, Scanning microscopy 8, 765 (1994).
4    W. Swiech, C. S. Rastonjee, R. Imbihl, J. W. Evans, A. M. Bradshaw, and E. Zeitler, Surf. Sci. 294, 297 (1993).
5    M. Mundschau, E. Bauer, W. Telieps, and W. Swiech, Phil. Mag. A 61, 257 (1990)
6    E. Bauer, Appl. Surf. Sci. 60/61, 350 (1992).
7    T. Duden and E. Bauer, Phys. Rev. Lett. 77, 2308 (1996).
8    B. T. Tonner, this proceedings.
9    C. Capasso, W. Ng, A.K. Ray-Chaudhuri, S.H. Liang, R.K. Cole, Z.Y. Guo, J. Wallace, F. Cerrina, J. Underwood, R. Perera, J. Kortright, G. De Stasio, and G. Margaritondo, Surf. Sci. 287/288, 1046 (1993).
10   L. Veneklasen, Ultramicroscopy 36, 76 (1991).
11   K.-D. Hermbecker, PHD Thesis (Clausthal 1979).
12   G.R. Zhao, J.F. Jia, and W.S. Yang, Phys. Rev. B 48, 5333 (1993).
13   G. LeLay, K. Hricovini, and J. E. Bonnet, Phys. Rev. B 39, 3927 (1989).
14   J. Stöhr, Y. Wu, B.D. Hermsmeier, M.G. Samant, G:R. Harp, S. Koranda, D. Dunham, and B.P. Tonner, Science 259, 658 (1993).
15   C. Koziol, T. Schmidt, M. Altman, T. Kachel, G. Lilienkamp, E. Bauer, and W. Gudat, unpublished.
16   E. Bauer, Ultramicroscopy 36, 52 (1991)
17   K.H. Herrmann and D. Krahl, Advances in Optical and Electron Microscopy Vol. 9, edit. by R. Barer and V.E. Cosslett (Academic Press; London, 1984).
18   W. Engel, this proceedings.

# Spectromicroscopy with Soft X-Rays at Hasylab

J. Voss, K. Berens von Rautenfeld, M. Fornefett, J. Friedrich, M. Pretorius,
M. Schroeder, H. Sievers, A. Ranck, M. Wachsmuth, V. Wedemeier

II. Institut für Experimentalphysik, Universität Hamburg, Germany
E-mail: voss@vxdesy.desy.de

**Abstract** The scanning soft X-ray microscope at Hasylab uses photoelectrons, luminescence, photodesorbed ions, reflected, scattered and transmitted photons as signals for imaging and spectroscopy. Mirror optics for grazing and normal incidence provides lateral resolution in the micron and submicron range for photon energies of 15 to 1500 eV. In this article we describe the design and operation of the microscope beamline, and present results obtained with solid state samples.

## 1 Introduction

Spectromicroscopy and microspectroscopy, the combinations of soft X-ray spectroscopy techniques with lateral resolution, have developed into powerful analytical methods to characterize structured and inhomogeneous samples. There is growing interest in material and surface science in laterally resolved information about elemental, chemical, electronic, magnetic and topographic features. The availability of synchrotron radiation sources has provided the possibility of developing more than 50 soft X-ray microscopes operating worldwide. These instruments with different characteristics are described in a series of conference proceedings including this volume.

## 2 The Microscope at Hasylab

The scanning microscope at Hasylab is designed to be operated in the energy range of the vacuum ultraviolet (VUV) and soft X-rays from 15 to 1500 eV. The used achromatic mirror optics provides tunability of the exciting synchrotron radiation without realignment, and a large working distance allowing detection of a variety of signals emitted by the surface of solid state samples. Some essential features of the microscope are summarized in Table 1. Details are published elsewhere [14].

### 2.1 Beamline and Focusing Optics

The light source for the microscope is a 32 period wiggler/undulator operated at the W1 beamline of the Doris storage ring at Hasylab. The Flipper plane grating monochromator supplies the microscope with linear polarised synchrotron radiation in the range of 15 to 1500 eV with an energy resolution of $E/\Delta E = 200\text{--}400$ [5]. The variable diameter exit pinhole of the monochromator defines the object of the microscope optics. It is demagnified by two different mirror optics used alternatively: an ellipsoidal ring mirror for grazing incidence and a Schwarzschild objective for normal incidence, both shown in Fig. 1. At present construction stage replacing the objectives cannot be carried out in situ, opening of the vacumm system is required.

**X-Ray Microscopy and Spectromicroscopy**
Eds.: J. Thieme, G. Schmahl, D. Rudolph, E. Umbach
© Springer-Verlag Berlin Heidelberg 1998

The ellipsoid mirror reflects radiation from 15 to 1500 eV with a minimum probe diameter of 1 μm (full width half maximum). The lateral resolution is limited by the surface figuring errors of the mirror [6]. The contrast depends on the photon energy, above 1000 eV it is increasingly reduced by roughness scattering. In the VUV below 30 eV a Pt coated Schwarzschild objective optimizes the imaging properties of the micro-scope. The objective consists of very small spherical mirrors with small radii of curvature as listed in Table 2. The mirrors have standard surface quality, alignment is performed manually with visible light. Despite the simple construction, resolutions at the diffraction limit of 100 nm at 20 eV have been obtained investigating test objects and solid state samples [7]. Details of the objective and applications are described in a separate contribution of this volume [8].

## 2.2 Scanning, Sample Preparation and Transfer

The scanning stage consists of three equal-type positioning devices [3] installed outside the ultra high vacuum chamber. Translatory movement is transfered by rods coupled close to the sample to form an orthogonal tripod. The usable image volume covers $(140 \ \mu m)^3$ using piezo elements and $(3 \ mm)^3$ with stepping motors.

Table 1. The parameters of the microscopy beamline at Hasylab.

| Beamline | DORIS wiggler/undulator W1 | |
|---|---|---|
| Polarisation | Linear | |
| Monochromator | Plane grating + focusing paraboloid | |
| Energy | 15 eV< hΩ<1500 eV | |
| Energy resolution | E/Ω    E~200-400 | |
| Optics | Ellipsoidal ring mirror | Schwarzschild objective |
| Grazing angle of incidence | $2^0$ | $80^0$ |
| Coating | Au | Pt |
| Microroughness (nm rms) | <2 | <0.5 |
| Energy range (eV) | 15 -1500 | 15 - 30 |
| Working distance (mm) | 30 | 10 |
| Intensity (photons/(s μm$^2$ 100 mA)) | $10^6 - 10^9$ | $10^6 - 10^7$ |
| Minimum probe Ø (FWHM) | 1 μm | 100 nm |
| Scanned volume | Piezos: $(140 \ \mu m)^3$ / Stepping motors: $(3 \ mm)^3$ | |
| Scanning resolution | 40 nm | |
| Orientation of scanning plane | Arbitrary | |
| Signals | Photoelectrons | |
| | Photoluminescence | |
| | Photodesorbed ions | |
| | Transmitted, reflected and scattered photons | |

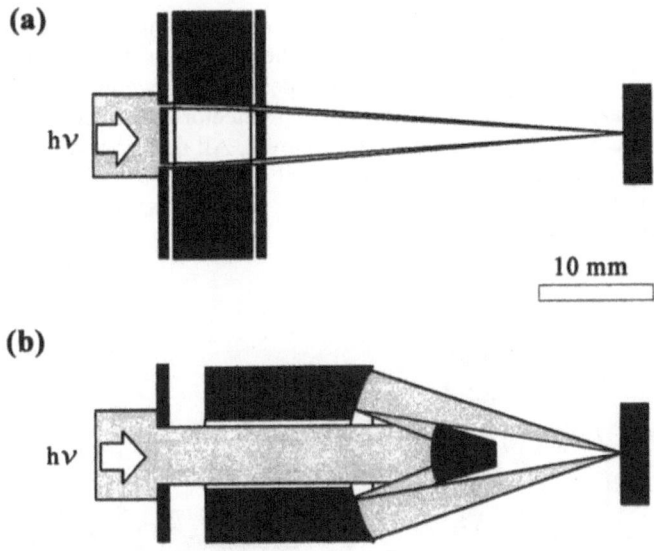

**Fig. 1.** Mirror objectives of the microscope at Hasylab. Both optics have symmetry of rotation. (a) Grazing incidence ellipsoidal ring mirror. (b) Schwarzschild objective.

The position of the sample can be monitored by optical encoders with a spatial resolution of 40 nm. The design offers the possibility to scan the sample in planes arbitrarily oriented. As discussed later this is necessary to measure specular reflectivity investigating multilayer defects and magnetic dichroism.

The microscope is equipped with a separate preparation chamber to sputter, anneal, and put in samples. Samples are transfered to the analysis chamber by means of a magnetically coupled transfer rod. Recently a Helium cryostat has been joined to the microscope chamber for investigations of in situ cooled samples. Transfer and focusing is monitored with an optical microscope.

**Table 2.** Parameters of the Schwarzschild objectives.

|  | Microscope | Spectrometer |
|---|---|---|
| Concave/convex mirror radius (mm) | 17.55/9.93 | 60/35 |
| Concave/convex mirror diameter (mm) | 17/6 | 60/21 |
| Magnification | 1:100 | 1:8 |
| Numerical aperture | 0.3 | 0.3 |
| Surface accuracy (peak to valley, $\lambda=632$nm) | $< \lambda/20$ | $\lambda/20$ |
| Microroughness (nm rms) | $< 0.5$ | 1 |
| Coating | Pt | $Al+SiO_2$ |

## 2.3 Detectors and Analyzers

Due to the large working distance of the mirror objectives used in the microscope, the free solid angle of the illuminated area of the sample is large enough to arrange detectors and analyzers for all conceivable types of signals emitted by the surface. A schematic diagram of the optics and the several detectors mounted permanently in the microscope vacuum chamber is shown in Figure 2. All detectors can be brought close to the illuminated sample area without mutual hindrance to detect different signals, for certain combinations simultaneously. Four of the possible detection modes are shown in the lower part of Figure 2.

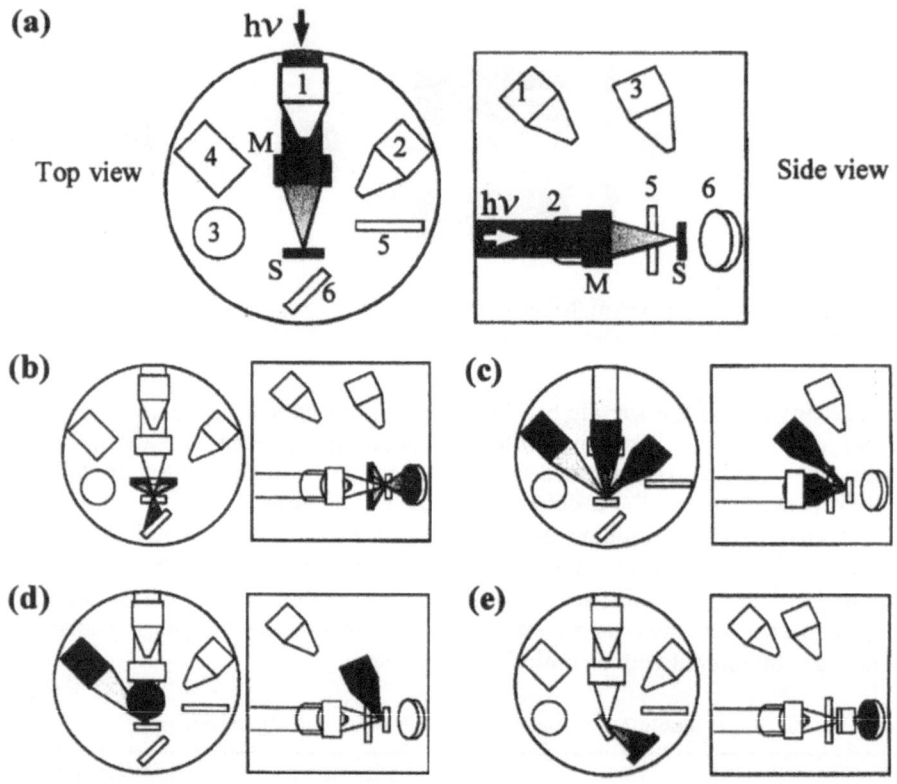

Fig. 2. The operating modes of the microscope at Hasylab. (a) Schematic representation of the optics and the detector arrangement in the microscope UHV–chamber. /M/ Focusing mirror objective, /S/ sample, /1/ time of flight (TOF) electron analyzer, /2/ hemispherical electron analyzer, /3/ TOF analyzer for photodesorbed ions, /4/ spectrometer for visible and UV luminescence, /5/ multichannelplate (MCP) with central hole, /6/ transmission MCP. (a) All detectors deactivated. (b) Center hole MCP and transmission detector activated.(c) Luminescence spectrometer and electron analyzers activated. (d) TOF ion analyzer and luminescence spectrometer activated. (e) Detector for reflected light in grazing incidence geometry activated.

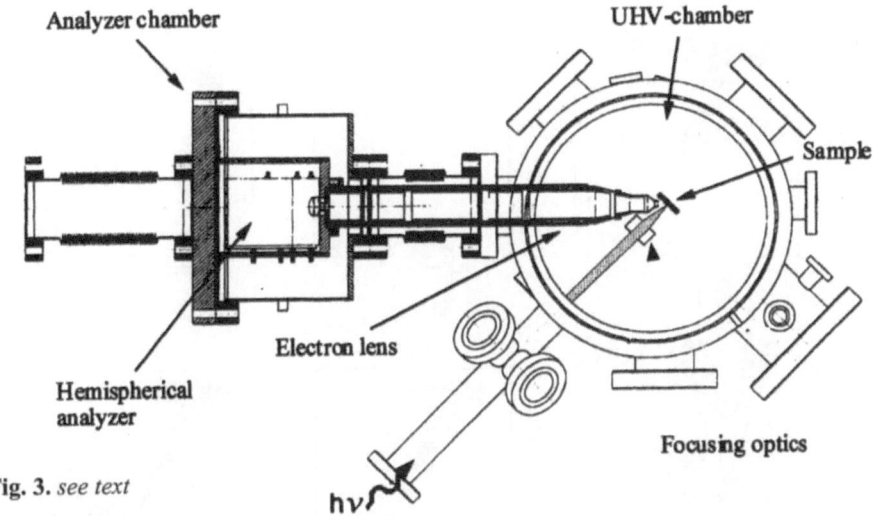

Fig. 3. *see text*

**Photoelectrons.** Different detectors and analyzers are used to measure photoelectrons. Detector /1/ in Fig. 2(a) consists of a multichannelplate (MCP) and a conical screening cover. With different voltage polarities this detector can be used to measure the total photoelectron yield or reflected and scattered soft X-rays. Using the pulsed time structure of the synchrotron radiation the drift distance of 11.5 cm inside the cone allows time-of-flight spectroscopy of the photoelectrons [9].

/2/ is a hemispherical photoelectron analyzer with a mean radius of 5 cm and a conical multielement electron lens. The layout of the analyzer and the vacuum system is shown in Fig. 3. To reach high acceptance and to avoid interference of further detectors the mounting permits translatory movement of 5cm with simultaneous high stability of the alignment. Details of the lens geometry, the analyzer and the alignment are descibed in a separate contribution of this volume [10].

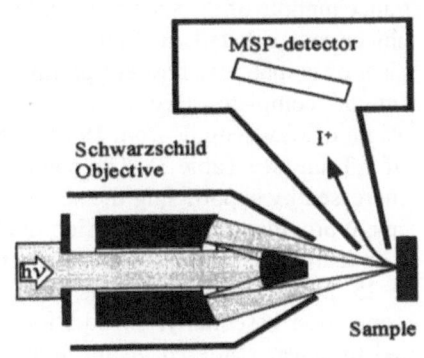

Fig. 4. Time of flight spectrometer for photodesorbed positive ions, MSP: multi-sphereplate detector.

**Photodesorbed Ions.** Device /3/ in Fig. 2(a) is a time-of-flight analyzer for photodesorbed positive ions (TOF). Masses up to 20 can be resolved with this device. With this analyzer it was managed for the first time to use desorbed ions excited by soft X-rays for imaging. In Fig. 4 the arrangement of the TOF-detector in combination with the Schwarzschild objective is illustrated. With an additional lens element the TOF-detector can also be used when the ellipsoid mirror focuses the incident radiation. Details of the detector design, mass spectra and ion micrographs are presented in a separate article of this volume [11].

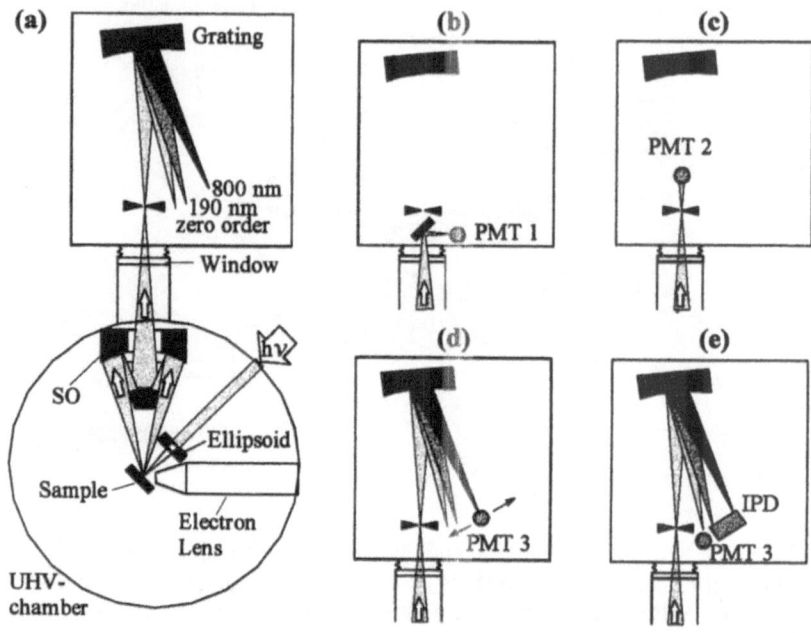

**Fig. 5.** The spectrometer for visible and UV photoluminescence. (a) Schematic diagram of the spectrometer, SO: Schwarzschild objective. (b) Non-confocal, total visUV yield mode, PMT: photomultiplier tube. (c) Confocal, total visUV yield mode. (d) Spectroscopy mode with PMT, used for imaging and excitation spectroscopy. (e) Spectroscopy mode with IPD (imaging photon detector), in this geometry PMT3 detects simultaneously zero order light.

**Photoluminescence.** In Fig. 2(a) /4/ is a spectrometer for visible and ultraviolet luminescence. It is illustrated in detail in Fig. 5. The self-luminous or scattering area of the illuminated sample area is imaged to the entrance pinhole of the spectrometer by a Schwarzschild preoptics (parameters of this objective are listed in Table 2). The spectrometer attached outside the vacuum consists of a spherical flatfield grating, a position sensitive detector (IPD) for readout of complete spectra and three photomultipliers for different detection modes. Light of wavelengths from 190 to 800 nm can be analyzed with a spectral resolution of 1.3 nm (see Table 3). The confocal arrangement has the advantage that contrast is increased by suppressing the scattered halo of the primary microfocus. In addition to this, it provides spatial resolution in all dimensions, in z-direction (parallel to incident optical axis) the resolution is 7 μm using a spectrometer entrance diaphragm of 25 μm in diameter. This feature simplifies focusing of the microscope during operation essentially. In addition to luminescence light the spectrometer can be used to detect scattered visible light to determine topographic structures of the sample. In this visible darkfield microscopy mode the sample is illuminated with zero order light of the monochromator. Further information about the spectrometer and applications are published e.g. in [12].

**Table 3**. Parameters of the luminescence spectrometer.

| | |
|---|---|
| Grating type | Spherical flat field |
| Grating radius | 138 mm |
| Lines per mm | 285 |
| Incidence angle | $5.6^0$ |
| Diffraction angles $\lambda=190/800$ nm | $8.7^0 / 19^0$ |
| Diffraction order | 1 |
| Spectral range | 190-800 nm |
| Spectrum length | 25 mm |
| Distance entrance-grating | 137.4 mm |
| Distance grating-spectrum | 130.9 mm |
| Spectral resolution | 1.3 nm (PMT3 with 50 µm slit) |
| | 6 nm (IPD) |

**Scattered, Transmitted and Reflected Soft X-Rays.** Fig. 2(a) /5/ is the symbol of a MCP with a central hole for the cone of focused light. This detector is mounted between the objective and the sample with the active side towards the sample. It is used to detect reflected and scattered light and emitted fluorescence. In addition this MCP works as a time-of-flight spectrometer for desorbed positive ions. /6/ is a MCP for detection of transmitted light and light reflected in grazing incidence geometry.

# 3 Applications

Our microscope is used for surface analysis of solid state samples. Due to the variety of usable signals investigations can be made with conductors, semiconductors and insulators. Luminescence powders and ceramics have been examined to determine luminescence efficiencies and homogeneities [12-14]. Among other things photoelectron microspectroscopy and spectromicroscopy has been performed with the hemispherical analyzer to use chemical shifts of binding energies to image elements in different chemical environments [15]. Charging effects, normally considered as hindering interferences, have been recognized as a new type of contrast mechanism in X-ray microscopy [9] and are planned to be used to image inhomogeneities in the conductivity of samples. Low concentration dopants of lithium niobate have been imaged with photoluminescence, scattered VUV-radiation and photoelectrons [16]. As an example Fig. 6(a) shows a VUV-straylight micrograph of a lithium niobate crystal doped with stripes of Ti. Simultaneous detection of photoluminescence and photodesorbed hydrogen ions from porous silicon proved the influence of H-passivation of nanostructured Si [17]. Fig. 6(b) illustrates the simultaneous degradation of both signals during illumination. Besides a wealth of spectral information the investigation of barium fluoride revealed the possibility of suppressing the slow, visible contribution to photoluminescence without increasing radiation sensitivity [16]. Fig. 6(c) and (d) show a total luminescence micrograph of a cleaved barium fluoride crystal and the corresponding excitation spectra of the total luminescence yield, the visible and UV cross-luminescence. Bragg reflection has been used to image microcrystallites in lminescence ceramics [10] and defects of multilayers. It turned out that detection of the total photoelectron yield is a promising

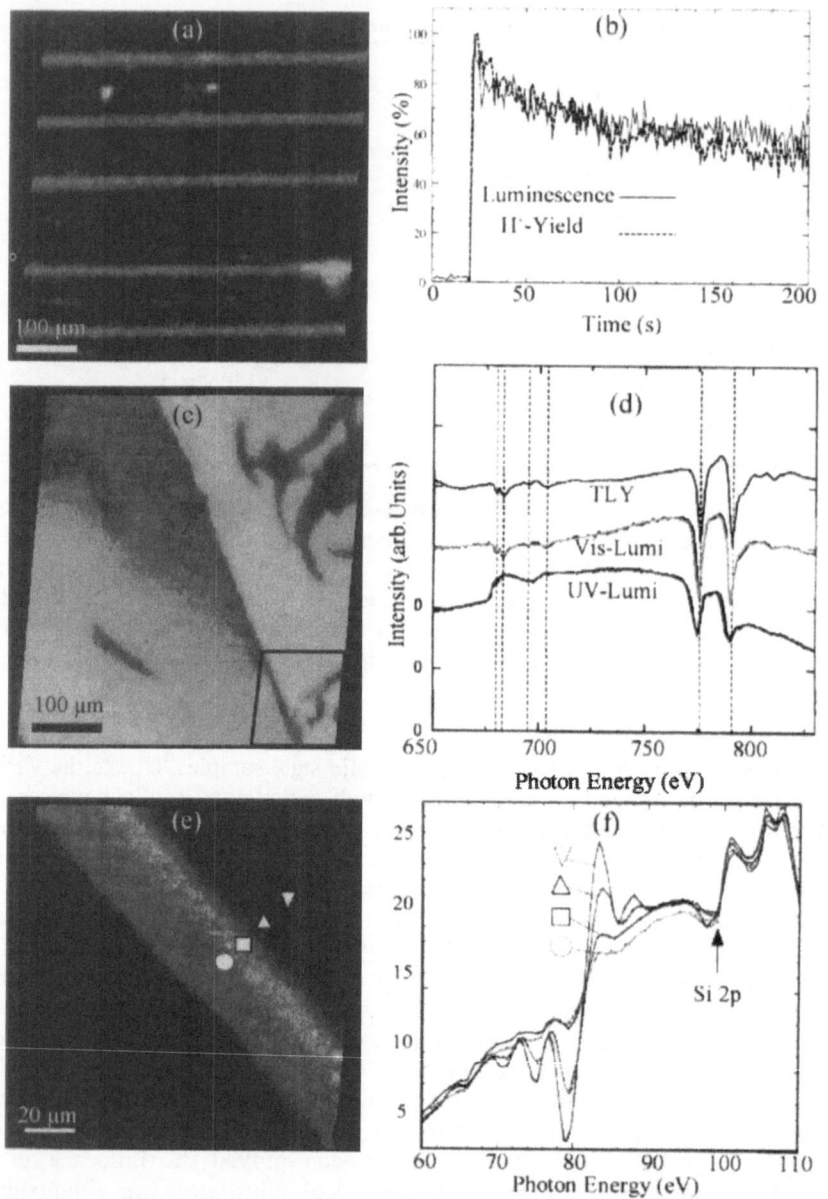

**Fig. 6.** Compilation of some results measured with the microscope at Hasylab. (a) VUV-stray-light micrograph of lithium niobate with Ti-doped stripes, hv = 40 eV. (b) Time dependence of the simultaneously measured luminescence and H⁺ signals from porous Si by illumination with primary zero order light [17]. (c) Total luminescence yield micrograph of a cleaved barium fluoride crystal, hv=683 eV. (d) Total, visible and UV luminescence excitation spectra of barium fluoride.(e) Photoelectron micrograph of an artificial defect on a Mo/Si multilayer, hv= 79 eV. (f) Total electron yield excitation spectra at the highlighted points in (e) [18].

method to analyse the homogeneity and quality of multilayer coatings for normal incidence optics used in EUV-lithography. Fig 6(e) and (f) show an electron micrograph of an artificial defect on a Mo/Si multilayer and the corresponding excitation spectra recorded from the highlighted points [18]. Finally linear magnetic dichroism in soft X-ray reflectivity will be used for element-specific imaging of samples with laterally inhomogeneous magnetization [19].

## 4 Summary

Our scanning soft X-ray microscope at Hasylab uses mirror optics for focusing of monochromatized synchrotron radiation to a microspot. The achromaticity of the grazing incidence optics provides full tunability for photon energies from 15 eV to 1500 eV with a lateral resolution in the micron range. An optionally usable normal incidence optics, restricted to photon energies below 30 eV, reaches a diffraction limited resolution of 100 nm.

A number of different signals can be used for spectromicroscopy and microspectroscopy. Photoelectrons can be detected in the total yield mode and spectrally analysed with time-of-flight and hemispherical analysers. Photodesorbed positive ions can be detected and analysed with multichannelplates and a time-of-flight spectrometer. Photoluminescence in the visible and UV are detected with several photomultipliers and spectrally resolved with a flat-field grating spectrometer. Specularly reflected, scattered and transmitted radiation is measured with different MCP detectors. The scanning plane is rotatable, the maximum scanning range has a volume of $(3mm)^3$ with a spatial resolution of 40 nm. In addition the defined pulse structure of the storage ring allows investigations with high temporal resolution.

Results of a variety of different samples have been presented to give an impression of the capabilities of our microscope.

## Acknowledgments

We thank the following persons for preparation of samples and for discussions: J. Becker (barium fluoride); S. Eisebitt (porous silicon); E. Louis, N. B. Koster, H.-J. Voorma, and F. Bijkerk (multilayers); W. Sohler and coworkers (lithium niobate).

This project is supported by the German Federal Minister of Education and Research (BMBF) under contract number 05 644 GUA 9.

## References

1    J. Voss, H. Dadras, C. Kunz, A. Moewes, G. Roy, H. Sievers, I. Storjohann and H. Wongel, *J. X-Ray Sci. Techn.* **3**, (1992) 85.

2    J. Voss, I. Storjohann, C. Kunz, A. Moewes, M. Pretorius, A. Ranck, H. Sievers, V. Wedemeier, M. Wochnowski and H. Zhang, *X-ray microscopy IV*. International Conference on X-Ray Microscopy, Chernogolovka/Russia, 1993 Bogorodski Pechatnik Publ. Comp., Chernogolovka Moscow Region, (1995).

3    J. Voss, C. Kunz, A. Moewes, and I. Storjohann, *Rev. Sci. Instrum* **63** (1), (1992) 569-573.

4    C.Kunz and J. Voss, *Fresenius J. Anal. Chem.* **335**, (1995) 494-498.

5    F. Senf, K. Berens v. Rautenfeld, S. Cramm, J. Lamp, J. Schmidt-May, J. Voss, C. Kunz, and V. Saile, *Nucl. Instrum. Methods* **A246**, (1986) 314.

6    C. Kunz and J. Voss, *Rev. Sci. Instrum.* **66**, (1995) 2021-2029.

7    J. Voss, M. Fornefett, C. Kunz, A. Moewes, M. Pretorius, A. Ranck, M. Schroeder and V. Wedemeier, *Journal of Electron Spectroscopy and Related Phenomena* **80**, (1996) 329-335.

8    M. Pretorius, M. Fornefett, J. Friedrich, C. Kunz, A. Ranck, K. Berens von Rautenfeld, M. Schroeder, V. Wedemeier and J. Voss, *X-ray microscopy V*. Proceedings of the International Conference XRM 96, this volume.

9    Ranck, M. Fornefett, J. Friedrich, M. Pretorius, K. Berens von Rautenfeld, M. Schroeder, V. Wedemeier and J. Voss, *X-ray microscopy V*. Proceedings of the International Conference XRM 96, this volume.

10   V. Wedemeier, K. Berens von Rautenfeld, M. Fornefett, J. Friedrich, M. Pretorius, A. Ranck, M. Schroeder, and J. Voss, *X-ray microscopy V*. Proceedings of the International Conference XRM 96, this volume.

11   M. Schroeder, M. Fornefett, J. Friedrich, M. Pretorius, A. Ranck, K. Berens von Rautenfeld, V. Wedemeier and J. Voss, *X-ray microscopy V*. Proceedings of the International Conference XRM 96, this volume.

12   H. Zhang, A. Föhlisch, C. Kunz, A. Moewes, M. Pretorius, A. Ranck, H. Sievers, I. Storjohann, V. Wedemeier and J. Voss, *Rev. Sci. Instrum.* **66** (6), (1995) 3513-3519.

13   Moewes, H. Zhang, C. Kunz, M. Pretorius, H. Sievers, I. Storjohann and J. Voss, *X-ray microscopy IV*. International Conference on X-Ray Microscopy, Chernogolovka/Russia, 1993 Bogorodski Pechatnik Publ. Comp., Chernogolovka Moscow Region, (1995).

14   Moewes, C. Kunz and J. Voss, *Nucl. Instrum. Meth.* **A 373**, (1996) 299-304.

15   Storjohann, C. Kunz, A. Moewes and J. Voss, in: *X-ray Optics and Microanalysis, Inst. Phys. Conf. Series, Bristol* **180**, (1993) 587 – 590.

16   J Voss, *X96 Proceedings of the International Conference, AIP Press*, to be published.

17   M. Fornefett, Diploma Thesis, Universität Hamburg, (1996).

18   J. Friedrich, K. Behrens v. Rautenfeldt, M. Fornefett, M. Pretorius, A. Ranck, M. Schroeder, H. Sievers, J. Voss, V. Wedemeier, and E. Louis, N. B. Koster, H.-J. Voorma, and F. Bijkerk, *X-ray microscopy V*. Proceedings of the International Conference XRM 96, this volume.

19   M. Pretorius, J. Friedrich, A. Ranck, M. Schroeder, V. Wedemeier, and J. Voss, to be published.

# X-Ray Magnetic Microspectroscopy
# Using the Circularly Polarized Undulator Radiation
# at the TRISTAN Accumulation Ring

Y. Kagoshima[1*], J. Wang[2**], T. Miyahara[1***], M. Ando[1], S. Aoki[2]

[1] Photon Factory, National Laboratory for High Energy Physics
Oho 1-1, Tsukuba, Ibaraki 305, Japan

[2] Institute of Applied Physics, University of Tsukuba
Tennodai 1-1, Tsukuba, Ibaraki 305, Japan

Present addresses:
* Faculty of Science, Himeji Institute of Technology, Kanaji 1479-1, Kamigoricho,
Akogun, Hyogo 678-12, Japan
E-mail: kagosima@sci.himeji-tech.ac.jp
** Nikon Corporation, Nishioi 1-6-4, Shinagawa, Tokyo 140, Japan
*** Faculty of Science, Tokyo Metropolitan University, Minamiohsawa 1-1, Hachioji,
Tokyo 192-03, Japan

**Abstract.** A scanning X-ray microscope was developed at beamline BL-NE1B of the TRISTAN accumulation ring. The microscope uses a Fresnel zone plate as a focusing element. The focused X-ray spot size has been evaluated to be 1 μm. The magnetic domains of a deposited nickel layer have been imaged using circularly polarized radiation produced by a helical undulator. The contrast arises from the effect of magnetic circular dichroism at the Ni $L_{2,3}$ photoabsorption edges. The photoabsorption spectra from a pair of domains having the opposite magnetization direction to each other have also been measured by probing the single domains with a focused X-ray spot. It was confirmed that the spectra exhibited features corresponding to the relative orientation between the direction of the circular polarization and that of the magnetization of domains. The microscope can offer element- and domain-specific X-ray magnetic spectromicroscopy/microspectroscopy on a sub-micrometer scale.

## 1 Introduction

An X-ray magnetic circular dichroism (MCD) measurement of the photoabsorption spectra has been a powerful experimental technique to investigate the magnetic properties of various magnetic materials [1-3]. The current MCD measurements give spectroscopic information mostly as an averaged value over the sample. It is well known that magnetic materials have magnetic domains without an external applied magnetic field. If a microscopic approach is introduced, differently oriented magnetic domains can be imaged by the contrast arising from MCD. This principle has been demonstrated by Stöhr *et al.* Using an electrostatic microscope with a spatial resolution of ~1μm [4, 5] and by Schneider *et al.* Using a commercially available hemispherical energy analyzer with a spatial resolution of 10 μm [6]. In addition to the observation, if a focused X-ray beam can be used as a probe, one can perform

**X-Ray Microscopy and Spectromicroscopy**
Eds.: J. Thieme, G. Schmahl, D. Rudolph, E. Umbach
© Springer-Verlag Berlin Heidelberg 1998

position-specific spectroscopy, which may be called microspectroscopy. The objective of our scanning X-ray microscope is not only to visualize the distribution of magnetic domains, but also to probe the magnetic properties inside single domains with an X-ray microbeam.

## 2 Beamline BL-NE1B at the TRISTAN AR

A plane view of the National Laboratory for High Energy Physics is shown at the left-bottom of Fig. 1. There are four electron/positron accelerators, a 2.5-GeV linear accelerator, a 2.5-GeV Photon Factory storage ring, a 8.0-GeV TRISTAN accumulation ring (AR) and a 30-GeV TRISTAN main ring (MR). The AR is an injection ring of the MR. During the intervals of MR-injection the AR is operated with an electron beam of 6.5 GeV as a synchrotron-radiation source. Typically, it takes one hour for MR-injection. The following two hours are used for synchrotron-radiation experiments. Principal parameters of PF, AR and MR have been given elsewhere [7].

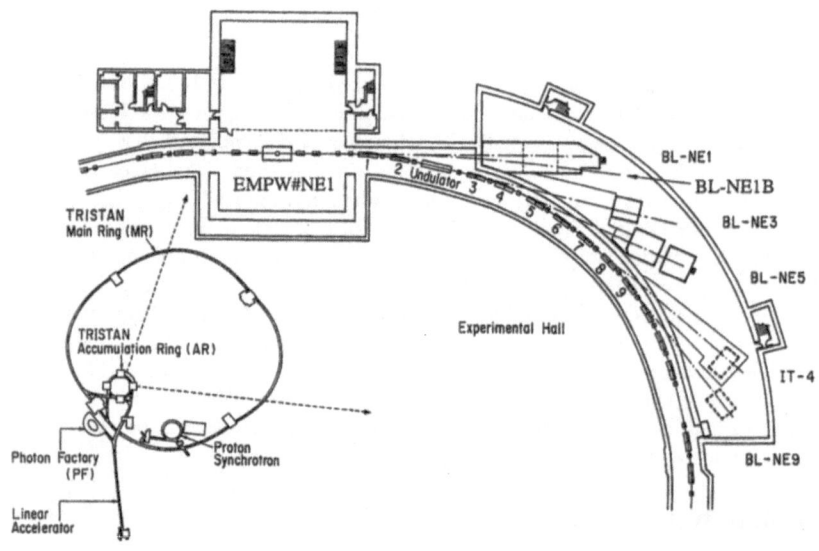

**Fig. 1.** A plane view of the National Laboratory for High Energy Physics

A quarter of the AR is used for the synchrotron radiation facility, as shown in Fig. 1. There is an insertion device, EMPW#NE1 [8], in one of the four long straight sections of the AR. The EMPW#NE1 comprises a combination of horizontal and vertical pairs of permanent-magnet arrays. The number of periods is 16 with a period length of 16 cm. It has two main operational modes: a wiggler mode and an undulator mode. In the undulator mode, we can select one of three configurations: a helical undulator (HU) an elliptic undulator (EU) or a linear undulator (LU). Beamline BL-NE1B [9] is for the undulator mode of the device. Under HU mode, soft X-rays with circular polarization ($P_c$) almost equal to 1 can be provided. Since the device provides on-axis highly brilliant radiation, we don't have to use off-plane radiation, such as from bending magnets. The source brilliance is about $1 \times 10^{16}$ phs/s/mm$^2$/mrad$^2$/0.1% b.w./50 mA. The BL-NE1B covers the photon energy range from 250 eV to 2 keV. Figure 2 shows the entire optical system of BL-NE1B. BL-NE1B is equipped with a grazing-incidence grating monochromator. It is of the inverse Vodar type, which is one of Rowland configurations, having two interchangeable gratings with groove densities of 1200 *lines/mm* (G1) and 2400 *lines/mm* (G2). A radius of curvature of the gratings ($R$) is 10.31 m. The monochromatized radiation is post-focused both vertically and horizontally by a toroidal mirror. The degree of circular polarization after the beamline is expected to be better than 90 % based on the results of an MCD measurement [3]. In the experiments mentioned in this paper the monochromator was operated with full-open slits, under which the monochromaticity ($E/\Delta E$) was estimated to be ~370 at a photon energy of 850 eV. At BL-NE1B, the MCD measurements [3], photoelectron diffraction experiments [10] and microscopy with zone plates [11-13] have been performed.

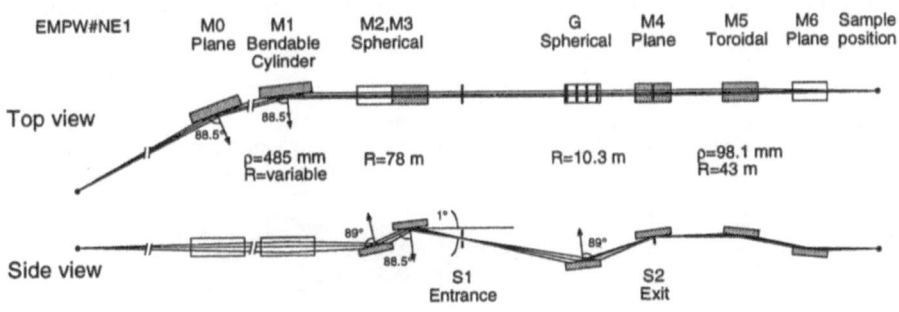

**Fig. 2.** Entire optical system of BL-NE1B

# 3 Scanning Microscope

## 3.1 Optical System and Apparatus

The optical system of the scanning microscope is shown in Fig. 3. There is a prepinhole (PPH) at the post-focused point to collimate the X-ray beam. The observed post-focused spot size was about 0.2 mm (vertical) and 0.8 mm (horizontal). Regarding PPH as a secondary light source, the X-ray beam is focused onto a sample by a demagnification optical system using a zone plate (ZP). The ZP was fabricated by NTT [14]. Ist diameter and outermost zone width ($\Delta r_N$) are 160 μm and 150 nm, respectively. The focal length is about 17 mm at the photon energy of 850 eV. The spatial resolution is determined by a size of PPH and the demagnification ratio of the optical system, because $\Delta r_N$ is enough small to be neglected in comparison to the geometrical spot size. The focused beam is incident obliquely onto the sample at an angle of 30° from the surface in order to effectively extract the MCD signal, because the magnetization direction of the sample is in-plane. The sample is scanned two-dimensionally, while simultaneously measuring the electron yield ($I$) from the sample and the incident intensity ($I_0$). The two-dimensional distribution of the normalized electron yield ($I/I_0$ ; relative absorption) is displayed as a microscopic image. The photoelectron yield is counted using a channel electron multiplier. A micro-channel plate is used to monitor the X-ray image for the optical alignment. The scanning unit consists of x-y coarse stages driven by pulse motors and x-y fine stages driven by piezoelectric actuators. The structure of a single fine stage is a monolithic parallel spring stage with flexure hinges and a magnification lever. The lever magnifies a motion of a piezoelectric actuator with the magnification ratio of 2. The fine stages can scan the sample two dimensionally with the smallest step of 10 nm. Ist specifications are summarized in Table 1. Figure 4 shows a schematic top view of the apparatus. All of the elements except for the scanning unit are mounted inside a vacuum chamber. The chamber, the scanning unit and an ion pump are fixed on the optical bench with an air spring vibration isolator. The sample is connected to the scanning unit through a flexible welded bellows. The position of an order sorting aperture (OSA) is fixed, while ZP can be aligned using the manipulator. The surface of the sample can be sputtered using an ion gun. A channel electron multiplier is connected to the linear motion drive.

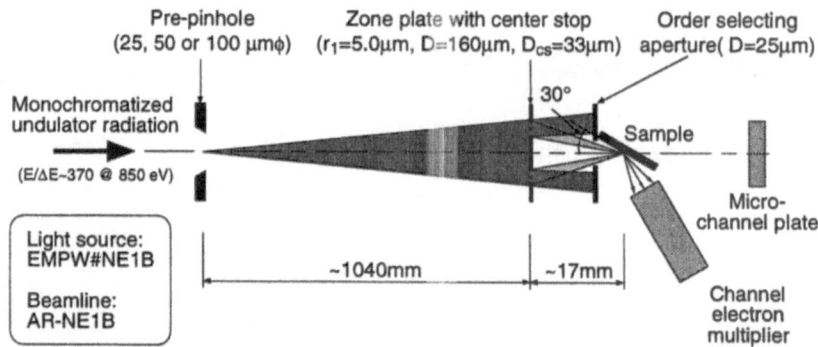

**Fig. 3.** The top view of the optical arrangement of a scanning X-ray microscope

**Table 1.** Specifications of the scanning unit

| Coarse stages | step width | Repeatability | Movable range |
|---|---|---|---|
| | 0.5μm | ± 1μm | ± 12μm |
| Fine stages | Resoluition (operated with linear decoder) | | Movable range |
| | 10nm | | ± 45μm |

## 3.2 Resolution Test

In order to evaluate the focused spot size, which corresponds to the spatial resolution of the microscope, an edge scan profile was taken as shown in Fig. 5. One side of a wire of a copper #2000 mesh was used as an edge sample. The scanning direction was vertical. The photon energy was tuned to be a copper $L_3$ absorption edge (933 eV). Dots were the raw data and two lines were the fitted results. The raw data were fitted to the error function (solid line) and its derivative, the Gaussian distribution, is shown (dashed line). The full width at a half maximum, defined here as the spot size, was evaluated to be 1.0 μm, which was almost equal to the geometrically expected size of 0.9 μm. It should be noted that the horizontal resolution was twice the vertical one, namely 2.0 μm, because the incidence angle of the beam to the sample was 30° from the surface.

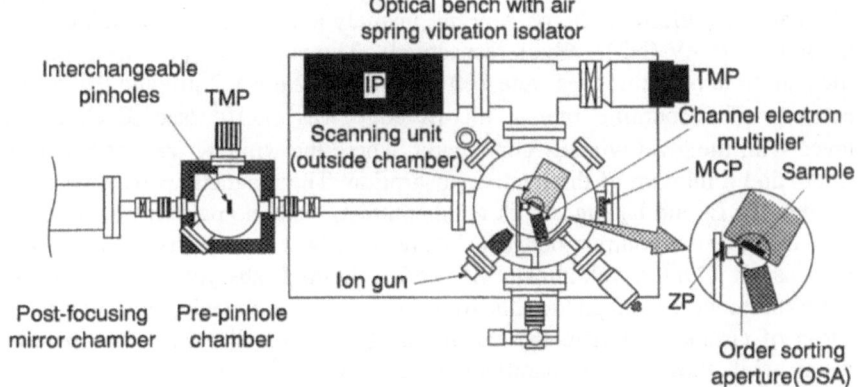

**Fig. 4.** A schematic top view of an apparatus of the scanning X-ray microscope

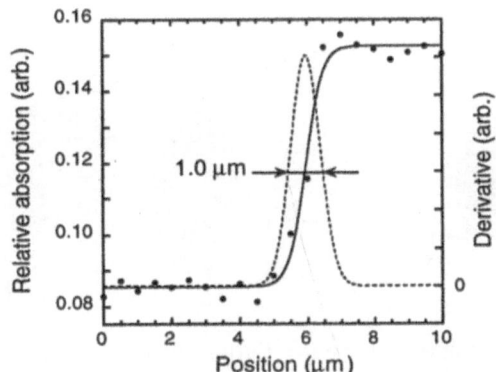

**Fig. 5.** Edge scan profile for evaluating the X-ray spot size . The dots , solid and dashed lines, represent the raw data, the fitted curve to the error function and its derivative, respectively. The full sidth at half maximum gave a spot size of 1.0 μm

## 4 Magnetic Spectromicroscopy and Microspectroscopy

### 4.1 Observation of Magnetic Domains as Spectromicroscopy

Figure 6 illustrates the experimental arrangement. The direction of circular polarization was fixed to the right one. The magnetic sample was a nickel layer with a thickness of 50 nm, deposited on a piece of commercially available 8 mm video tape. Before nickel was deposited on the tape, the magnetic pattern was drawn on it using an audio recording head. Therefore, the magnetization direction of the nickel was aligned subject to the originally drawn pattern in the tape during the depositing process. The pattern consisted of domains with an alternating in-plane magnetization direction. The width of each domain was 10 μm, while the length was much larger than the width. Therefore, the pattern looked like a grating. The experimental results are shown in Fig. 7. Three images are micrographs of the same area of the sample, but taken at three different photon energies, namely at (a) 844 eV, (b) 853 eV (Ni-$L_3$ edge) and (c) 870 eV (Ni-$L_2$ edge), respectively. The number of pixels, the pixel size and the gate time of the images were 100 x 40 pixels, 2 μm x 2 μm and 0.3 sec/pixel, respectively. The counting rate at the $L_3$ edge was ;$1x10^5$ /sec. Each image is displayed with the level window conversion, where the window was set between the maximum and minimum of the relative absorption. The magnetic pattern was clearly observed at the $L_3$ and $L_2$ edges with good contrast, while no pattern was observed at 844 eV. Further, the contrast between $L_3$ and $L_2$ edges was reserved. These results were consistent with the well known MCD effect in the L-absorption edges due to the magnetic moment of magnetic 3d transition metals. Namely, the photon spin (direction of circular polarization) and the magnetization direction were parallel in domain B, while they were antiparallel in domain A.

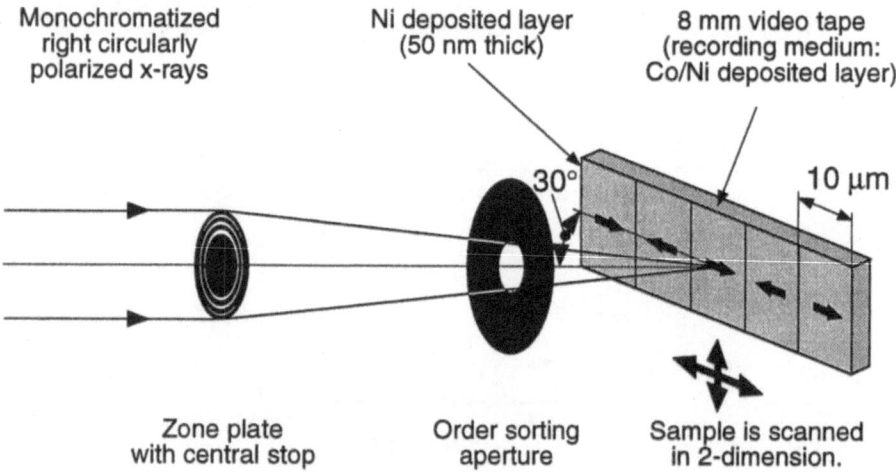

**Fig. 6.** Schematic arrangement of the scanning X-ray microscope. The thick black arrows indicate the magnitization direction drawn in the tape

## 4.2 Domain-Specific Absorption Spectra as Microspectroscopy

As a next step, fixing the position of the X-ray spot inside of a single domain (A or B), the photon energy was scanned in order to measure the absorption spectra. Figure 8 (a) shows the result. The gate time for counting at each photon energy was 1 sec. The solid and the dashed lines are the spectra from domain A and B, respectively. The absorption in A was smaller than that in B. at the $L_3$ edge, while it was larger at the $L_2$ edge, which is consistent with the images in Fig. 7. The difference between the two spectra (A-B = Antiparallel-parallel) is also shown in Fig. 8 (b). The energy scan step and the gate time for counting at each photon energy were 0.25 eV and 1 sec, respectively. The typical MCD spectrum in the L-absorption edges of the 3d transition metals was clearly observed. Since no cleaning process, such as ion-beam etching, was made on the surface of the sample, this technique is not too sensitive to the surface. This is because of the fact that the mean free pass of electrons is long relative to the thickness of the surface oxide layer. Therefore, the microscope can extract bulk-sensitive magnetic information, and this may be considered to be a merit compared to the technique using spin-polarized electrons [15].

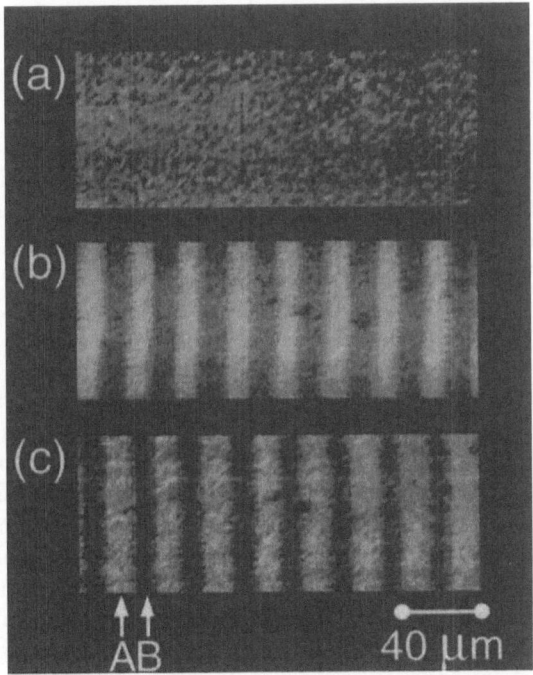

**Fig. 7.** Micrographs of the same microscopic area of the nickel-deposited layer taken at three photon energies: (a) 845 eV, (b) 854 eV (Ni-$L_3$ edge), and (c) 871eV (Ni-$L_2$ edge, respectively)

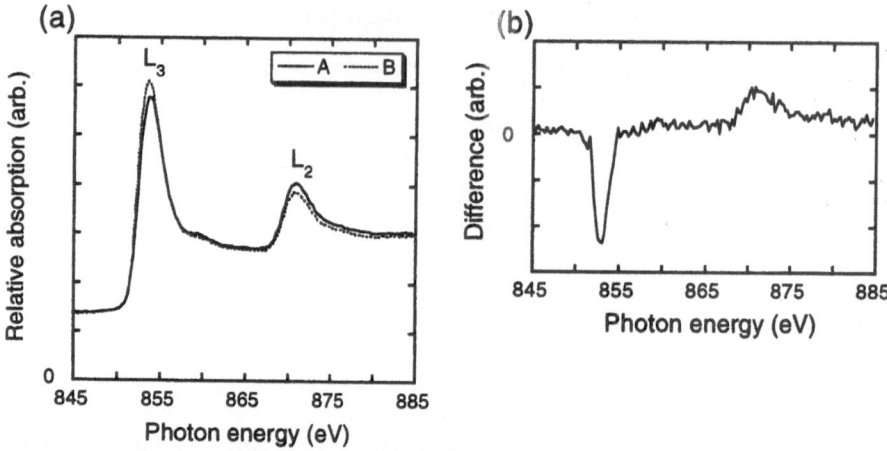

**Fig. 8.** (a) magnetic domain-specific absorption spectra taken by fixing the position of the X-ray spot inside a single domain (A or B), indicated by arrows in Fig. 7. The solid and dashed lines are the spectra from domains A and B, respectively. (b) The difference between the two spectra (A-B = antiparallel-parallel)

## 5 Summary and Future Prospects

Since the outermost zone width of the zone plate was 150 nm, we could obtain a resolution of 180 nm according to Rayleigh's criterion. To achieve this, the zone plate had to be coherently illuminated. Since the available coherent flux was deficient in taking images with a reasonable signal-to-noise ratio within a reasonable acquisition time, we chose a 50 μm pinhole, which limited the resolution to 1.0 μm. As we already have a zone plate with the outermost zone width of 50 nm developed by NIT [16], we can achieve a resolution of 60 nm at any third-generation synchrotron radiation facility.

Spectroscopy using X-rays provides elemental specificity by tuning the photon energy to each characteristic absorption edge of the concerned element. By combining all popular X-ray spectroscopic studies with our microscope, we can perform them on a sub-micrometer scale. For example, micro-XANES (X-ray absorption near-edge structure) is possible. It could detect the chemical state-specific magnetic circular dichroism effects, which would reflect the magnetic properties in a particular chemical state. Furthermore, if an electron energy analyzer is introduced for collecting photoelectrons emitted from a sample, micro-XPS (X-ray photoemission spectroscopy) with magnetic sensitivity would also be possible. It would be a new experimental technique for investigating the magnetism of surfaces and thin films.

## Acknowledgment

This work has been carried out under an approval of Photon Factory Program Advisory Committee (proposal number: 93G126).

# References

1   L. Baumgarten, C.M. Schneider, H. Petersen, F. Schäfers and J. Kirschner, Phys. Rev. Lett. **65**, 492 (1990).

2   C.T. Chen, F. Sette, Y. Ma and S. Modesti, Phys, Rev. B **42** 7262 (1990).

3   T. Miyahara, S.-Y. Park, T. Hanyu, T. Hatano, S. Muto and Y. Kagoshima, Rev. Sci. Instrum. **66**, 1558 (1995).

4   J. Stöhr, Y. Wu, B.D. Hermsmeier, M.G. Samant, G.R. Harp, S. Koranda, D. Dunham and B.P. Tonner, Science **259**, 658 (1993).

5   Y. Wu, S.S.P. Parkin, J. Stöhr, M.G. Samant, B.D. Hermsmeier, S. Koranda, D. Dunham and B.P. Tonner, Appl. Physl Lett. **63**, 263 (1993).

6   C.M. Schneider, K. Holldack, M. Kinzler, M. Grunze, H.P. Oepen, F. Schäfers, H. Petersen, K. Meinel and J. Kirschner, Appl. Phys. Lett. **63**, 2432 (1993).

7   M. Ando and Y. Kagoshima, in *X-ray Microscopy III*, edited by A.G. Michette, G.R. Morrison and C.J. Buckley (Springer, Berlin, 1992), p. 23.

8   S. Yamamoto, H. Kawata, H. Kitamura, M. Ando, N. Sakai, N. Shiotani, Phys. Rev. Lett. **62**, 2672 (1989).

9   Y. Kagoshima, T. Miyahara, S. Yamamoto, H. Kitamura, S. Muto, S.-Y. Park, and J.-D. Wang, Rev. Sci. Instrum. **66**, 1696 (1995).

10   H. Daimon, T. Nakatani, S. Imada, S. Suga, Y. Kagoshima and T. Miyhara, Jpn. J. Appl. Phys. **32**, L1480 (1993).

11   J.-D. Wang, Y. Kagoshima, T. Miyahara, M. Ando, S. Aoki, E. Anderson, D. Attwood, D. Kern, these proceedings.

12   Y. Kagoshima, T. Miyahara, M. Ando, J. Wang and S. Aoki, Rev. Sci. Instrum. **66**, 1534 (1995).

13   Y. Kagoshima, T. Miyahara, M. Ando, J. Wang and S. Aoki, J. Appl. Phys. **80**, 3124 (1996).

14   M. Sekimoto, A. Ozawa, T. Ohkubo, H. Yoshihara, M. Kakuchi and T. Tamamura, in *X-ray Microscopy II*, edited by D. Sayer, M. Howells, J. Kirz, and H. Rarback (Springer, Berlin, 1988), p. 178.

15   K. Koike and K. Hayakawa, Jpn. J. Appl. Phys. **23**, L187 (1984).

16   A. Ozawa, T. Tamamura, T. Ishii, H. Yoshihara and Y. Kagoshima, to be published in the proceedings of International Conference on *Micro- and Nanoengineering 96*.

# Concept and Design
# of the SMART Spectromicroscope at BESSY II

W. Engel, R. Degenhardt, A. M. Bradshaw, W. Erlebach, K. Ihmann,
H. Kuhlenbeck, R. Wichtendahl, H.-J. Freund, R. Schlögl[1],
D. Preikszas, H. Rose, R. Spehr, P. Hartel[2],
G. Lilienkamp, Th. Schmidt, E. Bauer[3],
G. Benner[4],
R. Fink, M. R. Weiss, E. Umbach[5]*

[1] Fritz-Haber-Institut der Max-Planck-Gesellschaft,
Faradayweg 4-6, D-14195 Berlin, Germany
[2] Technische Hochschule Darmstadt,
Angewandte Physik, Hochschulstraße 6, D-64289 Darmstadt, Germany
[3] Technische Universität Clausthal,
Leibnizstraße 4, D-38678 Clausthal-Zellerfeld, Germany
[4] LEO Elektronenmikroskopie GmbH, D-73446 Oberkochen, Germany
[5] Universität Würzburg, Experimentelle Physik II,
Am Hubland, D-97074 Würzburg, Germany

**Abstract.** The concept of a new spectromicroscope (SMART = spectromicroscope for all relevant techniques) currently under construction for an undulator beam line at BESSY II is discussed. The design of the optical system is described as well as the modes of operation and the experiments that can be performed. Monochromatic XUV-radiation (tunable within the energy range 20 to 2000 eV) will be provided by a plane-grating monochromator and focussed onto the sample by means of an ellipsoidal mirror. In addition, an electron gun will be installed allowing LEEM, MEM and other forms of microscopy as well as small spot LEED to be performed. With an aberration corrector we expect to achieve a lateral resolution better than 5 nm and to increase considerably the transmission of the optical system compared to previous instruments. An imaging band pass filter corrected to second order will select the energy and the energy band width for the image-forming electrons. The instrument will also allow spectroscopy of photoelectrons from selected small areas of the sample (5–500 nm) with an energy resolution of 0.1 eV.

## 1 Introduction

Photoelectron microscopy with energy selected electrons (termed spectromicroscopy) and spectroscopy of photoelectrons from selected small areas of the sample (termed microspectroscopy) rely on the availability of high brilliance soft X-ray undulators on modern synchrotron radiation sources [1-6]. Some of the

---

* Project coordinator

**X-Ray Microscopy and Spectromicroscopy**
Eds.: J. Thieme, G. Schmahl, D. Rudolph, E. Umbach
© Springer-Verlag Berlin Heidelberg 1998

so-called third generation sources (ALS in Berkeley, ELETTRA in Trieste) are already in operation; others (BESSY II in Berlin) will be in operation within a short time. Research groups of five German institutions have decided to develop jointly a versatile and high-performance spectromicroscope for an undulator beam line at BESSY II [7]. The aim is to achieve a spatial resolution of about 2 nm and an energy resolution of 0.1 eV for the spectroscopy of photo- and Auger electrons. The photon energy will be tunable over a wide range in the XUV giving the possibility of optimising the experimental conditions for particular measurements and also allowing X-ray absorption spectroscopy to be performed. It is intended that image acquisition is so fast that chemical and physical processes at surfaces can be followed with a temporal resolution of about 20 ms.

## 2   General Concept

A high-performance spectromicroscope is necessarily a parallel imaging device and cannot be realised with a scanning-type instrument. In the latter the image is constructed in a serial way, making image formation far too slow for the observation of processes at a sample surface in real time. Further, the lateral resolution of soft X-ray scanning microscopes is determined by the size of the irradiated spot, for which the diffraction limit of the imaging optics will be larger than the desired lateral resolution. In principle, electrons have the capability of forming more highly resolved images than photons of the same energy: the electron mass results in a much shorter wavelength. In order to take advantage of this fact in an electron microscope, however, lens aberrations, in particular the spherical and the axial chromatic aberrations of the objective lens have to be sufficiently small. To compensate for at least the third-order spherical aberration and the second-rank axial chromatic aberration of the objective lens rather sophisticated electron optics is required. The corrected imaging system proposed by Rose and Preikszas [7] will be used in the instrument described here. In addition, an imaging energy band pass filter is needed to select the kinetic energy and the energy bandwidth for the image-forming electrons. This feature will also allow photoelectron spectroscopy to be performed.

The correction of spherical and chromatic aberrations allows the beam aperture and thus the transmission of the objective lens to be considerably enlarged which may be even more important than the improvement of the lateral resolution limit. Although the photon flux density obtainable at BESSY II will be extremely high, lack of intensity will remain a problem in photoemission microscopy, in particular if core electrons are used for imaging. The photoionisation cross sections for core electrons are low and the estimated intensities for typical samples will just allow 2 nm spatial resolution to be achieved [8]. Reaching simultaneously the limits of both the spatial and the energy resolution will, in general, not be possible. The problem of transmission is illustrated in Fig. 1 which shows a section through a quadrant of momentum space. The total photoelectrons emitted from the sample fall in a region of momentum space lying within a sphere the radius of which corresponds to the maximum in the photo-

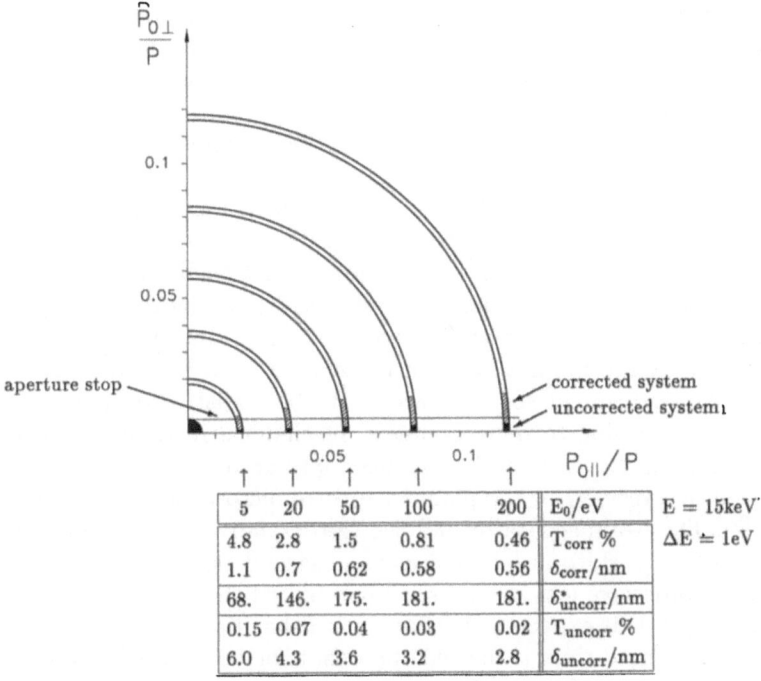

| | | | | | $E_0$/eV | $E = 15\text{keV}$ |
|---|---|---|---|---|---|---|
| 5 | 20 | 50 | 100 | 200 | $E_0$/eV | $E = 15\text{keV}$ |
| 4.8 | 2.8 | 1.5 | 0.81 | 0.46 | $T_{\text{corr}}$ % | $\Delta E \doteq 1\text{eV}$ |
| 1.1 | 0.7 | 0.62 | 0.58 | 0.56 | $\delta_{\text{corr}}$/nm | |
| 68. | 146. | 175. | 181. | 181. | $\delta^*_{\text{uncorr}}$/nm | |
| 0.15 | 0.07 | 0.04 | 0.03 | 0.02 | $T_{\text{uncorr}}$ % | |
| 6.0 | 4.3 | 3.6 | 3.2 | 2.8 | $\delta_{\text{uncorr}}$/nm | |

**Fig. 1.** Section through a quadrant of momentum space illustrating the small percentage of emitted electrons that can be used for imaging in order to achieve the resolution limits, $\delta_{\text{corr}}$ and $\delta_{\text{uncorr}}$, for the corrected (hatched areas) and uncorrected (black areas) objective lenses, respectively. The corresponding values for the transmission are $T_{\text{corr}}$ and $T_{\text{uncorr}}$. $\delta^*_{\text{uncorr}}$ is the resolution of the uncorrected system calculated for $T_{\text{corr}}$. Data are taken from ref. [9] or derived from data therein. ($p_{0\perp}$, $p_{0\parallel}$ components of the initial momentum perpendicular and parallel to the optic axis, $E_0$ initial kinetic energy, $p$ and $E$ momentum and kinetic energy of image-forming electrons, respectively, $\Delta E$ energy bandwidth; see text.)

electron kinetic energy. Only a small part of the total volume, however, can be used for imaging, as indicated by the hatched areas in Fig. 1. In order to limit the influence of the remaining chromatic aberration only a small volume between two concentric spheres has to be selected by means of the band pass filter. It determines the mean radius of the spheres (i. e. the initial kinetic energy) and the separation of the spheres (i. e. the width of the energy band for the electrons which pass through the filter and form the image). The remaining spherical aberration requires an aperture stop that limits the volume to the vicinity of the optic axis. The decrease in transmission with increasing kinetic energy, dictated by the condition of optimal lateral resolution, is obvious from the table in Fig. 1. A transmission close to 100% is attained only when photoemission microscopy is carried out in its classical form as a threshold microscopy using photons with energies slightly higher than the work function. The transmission obtained at the resolution limit for the uncorrected system is extremely small which may be the reason that these calculated resolution limits have never been achieved experimentally. The transmission at the resolution limit for the corrected system, however, is reasonable and represents the more realistic experimental conditions achieved in existing uncorrected instruments. The resolution $\delta_{uncorr}^{*}$ that can be obtained theoretically with the increased transmission of the corrected system, but with the corrector switched off, is also given in Fig. 1. It is interesting to note that these values correspond to the lateral resolution routinely attained in uncorrected systems. The best lateral resolution that has been achieved so far is about 40 nm [1]. Any improvement in lateral resolution by the corrector on the basis of increased transmission would represent important progress even though the theoretical resolution limit may not be achieved. The expected value of 2 nm for the present instrument seems, however, to be realistic because the deterioration in resolution due to misalignments, mechanical instabilities, alternating magnetic fields etc. have already been taken into account.

A schematic overview of the planned spectromicroscope (SMART) is shown in Fig. 2 [7]. The beam separator and the tetrode mirror, which simultaneously compensates for the third-order spherical and second-rank axial chromatic aberrations of the objective lens, have been investigated theoretically but so far not been constructed. It is expected that for the construction and experimental tests of these devices more time is needed than that for the rest of the microscope. Thus, in the first stage of the project, the spectromicroscope will not be equipped with these components; an intermediate lens will bridge them. When the beam separator and mirror are incorporated, a field emission electron gun will also be installed opposite the mirror, allowing low energy electron microscopy (LEEM), mirror microscopy (MEM), secondary electron emission microscopy (SEEM), Auger electron emission microscopy (AEEM) [2], and small spot low energy electron diffraction (LEED) to be performed. The imaging band pass filter, an UHV-compatible version of the omega filter developed by Lanio et. al. [10], will be installed in the first stage of the project. The omega filter is fully corrected to second order.

Recently, Rempfer et. al. have reported the simultaneous correction of the spherical and the axial chromatic aberrations of a probe forming objective lens by means of a mirror operated on an optical bench [11]. This represents a first experimental success towards a corrector for the Oregon photoemission microscope [12]. Band pass filters with a lower level of aberration correction are used in the Clausthal instrument [13,14] and in the spectromicroscope called PRISM which has recently been set up for use at the ALS [6].

## 3   Description of the Major Components

The SMART spectromicroscope will be installed on an U49 undulator beam line at BESSY II in Berlin. A plane-grating, grazing-incidence monochromator with the designation PM-6 will provide soft X-radiation in the spectral range from

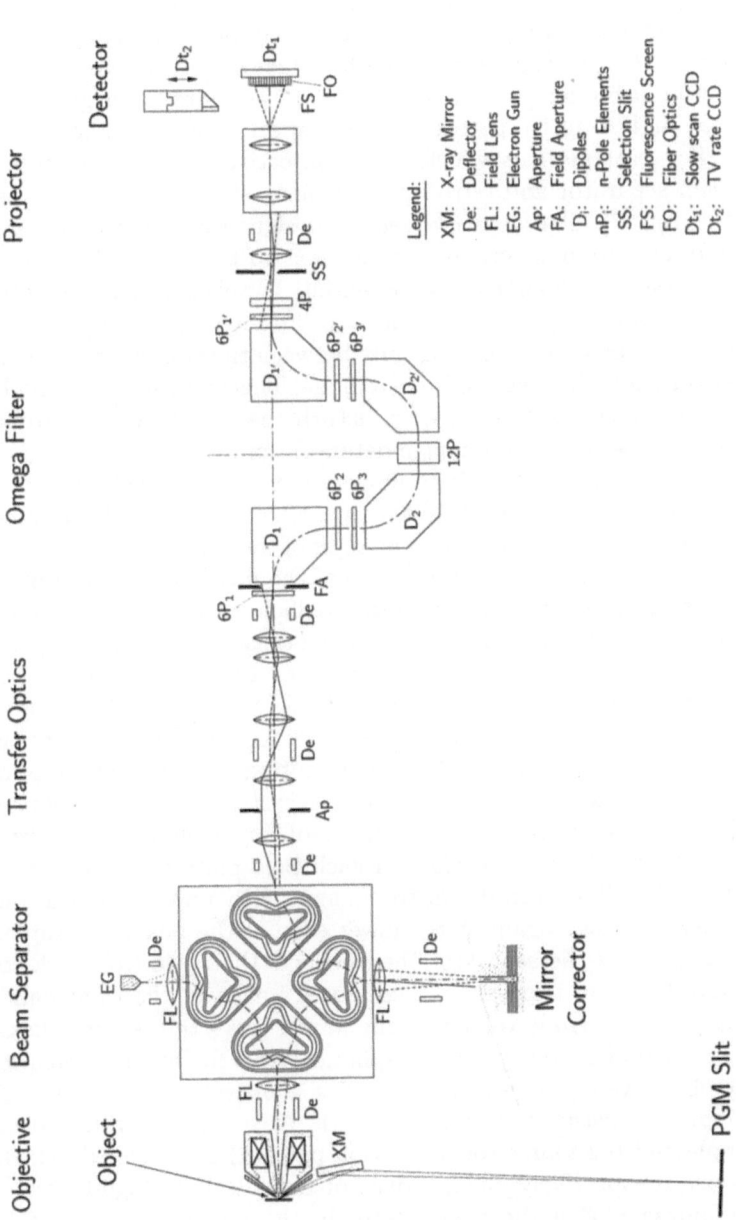

**Fig. 2.** Schematic layout of the SMART spectromicroscope currently under construction for a BESSY II undulator beamline

about 20 eV to 2000 eV. The maximum of its spectral resolving power is about $\nu/\Delta\nu = 10^4$. The exit slit of the monochromator is imaged onto the sample by means of a grazing-incidence ellipsoidal mirror with a demagnification ratio of 10:1, thus irradiating a sample area of $5\mu$m $\times$ $10\mu$m. At a photon energy of 400 eV and a bandwidth of 60 meV the flux density at the sample is estimated to be $4 \times 10^9$ photons $\mu$m$^{-2}$s$^{-1}$. This can be increased by a factor of 10 at the expense of energy resolution by using a second, low-resolution monochromator grating.

The sample is at high negative potential with respect to ground and is the cathode of a so-called cathode lens which is the objective of the microscope. Objective lens, beam separator and tetrode mirror form the corrected imaging system described in ref. [7]. The objective lens is designed as an electrostatic magnetic compound lens. Its magnetic pole piece next to the sample is electrically isolated and can be set at high negative potential. The electric field strength at the sample is then reduced and can even be zero if desired. The acceleration of the emitted photoelectrons then occurs mainly between the two magnetic pole pieces which are at the same time electrodes of a diode. This feature allows rough surfaces to be imaged which would otherwise disturb the electric field in front of the surface and give rise to a severe image deterioration.

The cathode lens forms an intermediate image of the sample surface at the entrance of the beam separator where a field lens is located which sets the diffraction plane, i. e. the source image, at infinity. The signs of the third-order spherical and second-rank chromatic aberration coefficients of the proposed electron mirror are opposite to the signs of the corresponding coefficients of the rotationally symmetric magnetic and electrostatic lenses which can thus be compensated [15]. The incident and reflected beams have to be separated by means of magnetic dipole fields which, in general, cause dispersion and second order aberrations. The proposed beam separator, however, does not affect the transferred image by dispersion and second order aberrations, thus representing a remarkable step in the development of mirror-based aberration correction systems. It consists of two quadratic plane pole plates separated by a distance of 7 mm. Four coil triplets are placed into grooves on the inner surface of each pole plate to produce the necessary magnetic fields. The tetrode mirror images with unit magnification and places the corrected image again at the lower edge of the beam separator. The lower right quadrant of the separator then deflects the optic axis back to that of the objective lens and transfers the corrected image to the right hand edge of the separator. The tetrode version of the mirror gives the required flexibility to compensate simultaneously for the spherical and the axial chromatic aberrations of the objective lens in almost all modes of operation.

The transfer optics consists of five electrostatic einzel lenses which form images of the sample and the source (or diffraction pattern) at two fixed planes $z = z_{E2}$ and $z = z_{E1}$, respectively, independent of the selected magnification. This meets the requirements of the omega filter in the imaging mode of the microscope. The plane $z = z_{E1}$ of the source image is in the centre of the last lens, and an intermediate sample image is positioned in the centre of the penultimate lens, giving an arrangement that allows focussing of the two images almost independently. Further, the position of the sample image and the source image (or diffraction pattern) can be exchanged so that the instrument is also capable of producing energy-filtered LEED pattern. Third-order geometric and second-rank chromatic aberrations are negligible for all the planned modes of operation [16]. The first lens of the system places a source image or diffraction pattern at a plane $z = z_{Ap}$ which is fixed for all modes of operation and contains a set of interchangeable diaphragms that are to be used to limit the beam aperture. Such a set is also available in the plane $z = z_{E2}$ at the entrance of the omega filter which will be used to limit the field of view.

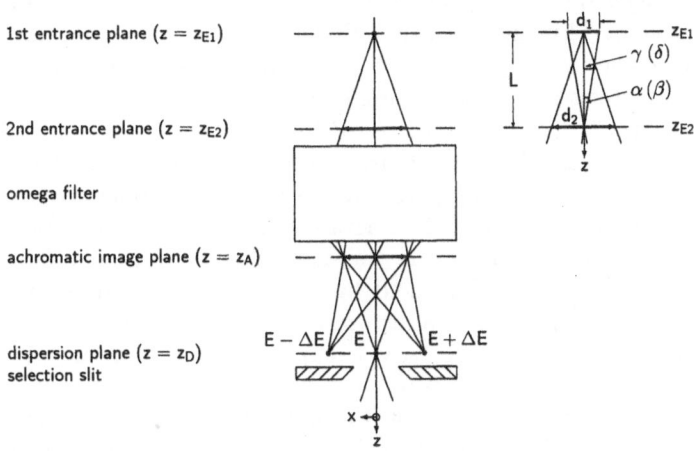

**Fig. 3.** Conjugate planes $(z_{E1}, z_D)$ and $(z_{E2}, z_A)$ for stigmatic focussing of the omega filter and definition of the ray parameters $\alpha(\beta)$ and $\gamma(\delta)$. $d_1$ defines the diameter of the source image and $d_2$ the diameter of the field of view in front of the energy filter. $L$ is the Helmholtz length.

The omega filter consists of four sector magnets each deflecting the optic axis by 90° in such a way that its form resembles the Greek capital letter omega. It images the entrance planes $z = z_{E1}$ and $z = z_{E2}$ stigmatically with unit magnification into the dispersion plane $z = z_D$ and the achromatic image plane $z = z_A$, respectively, as shown schematically in Fig. 3. The spatial dispersion $D$, which vanishes at the achromatic image plane, has a value of $D = 35$ $\mu$m/eV at the dispersion plane for a pass energy of 15 keV. The intermediate sample image (LEED pattern) is formed at $z = z_A$, while the spectrum appears in the plane $z = z_D$. Magnetic energy filters possess 18 linearly independent second-rank aberration coefficients [17]. The aberrations are proportional to $\alpha^k \beta^l \gamma^m \delta^n \kappa^p$ with k+l+m+n+p=2 where $\alpha, \beta, \gamma, \delta$ are the geometrical ray parameters as defined in Fig. 3 and $\kappa = \Delta E/E$ takes into account the energy deviation $\Delta E$ from the pass energy E. With the exception of the axial chromatic aberrations proportional to $\alpha\kappa$ or $\beta\kappa$ in the plane $z = z_A$ and to $\gamma\kappa$ or $\delta\kappa$ in the plane $z = z_D$ as well as the second order dispersions (proportional to $\kappa^2$) in both planes, all other second-rank aberration coefficients vanish due to the mid-plane symmetry or due to appropriate excitation of the sextupole elements [18]. A sextupole field in the mid-plane is generated by a dodecapole which in addition can produce quadrupole fields used to correct for astigmatism due to misalignments. The feasibility of the aberration correction has been demonstrated experimentally by Krahl [19]. The axial chromatic aberrations and the second order dispersion in the achromatic image plane do not affect the resolution limit if the magnification of the intermediate image in front of the omega filter is larger than 310 (even if an energy bandwidth of 5 eV has been chosen).

Of particular importance is the correction of second-order aperture aberrations in the dispersion plane, because it guarantees isochromatic filtering in a large field of view. This means in turn that the pass energy does not depend on the position of the image point which is absolutely necessary for unambiguous elemental characterization. The combined effect of the dispersion and the axial chromatic aberration (proportional to $\gamma\kappa$) in the energy selection plane is a tilt of this plane which can result in a blurring of spectra at the detector. This effect can be avoided by readjusting three of the seven sextupole strengths leading to a simultaneous correction of the relevant axial chromatic aberration and the second-order aperture aberration in the energy selection plane. The achievable energy resolution is then determined by the size of the source image and its blurring due to the third-order aperture aberrations in the plane $z = z_D$. An energy resolution of less than 0.1 eV is expected to be obtained for all initial energies $E_0$.

The projector system consists of 3 electrostatic lenses which project either the sample image (diffraction pattern) at the achromatic image plane or the spectrum at the selection plane onto the image detector. In the first case, the magnification can be varied between 16 and 150. The total magnification of the microscope can then be varied between 500 and 100 000. In the second case, the spectroscopy mode, the magnification can be set between 15 and 60. This is sufficient on the one hand to simultaneously record the transferable spectrum of 35 eV and, on the other hand, to ensure that the energy resolution is not affected by the spatial resolution of the detector. Focussing perpendicular to the dispersive direction can be lifted by means of a quadrupole. A line spectrum is then produced which increases the dynamic range detectable with a CCD detector.

A TEM-1000n slow scan CCD camera and image acquisition system from Tietz Video and Image Processing Systems GmbH (Gauting, Germany) will be used for the recording of high-resolution images and spectra. A second detector system consists of a channel-plate image intensifier combined with a TV-rate CCD camera that can be inserted for alignment purposes and for the observation of relatively fast processes at the sample surface.

## 4    Modes of Operation

The various modes of operation include microscopy, spectroscopy and small spot low energy electron diffraction. In all these cases the pass energy of the omega filter has to be kept constant because a change in the excitation of its sector magnets requires a readjustment of all multipole correctors and realignment. Setting up the instrument therefore begins with the appropriate excitation of the omega filter which determines the pass energy $E$ (see Fig. 4). The applied acceleration voltage $U$ between sample (cathode) and anode then determines the initial kinetic energy $E_0 = E - eU$ of the electrons at the sample surface which can pass through the omega filter and form the image. $E_0$ can be varied between 0 and 2000 eV and, in general, will be kept constant during an experiment because a change of $E_0$ requires refocussing of the objective lens and a readjustment of the mirror corrector. Some exceptions will be discussed below.

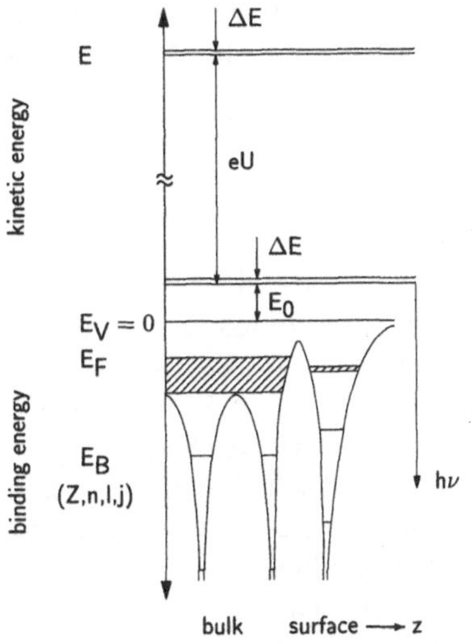

**Fig. 4.** Schematic energy level diagram for the spectromicroscope ($E_P$ pass energy, $\Delta E$ energy window, $E_0$ initial energy, $U$ acceleration voltage, $E_B$ binding energy, h$\nu$ photon energy, $E_F$ Fermi level, $E_V$ vacuum level).

## 4.1   High-Resolution Microscopy with Electrons of Selected Energy

LEEM, MEM, SEEM, AEEM as well as UV- and X-ray-induced photoemission electron microscopy (PEEM) can be performed with the instrument. (X)PEEM is of main interest because core level photoelectrons can be excited, thus providing a chemical contrast mechanism. There are three different ways of imaging the distribution of a particular element, i.e. of performing elemental mapping. Firstly, core electrons of the element of interest with a binding energy $E_B$ are used. The photon energy h$\nu$ has to be set within the interval $E_0 + E_B <$ h$\nu <$ $E_0 + E_B + \Delta E$. Optimally, $E_0$ should be within the energy range 30 – 100 eV for several reasons: (i) $E_0$ then lies above the large peak of secondary electrons which considerably improves the signal-to-background ratio; (ii) at photon energies near the ionization threshold the cross section for exciting the core electron is close to its maximum; (iii) the experiment is very surface-sensitive because the inelastic mean free path for electrons with kinetic energies in this range is rather small; (iv) the transmission of the objective lens, which decreases with increasing $E_0$, is still reasonably high. Secondly, Auger electrons characteristic of the element of interest are imaged by adjusting $E_0$ to the corresponding kinetic energy $E_A$ of an Auger electron. A considerably higher photon bandwidth can be used in this mode, thus increasing the photon flux density at the sample surface. This method is also very surface-sensitive. Thirdly, secondary electrons can also be used for elemental mapping by taking advantage of the fact that the intensity of secondaries rises steeply when the photon energy exceeds the binding energy of a particular core level. Good contrast is obtained if two images at slightly different photon energies, one above and the other below the binding energy, are taken and the difference is plotted. In order to obtain high intensities, an $E_0$ value at the intensity maximum of secondaries between 2 and 3 eV is selected. In contrast to the photoelectron and Auger modes, bulk information is also obtained because photoionised atoms in deeper layers contribute substantially to the emission of secondary electrons.

## 4.2   Spectroscopy in the Microscopy Mode

In this mode of operation a magnified image of the sample is always produced at the final image detector. The pass energy $E$ of the omega filter and the initial kinetic energy $E_0$ that is accepted by the optics are fixed, while the photon energy $h\nu$ is tuned in steps through the energy range of interest and an image recorded after each step. Spectra can be recorded simultaneously for all pixels the size of which can be as small as $2 \times 2$ nm$^2$, if sufficient intensity is available. In many cases, however, recording of spectra from some specific areas of interest will be sufficient. Number, position and size of these areas can be selected by the detector system. Various spectroscopies can be performed this way depending on the selected value of $E_0$.

(i) If $E_0$ has been chosen far beyond the secondary peak, photoelectron spectroscopy of valence or core levels can be performed in the so-called constant final state (CFS) mode where the final state energy of the detected electrons is fixed. Energy resolution results from a convolution of the transfer functions of both the X-ray monochromator and the omega filter. The latter, however, effectively determines the resolution limit of about 0.1 eV. A spatially resolved chemical analysis with high surface sensitivity can be carried out by choosing optimal settings for $E_0$ and tuning the photon energy from $E_0$ to 2000 eV, which causes a brightening in the image at the location of elements when their core electrons fulfill the energy condition for passing through the omega filter. The binding energies of all constituent elements with $E_B < 2000$ eV $- E_0$ can in principle be identified this way.

(ii) X-ray absorption spectroscopy at a surface (XANES, NEXAFS, SEXAFS) is performed by recording the signal of a secondary effect (e.g. secondary electron or Auger emission) induced by the primary core electron excitation as the photon energy is tuned through the energy range of interest. In the case of secondary electrons, $E_0$ will be set at the intensity maximum of the secondaries. By tuning the photon energy through the available energy range a chemical analysis giving bulk and surface information can also be performed. Micro-XANES measurements with secondary electrons were first performed by Tonner and coworkers [20]. A significantly improved signal-to-noise ratio and an increase in the surface sensitivity are achieved when $E_0$ corresponds to an Auger transition in the element of interest. It should be noted that the energy resolution in the X-ray absorption spectroscopy mode is solely determined by the energy resolution of the X-ray monochromator. The energy window of the omega filter is thus a parameter that can be varied in order to optimize the imaging conditions which, in general, will involve a trade-off between contrast, intensity and resolution.

(iii) Photoelectron spectroscopy in the classical sense (ESCA, or XPS, and AES) using photons with fixed energy is also possible in the microscopy mode if $E_0$ is tuned through the energy range of interest and automatic refocussing of the image is performed. The aberration corrector, however, complicates matters because refocussing necessitates a readjustment of the corrector. This kind of spectroscopy will be practicable only if refocussing and readjustment of the aberration corrector are completely computer-controlled.

(iv) Photoelectron spectroscopy in the CIS (constant initial state) mode is also possible in a computer-controlled instrument by recording images while both $E_0$ and $h\nu$ are tuned stepwise in such a way that $h\nu - E_0 = E_B$ remains constant.

### 4.3    Spectroscopy of Electrons from Selected Small Sample Areas (Microspectroscopy)

In this mode the spectrum is projected onto the final image detector which allows parallel recording of spectra within an energy window of 35 eV at an energy resolution of 0.1 eV. Sample areas of interest as small as 5nm in diameter can be selected by the field aperture at the entrance of the omega filter.

### 4.4    Low Energy Electron Diffraction (LEED)

In order to perform small spot LEED experiments the transfer optics projects the source image rather than the sample image onto the second entrance plane in front of the omega filter. An energy-filtered diffraction pattern which is free of secondary and inelastically scattered electrons is thus projected onto the final image detector. Selectable apertures in the electron gun system allow the size of the illuminated sample area to be reduced to 1 $\mu$m.

### Acknowledgement
The project is supported by the Federal German Ministry of Education, Science, Research and Technology (BMBF) under contract no. 05644WWA9.

# References

1. G. Lilienkamp, Th. Schmidt, C. Koziol, and E. Bauer, BESSY Annual Report 1994, p. 469.
2. E. Bauer, T. Franz, C. Koziol, G. Lilienkamp, and T. Schmidt, in *Chemical, Structural and Electronic Analysis of Heterogenous Surfaces on the Nanometer Scale,* edited by R. Rosei (Kluwer Academic Publishing; Dordrecht), in press.
3. J.D. Denlinger, E. Rotenberg, T. Warwick, G. Visser, J. Nordgren, J.-H. Guo, P. Skytt, S.D. Kevan, K.S. McCutcheon, D. Shuh, J. Bucher, N. Edelstein, J.G. Tobin, and B.P. Tonner, Rev. Sci. Instrum. **66**, 1342 (1995).
4. J. Welnak, Z. Dong, H. Solak, J. Wallace, F. Cerrina, M. Bertolo, A. Bianco, S. Di Fronzo, S. Fontana, W. Jark, F. Mazzolini, R. Rosei, A. Savoia, J.H. Underwood, and G. Margaritondo, Rev. Sci. Instrum. **66**, 2273 (1995).
5. E. Umbach, Physica **B 208&209**, 193 (1995).
6. B.P. Tonner, D. Dunham, T. Droubay, J. Kikuma, J. Denlinger, E. Rotenberg, and A. Warwick, J. Electron Spectrosc. Rel. Phen. **75**, 309 (1995).
7. R. Fink, M.R. Weiss, E. Umbach, D. Preikszas, H. Rose, R. Spehr, P. Hartel, W. Engel, R. Degenhardt, R. Wichtendahl, H. Kuhlenbeck, W. Erlebach, K. Ihmann, R. Schlögl, H.-J. Freund, A.M. Bradshaw, G. Lilienkamp, Th. Schmidt, E. Bauer, and. G. Benner, J. Electron Spectrosc. Rel. Phen. **84**, 231 (1996).
8. H. Rose and D. Preikszas, Optik **92**, 31 (1992).
9. E. Bauer, Ultramicroscopy **36**, 52 (1991).
10. D. Preikszas, Ph. D. Dissertation D 17, TH Darmstadt (1995).
11. S. Lanio, H. Rose, and D. Krahl, Optik **73**, 56 (1986).
12. G.F. Rempfer, D.M. Desloge, W.P. Skoczylas, and O.H. Griffith, Microscopy and Microanalysis **1** (1997), in press.
13. W.P. Skoczylas, G.F. Rempfer, and O.H. Griffith, Ultramicroscopy **36**, 252 (1991).
14. L.H. Veneklasen, Rev. Sci. Instrum. **63**, 5513 (1992).
15. E. Bauer, Rep. Prog. Phys. **57**, 895 (1994).
16. H. Rose and D. Preikszas, Nucl. Instrum. Methods **A363**, 301 (1995).
17. R. Degenhardt and W. Engel, to be published.
18. H. Rose, Optik **51**, 15 (1978).
19. S. Lanio, Ph. D. Dissertation D 17, TH Darmstadt (1986).
20. H. Rose and D. Krahl, in *Energy-filtering Transmission Electron Microscopy,* edited by L. Reimer (Springer; Berlin, 1995), p. 43.
21. B.P. Tonner and G.R. Harp, J. Vac. Sci. Technol. **A7**, 1 (1989).

This article was processed using the LaTeX macro package with LLNCS style

# Microchemical Analysis of Boron in Rat Brain Tumor: A Spectromicroscopy Study with MEPHISTO

Gelsomina De Stasio

Institut de Physique Appliquée, Ecole Polytechnique Fédérale, PH-Ecublens,
CH-1015 Lausanne, Switzerland
and
Istituto di Struttura della Materia del CNR, Via Enrico Fermi 38,
I-00044 Frascati, Roma, Italy

**Abstract.** The boron microscopic distribution in brain tissue sections is an extremely important issue for the success of boron neutron capture therapy for cancer. We present the first results of a new approach to assess this distribution and its homogeneity: photoelectron spectromicroscopy with synchrotron radiation enabled us to detect boron in rat brain tissue specimens with a lateral resolution of 0.2 µm and a detection sensitivity of a few ppm. These experiments were performed by the newly commissioned MEPHISTO spectromicroscope (from the French acronym "Microscope à Emission de Photoélectrons par Illumination Synchrotronique de Type Onduleur", or "Photoelectron Emission Microscope by Synchrotron Undulator Illumination"). The specimens were brain tissue sections from two rats injected with BPA (boronophenylalanine) or BSH (dodecahydro-dodecaborate). We found boron only in cancer-related structures, and with a higher concentration in the BPA-treated specimens than in the BSH-case. These results are in good agreement with the quantitative results obtained by inductively coupled plasma atomic emission spectroscopy.

## 1 Introduction

One of the fundamental issues for the success of boron neutron capture therapies (BNCT) [1] is the availability of analytical techniques to study the microscopic boron distribution in tissues [2]. In fact, techniques like the widely used Inductively Coupled Plasma Atomic Emission Spectroscopy (ICP-AES) offer no spatial resolution [3, 4]. We present a feasibility test to solve this fundamental problem, based on synchrotron radiation photoelectron spectromicroscopy [5, 6]. This approach is an alternative to electron energy loss spectroscopy combined with transmission electron microscopy (EELS-TEM) [2], immunohistochemical methods [7], boron neutron capture autoradiography [8], or SIMS [9] also used for the same purpose.

Spectromicroscopy does not require ultra-thin tissue sections (<60 nm for EELS-TEM) and can analyze tissue sections 5–7 µm thick. It can detect trace concentrations of boron of the order of a few ppm. EELS-TEM can detect the presence of boron only if its local concentration is higher than 100 ppm [10].

On the other hand, EELS-TEM has a spatial resolution of the order of a few Å. At present, we are limited to 200 nm; at the end of the commissioning period of our spectromicroscope MEPHISTO we should be able to reach the 50 nm level. MEPHISTO's

X-Ray Microscopy and Spectromicroscopy
Eds.: J. Thieme, G. Schmahl, D. Rudolph, E. Umbach
© Springer-Verlag Berlin Heidelberg 1998

resolution is, and will always be, much worse than that of EELS-TEM, but still better than other techniques based on optical microscopy such as neutron capture autoradiography or immunohistological methods, or with respect to SIMS [9].

Compared to immunohistochemical or neutron capture radiography techniques, spectromicroscopy with MEPHISTO can detect *any* element present in the specimen, with no need for staining or labeling. The elements are not required to have a high neutron absorption cross section to be detectable, and the analysis is not limited by the availability of the proper enzyme to stain specific biological targets, with a limited number of elements for staining and limited histological parts detectable.

Another convenient characteristic of spectromicroscopy is its dynamic range: the field of view of the micrographs obtained with MEPHISTO may be quickly varied between 5 μm and 500 μm, enabling the microchemical analysis of a single cells or the whole region of tissue surrounding it.

One of the main limitations of spectromicroscopy when compared to the other techniques is that it requires a synchrotron source of photons.

In the present work we used MEPHISTO to analyze the boron distribution in specimens of brain tissues of two rats with experimental cerebral tumors, previously injected with $^{10}$B enriched BPA (borono-phenylalanine) or BSH (dodecahydro-dodecaborate) two hours prior to sacrifice. We observed strong spectroscopic signal from boron in the experimental glioma of the BPA and BSH injected rats, and we determined the boron spatial distribution with a lateral resolution of 0.2 μm.

One critical point in BNCT is the boron distribution, which is required to be concentrated selectively in cancer cells. The results presented here, obtained for the BPA- and BSH-treated cases, revealed indeed that boron is localized in the specimen structures that are related to cancer. By contrast, no boron signal was observed outside these structures.

## 2 Sample Preparation

The specimens were brain tissue sections from rats that had been injected with $4 \times 10^4$ cancer cells (C6 glioma cell line # CCL107) twelve days before the experiment. These cells were injected in the striatum at a depth of 5.5 mm. The rats were intraperitoneally injected with 150 mg of BPA in 2 ml of isotonic solution or 150 mg of BSH suspended in 3 ml of isotonic solution. Two hours after the injection, the rats were sacrificed by decapitation and the brains extracted. The brain tissues were cryofixed by quick immersion in isopentane kept at liquid nitrogen temperature; this quick freezing prevents the formation of large ice crystals and the consequent microscopic damage to the tissue structures. Then, the tissue was cut in 5 μm thick sections with a cryostat. The morphology of individual sections was examined after immunohisto-chemical staining and tumoral tissue identified.

Normal brain or tumoral cryocut specimens were put on silicon wafer substrates, and then ashed with a cold plasma (150 °C, Plasma-Processor 300E, Techn. Plasma GmbH, München) in the presence of oxygen for ≥24 hours. The ashing removes most of the tissues' carbon; the absence of boron from substrate areas indicates that the ashing process does not promote boron diffusion or spread through the specimen [11].

# 3 Experimental Techniques

## 3.1 Spectromicroscopy

The specific experimental technique used for microchemical analysis was the photo-electron spectromicroscopy approach known as x-ray secondary electron (emission) microscopy [5], using our newly commissioned MEPHISTO system. This technique uses monochromatized x-ray photons emitted by a synchrotron source, in our case the 6-m Toroidal Grating Monochromator (TGM) beamline of the Aladdin ring at the Wisconsin Synchrotron Radiation Center. The photons stimulated the emission of primary and mostly secondary photoelectrons, through intermediate energy-loss steps.

Intensity vs photon energy spectra taken in this way (total yield) correspond to the x-ray optical absorption coefficient. [12] The secondary electrons emitted by the sample interacted with an electron-optical magnifying system [5] yielding images of the specimen's geometric features and spectra of microscopic areas. The electron optics are basically composed by a cathode lens (the sample is kept at -20 kV, the objective lens' intermediate element is kept at <20 kV), an aperture (70, 40 or 25 μm pinhole) in the back focal plane of the objective lens, and intermediate lens, a projective lens, a series of two microchannel plates to intensify the electron magnified image produced by the optical system and a phosphor screen to convert this electron image into a visible image, which is finally collected by a video acquisition system.

Extensive tests were systematically performed to assess the possible presence of problems such as radiation damage or sample charging, with negative results. Specifically, samples exposed for hours to the monochromatized x-ray beam (whose size was of the order of 0.3 x 0.3 mm$^2$) did not exhibit any detectable evidence of damage – such as spectral or morphological changes. Substantial damage was observed instead after several minutes of exposure to a much more intense unmonochromatized beam (which was never used in our present experiments).

## 3.2 ICP-AES

ICP-AES, allows to perform quantitative chemical analysis on liquid samples. An aerosol of the sample solution and argon is heated up to a cold-plasma temperature (8000 °C) by an electromagnetic field. At this temperature, each element emits a specific near-ultraviolet light wavelength, that can be accurately identified.

Measuring the emission intensity at each wavelength, and comparing it to a standard reference, make it possible to determine the concentration of each element (boron, in the present case) in the solution. The sensitivity reaches 1 ppb, evidently with no spatial resolution: this, in fact, is a space-averaged technique. The conceptual background of the ICP-AES technique can be found, for example, in Refs. 13 and 14.

Samples for ICP-AES analysis were not ashed. For these experiments tissue samples were taken from the tumor volumes or the contra-lateral hemispheres, for comparison, from both rats treated with BSH and BPA. After quick freezing in isopentane, the tissue portions were weighted, then added to 2 ml HNO$_3$ and sonicated until the solution appeared homogeneous.

Thirteen replicates were taken for each ICP-AES measurement, and the blank HNO₃ was repeatedly tested to rule out the presence of boron after each experiment on different parts of the rat brain.

## 4 Results and Discussion

Figures 1-7 show typical sets of experimental data, extracted from a much larger body of results. Specifically, Fig. 1 presents a photoelectron microimage taken at low magnification, and Fig. 2 presents a set of spectra from the correspondingly labeled microscopic areas of Fig. 1.

The micrograph corresponds to an ashed section of cryocut brain cancer tissue. We note that most of the image exhibits a reticulate pattern of fine corrugations with limited vertical protrusion. There are, however, more protruding, elongated structures (appearing as clear areas in the image, corresponding to higher photoelectron yield). The analysis of the specimens indicates that these elongated structures are typical of the cancer tissue, since none was observed in non-cancer areas.

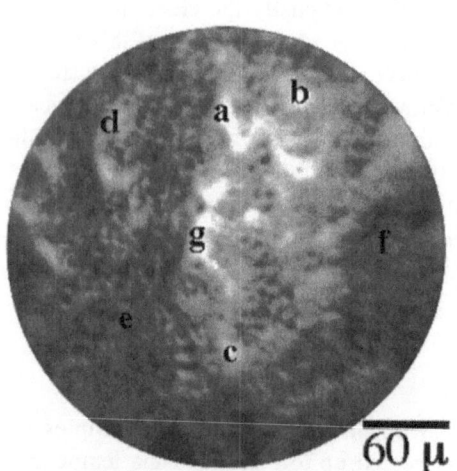

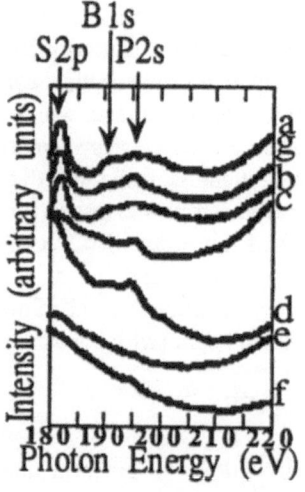

**Fig. 1 .** An X-ray secondary pho-toelectron emission micrograph taken with the MEPHISTO system on a section of extra-cranially grown rat brain cancer tissue, after ashing. The rat had been treated intraperitoneally with BPA.

**Fig. 2.** X-ray secondary photoelectron emission spectra taken with MEPHISTO in the labeled areas of Fig. 1. Note that the B signal is confined to the elon-gated bright areas.

The spectra of Fig. 2 indicate the presence of three absorption structures, due to the elements boron, sulfur and phosphorous, of which the last two are naturally present in the brain tissue. It is evident from Fig. 2 that extreme care must be used in distinguishing from each other the boron and phosphorous signals, which fall in the same spectral region.

The key point in discriminating between P and B was the analysis of control specimens extracted from normal tissue areas. Figure 3 shows one of the spectra obtained on this normal tissue: it is quite clear that the P and S signals are still present, whereas no B signal can be observed. This conclusion is supported by an analysis of the first and second derivatives of the spectra (data not shown). Quite to the contrast, the absence of the B signal corroborates the identification of the corresponding spectral structure in Fig. 2 as due, indeed, to this element.

Having solved this crucial issue, we can now analyze some interesting points in Figs. 1 and 2. The relative intensity of the B-related spectral feature at 190 eV is not constant from place to place in the examined areas. On the contrary, it disappears in some points and is rather weak in others, indicating a strongly inhomogeneous distribution of boron.

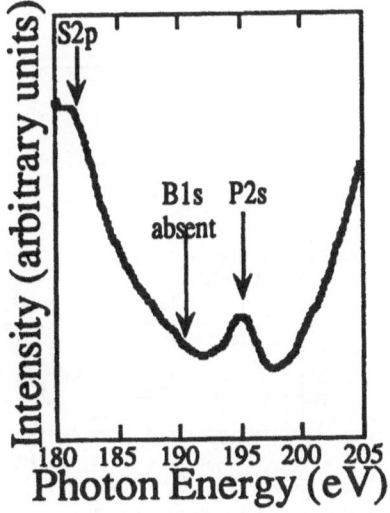

**Fig. 3.** Spectrum from a non-cancer control specimen: note the absence of the B signal.

More specifically, we found very strong B signal in areas corresponding to the elongated and protruding structures in the cancer specimens – see, for example, curves (a), (b), (d) and (g) in Fig. 2. Outside the elongated structures, on the reticulate tissue, no B signal was observed (curves (e) and (f)). These specific results are representative of a much larger body of consistent data, including about 100 spectra and 20 images acquired on 3 different sections of the same BPA treated tumor.

The strong localization of boron in the elongated structures is indicated by the results of Figs. 4 and 5, which illustrate a "zoom" test. Figure 4 shows micrographs taken in the same specimen area with increasing magnification; the area is centered around point (g) in Fig. 1. Spectra were taken in microscopic regions of the higher-magnification images of Fig. 4 with the following two relevant results.

First of all, a fine analysis of the reticulate pattern regions close to, but outside, the elongated protruding structures always failed to produce any evidence of boron. Second, spectra taken in different areas within the elongated structures give different relative intensities of the B signal.

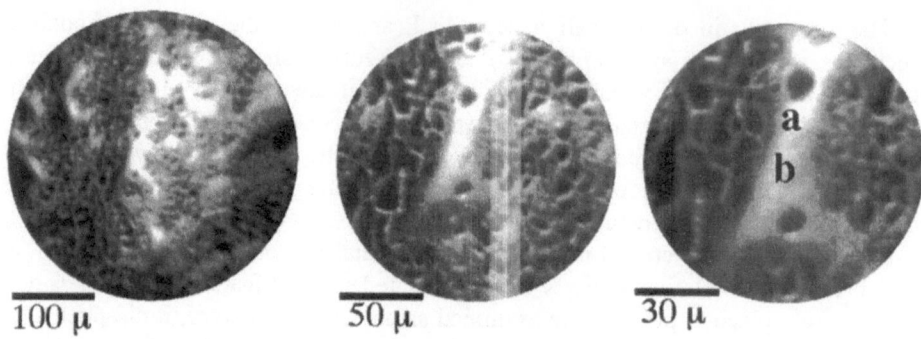

100 μ          50 μ          30 μ

**Fig. 4.** Three different images centered at point (g) in Fig. 1, taken with increasing magnification.

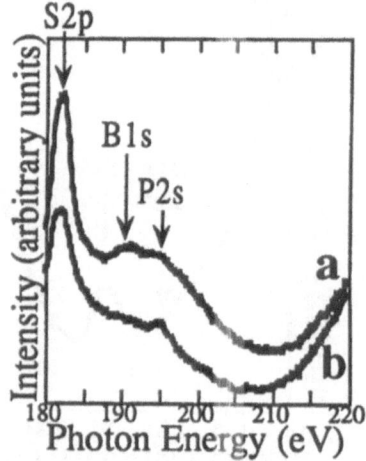

**Fig. 5.** Spectra taken from the areas labeled as (a) and (b) in Fig. 4 and (g) in Fig. 1, but much smaller than for the spectrum (g) in Fig. 2. Note that the relative boron intensity is higher in curve (a) and lower in curve (b) than in the spectrum (g) of Fig. 2, indicating that this element is not homogeneously distributed within the elongated structures.

An example is shown in Fig. 5; the spectra (a) and (b) were taken at the same point as region (g) in Fig. 2a, but from two much smaller regions (approximately 7 x 7 $\mu m^2$, whereas the acquisition area (g) was approximately 12 x 20 $\mu m^2$): the relative B intensity in curve (a) is higher than in the (g) spectrum of Fig. 2, and lower in the curve (b).

These results demonstrate that (1) boron is strongly localized and confined to the cancer-related elongated protruding structures; (2) the distribution of boron within such structures is not homogeneous. This confirms the need of a high-spatial-resolution technique for microchemical analysis in this type of problems.

At the present time it is not clear to what cellular/tissue structures the reticulate pattern and the elongated protruding structures of Figs. 1 and 4 correspond. It is tempting to speculate that the bright reticulate edges might be delimiting individual cells (5–20 mm size) and could represent the extracellular space.

The protruding structures, on the other hand, may originate from tumor specific blood vessels. Neovascularization (or angiogenesis), in fact, is an important component of tumor growth. But these structures could as well possibly originate from higher density areas characteristic of the tumor tissue, which "resist" ashing more than the surrounding areas, and are therefore thicker after ashing. Further investigation and histochemical analysis will be necessary to univocally identify such structures.

We note that however these protruding structures are interpreted, our technique could be detecting relative tumor tissue vs normal tissue B uptake, or vascular structure vs tumor cell uptake. In this latter case the spectromicroscopical information retrieved would also be valuable. If we are detecting boron in blood vessels, it means that BNCT would destroy them, possibly leading to life threatening hemorrhages in the brain.

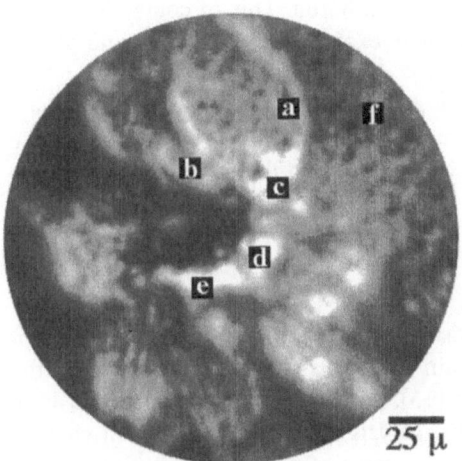

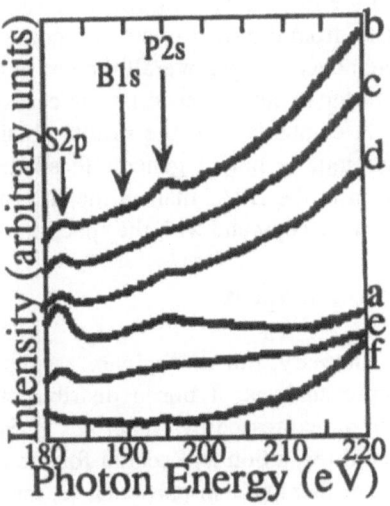

**Fig. 6.** MEPHISTO micrograph of a section of intracranial rat brain cancer tissue, after ashing. The rat had been treated intraperitoneally with BSH.

**Fig. 7.** Spectra taken in the labeled areas of Fig. 6. Note that the B signal is also in this case confined to the elongated bright areas, but it's corresponding signal is much lower than the one found treating the rat with BPA (spectra of Fig. 2).

Figures 6 and 7 show spectromicroscopy results similar to those of Figs. 1 and 2, obtained on tumor tissue from a rat treated with BSH. Note that boron is barely visible in the spectra of Fig. 7. First and second derivative analysis of these spectra confirms that the spectral feature at 190 eV can indeed be overlapped to the corresponding signal of Fig. 2, only much less intense. Also in this case, boron, when detectable, is confined to the protruding cancer-related structures, and absent from the reticulate tissue. In this BSH-case too, the control sample obtained from the contra-lateral (with respect to cancer) hemisphere did not show any boron signal.

As already mentioned, we also performed ICP-AES analysis of boron, with no spatial resolution, of tissue samples extracted from the same two rats used for spectromicroscopy experiments.

**Table 1.** ICP-AES results on boron concentration in the various rat brain tissue parts. The results for B concentration are normalized to the tissue volume (10–50 mg). The errors shown are the standard deviation calculated on the 13 different measurements performed on each sample.

| Sample from | B - compound injected | Boron in tissue(ppm) |
|---|---|---|
| HNO$_3$ | - | $0.0 \pm 0.3$ |
| normal brain tissue | BSH | $1.3 \pm 0.5$ |
| intracranial tumor | BSH | $1.9 \pm 0.8$ |
| normal brain tissue | BPA | $0.7 \pm 0.3$ |
| extra-cranial tumor | BPA | $5.7 \pm 0.3$ |

Table 1 shows the results of the quantitative ICP-AES analysis. Note, in particular, that the BPA-treated rat tumor tissue contains about 6 ppm boron, whereas the cancer tissue from the BSH treated rat contains about 2 ppm boron. This of course is an average concentration overall the entire tissue volume examined (between 10 and 50 mg of tissue in the various different cases).

A comparison of the results obtained by a space-resolved and a space-integrated technique is not in general feasible. Nevertheless, we note that we detected more boron in the BPA- than in the BSH-treated case, both with the space-resolved spectromicroscopy and with the space-integrated ICP-AES technique.

## 5 Conclusions

In summary, our preliminary results demonstrate the feasibility of a spectromicroscopic analysis of boron distribution in brain tumor tissues. We concur with the opinion expressed by Gabel [2, 7], Setiawan [15], and many other authors, that high spatial resolution is essential for the complete analysis of boron in tissues, in view of cancer therapy – in particular as far as the assessment of spatial distribution is concerned for a complete destruction of cancer cells. Our approach provides a satisfactory and practical method to answer to this need.

## Acknowledgments

Work supported by the Fonds National Suisse de la Recherche Scientifique, by the Istituto di Struttura della Materia del CNR and by the EPFL. We are indebted to Giorgio Margaritondo and Paolo Perfetti for their constant support, to Brian P. Tonner for his help and encouragement during the spectromicroscopy experiments, and for the construction of the electron optics for MEPHISTO. We thank Erwin Van Meir, Marie-France Hamou, C. Jayet, B. Ess, Gian Francesco Lorusso, Jose Redondo, Delio Mercanti and Maria Teresa Ciotti for rat treatments and sample preparation, Didier Perret and Benjamin Gilbert for performing the ICP-AES experiments. We also thank Mario Capozi, Sandro Rinaldi, and Mary Severson, Tom Nelson and the entire staff of the Wisconsin Synchrotron Radiation Center (a national facility supported by the NSF) for their expert professional help. We are grateful to Borje Larsson for his encouragement and advise.

# References

1    "Advances in Neutron Capture Therapy", A. H. Soloway, R. F. Barth and D. E. Carpenter Eds., Plenum Press, New York, pp. 829 (1993).

2    See for example B. Otersen et al., Seventh International Symposium on Neutron Capture Therapy for Cancer, Zurich 1996, Proceedings to be published by Elsevier Science, Excerpta Medica, International Congress Series 1132, submitted.

3    W. F. Bauer, P. L. Micca and B. M. White, in "Advances in Neutron Capture Therapy", A. H. Soloway, R. F. Barth and D. E. Carpenter Eds., Plenum Press, New York (1993), p. 403.

4    N. Hotz and W. Bauer, ibid, p. 439.

5    Tonner, B. P., and Harp, G. R., Rev. Sci. Instrum. 59, 853-858 (1988); Tonner, B. P., and Harp, G. R., J. Vac. Sci. Technol. 7, 1-4 (1989); Tonner, B. P., Harp, G. R., Koranda, S. F., Zhang, J., Rev. Sci. Instrum . 63, 564-568 (1992).

6    De Stasio, Gelsomina, Hardcastle, S., Koranda, S. F., Tonner, B. P., Mercanti, D., Ciotti, M. Teresa, Perfetti, P., and Margaritondo, G., Phys. Rev. E47, 2117-2121 (1993); Gelsomina De Stasio, Journal de Physique IV 4, C9-287-292 (1994); G. Margaritondo, G. De Stasio, C. Coluzza, J. Electron Spectrosc. 72, 281-287 (1995); Gelsomina De Stasio et al., Synchrotron Radiation News 7, 18-21 (1994).

7    see D. Gabel et al., Seventh International Symposium on Neutron Capture Therapy for Cancer, Zurich 1996, Proceedings to be published by Elsevier Science, Excerpta Medica, International Congress Series 1132, submitted.

8    See, for example, Thellier, M., Ripoll, C., Quintana, C., Sommer, F., Chevallier, P., and Dainty, J., Methods in Enzymology 227, 535-586 (1993).

9    V. K. F. Chia, R. J. Blieler, D. B. Sams, et al., in "Advances in Neutron Capture Therapy", A. H. Soloway, R. F. Barth and D. E. Carpenter Eds., Plenum Press, New York (1993) p. 409.

10   R. D. Leapman and D. E. Newbury, Anal. Chem. 65, 2409 (1993); H. Raether, in "excitation of Plasmons and Interband Transitions by Electrons", Springer-Verlag, Berlin 1980.

11   Gelsomina De Stasio et al., Seventh International Symposium on Neutron Capture Therapy for Cancer, Zurich 1996, Proceedings to be published by Elsevier Science, Excerpta Medica, International Congress Series 1132, submitted.

12   Gudat, W. and Kunz, C., Phys. Rev. Lett. 29, 169-173 (1972).

13   A. Varma, Hanbook of Inductively Coupled Plasma Atomic Emission Spectroscopy, CRC, Boca Raton, 380 pp. (1991); R. M. Barnes, Chemia Analityczna 28, 179 (1983); S. Caroli, E. Coni, A. Alimonti, E. Beccaloni, E. Sabbioni and R. Pietra, Analusis 16, 75 (1988); L. Bourrier-Guerin, Y. Mauras, J. L. Truelle and P. Allain, Trace Element in Medicine 2, 88(1985); C. De Martino, S. Caroli, A. Alimonti, F. Petrucci, G. Citro and A. Nista, J. Exp. Clin. Cancer Res. 10, 1 (1991). A. Alimonti, S. Caroli, L. Musmeci et al., Sci. Total Environm. 71, 495-500 (1988).

14   E. Sabbioni, G. R. Nicolini, R. Pietra et al., Biol. Trace Elem. Res. 26/27, 757-768 (1990).

15   Setiawan, Y., Halliday, G. M., Harding, A. J., Moore, D. E., and Allen, B. J., Cancer Res. 55, 874-877 (1995).

Part IV

# X-Ray Optics

# Zone Plates in Nickel and Germanium
# for High-Resolution X-Ray Microscopy

T. Schliebe and G. Schneider

Georg-August-Universität Göttingen,
Forschungseinrichtung Röntgenphysik,
Geiststraße 11, D-37073 Göttingen, Germany

**Abstract.** New manufacturing processes for highly efficient high resolution phase zone plates in nickel and germanium are described. A high resolution cross-linked PMMA resist has been synthesized and optimized for the generation of zone plate patterns with smallest zone width down to 19 nm by e-beam lithography. The resist shows an increased resolution compared to conventional PMMA for periodic structures with line to space ratio of 1:1. For the pattern transfer into the zone plate material a cross-linked polymer has been developed. This polymer can be structured by reactive ion etching (RIE) with oxygen, thereby aspect ratios of up to 6:1 can be obtained using a tri-level process. The polymer structures can be used either as galvanoform for electrodeposition of nickel or as an etching mask for structuring germanium zone plates. Electroplated nickel zone plates with outermost zone width of 40 nm and 30 nm have been fabricated and achieve 15% and 11% diffraction efficiency. For germanium zone plates with outermost zone width of 40 nm, 30 nm and 19 nm diffraction efficiencies of 14%, 10% and 4% were measured at 2.4 nm wavelength.

## 1 Introduction

Highly efficiency zone plates with high spatial resolution are important optical elements in X-ray microscopy. Since the resolution of zone plates is mainly determined by the smallest zone width, processing of structures with widths down to 20–30 nm is necessary. Therefore, one task of our work is the generation of periodic structures down to 20 nm linewidth and line to space ratio of 1:1.

Germanium and especially nickel are well suited materials for phase zone plates in the waterwindow wavelength region between the K-absorption edges of oxygen at 2.34 nm and carbon at 4.38 nm wavelength. Figure 1 shows the first order diffraction efficiency at 2.4 nm wavelength for the materials silicon, germanium and nickel. This theoretical data allows to characterize the quality of the stucturing process by comparison with the experimentally derived efficiency data of zone plates. The X-ray optical properties of nickel and germanium allow the combination of high diffraction efficiency and high resolution, because the aspect ratios, which is the ratio of zone height to zone width, of the outermost zones are small compared to other materials. However, structures with linewidth of 20 nm to 30 nm and aspect ratios of up to 5–10 have to be generated.

**X-Ray Microscopy and Spectromicroscopy**
Eds.: J. Thieme, G. Schmahl, D. Rudolph, E. Umbach
© Springer-Verlag Berlin Heidelberg 1998

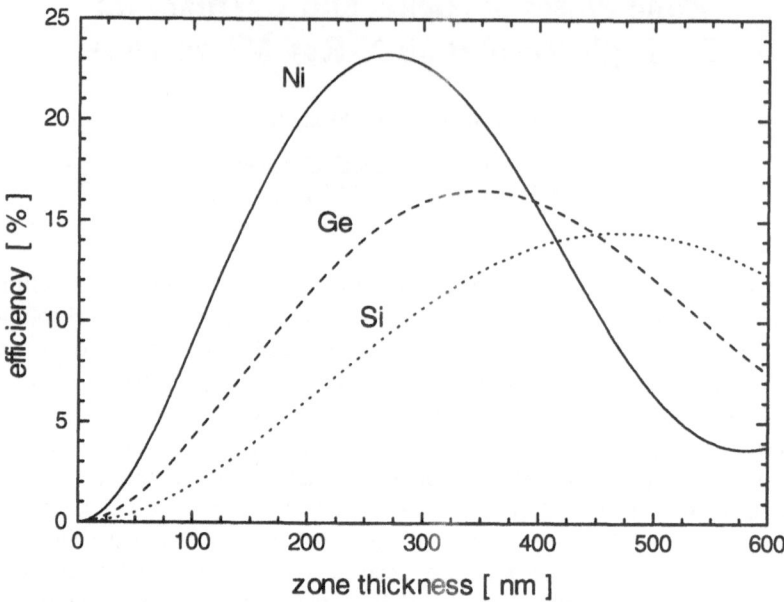

**Fig. 1.** First order diffraction efficiencies at 2.4 nm wavelength of nickel, germanium and silicon zone plates with line to space ratio of 1:1 and rectangular profile, unslanted zones and imaging magnification of 1000× as a function of the zone thickness for 20 nm structure width [1].

## 2 Cross-Linked PMMA Electron Beam Resist

Although PMMA is suitable to record single lines with widths down to 8-10 nm [2], our experiments on the generation of periodic structures with line to space ratios of 1:1 showed that the tolerable dose range becomes ever smaller for linewidths below 30 nm. Furthermore, the etch resistance of the PMMA structures during a RIE process is degraded with decreasing linewidth. This becomes clear, because theoretical considerations show that after e-beam exposure using 40 keV electrons and an exposure dose of 170 µC/cm$^2$ , the linear PMMA chains are broken into fragments of an average length of 10 -15 nm. This is in the region of the linewidth of the generated structures, so the selectivity of the developing process becomes lower with decreasing linewidth. The idea to overcome this problem is to cross-link the linear PMMA chains, so that the resist consists of a three dimensional cross-linked giant molecule of infinite molecular weight. Then only the exposed areas, where the linear chains and the crosslinking bridge are broken, become soluble in the developer. Figure 2 illustrates this effect for conventional PMMA and for cross-linked PMMA.

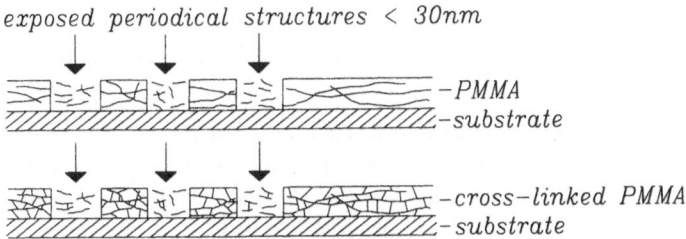

*exposed periodical structures < 30nm*

**Fig. 2.** Illustration of the effect of cross-linking PMMA on the generation of smallest periodic structures by electron beam lithography.

The cross-linked resist consists of the monomers methyl methacrylate (MMA) and triethylene glycol dimethacrylate (TEGDMA) [2], which are polymerized by radical reactions. The scheme for the chemical reactions is shown in figure 3. These reactions are initiated with benzoyl peroxide, which was dehydrated by phosphorus pentoxide in a dessicator. The ratio R, of MMA to TEGDMA determines the degree of cross-linking and strongly influences the resolution and sensitivity of the resist. For R = 100:1, high resolution and acceptable exposure times can be combined. For the synthesis of the cross-linked resist, 0.2 g of the radical starter were dissolved in 6 ml MMA and 0.17 ml TEGDMA (R = 100:1) and heated to 90°C.

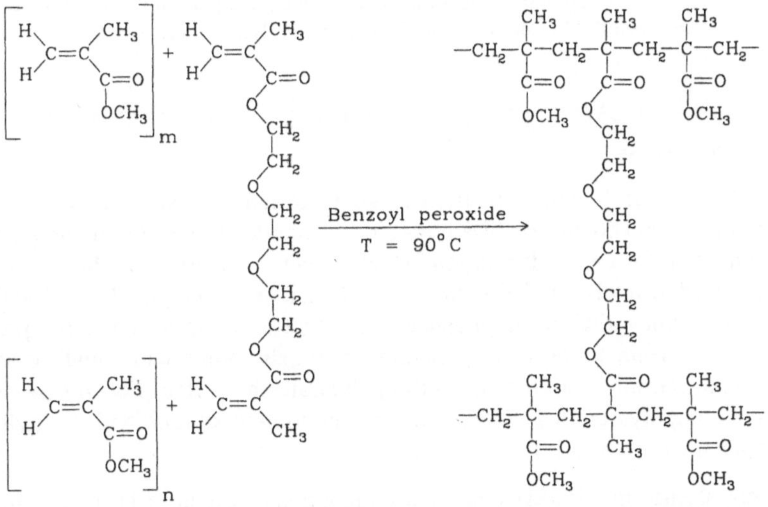

**Fig. 3.** Synthesis of the cross-linked PMMA by radical reactions of methyl methacrylate and triethylene glycol dimethacrylate initiated with benzoyl peroxide at 90 °C.

After about 5 minutes, the polymerization is stopped by cooling down to 20°C. Then the polymer is dissolved in 2-methoxyethyl acetate to obtain the desired thickness of the resist layer after spin coating. Finally the resist is polymerized again at 90°C for 24 hours.

Figure 4 shows a differential interference contrast image of a zone plate pattern with outermost zone width of 19 nm, which is recorded in cross-linked PMMA and transferred into a 90 nm thick polymer layer by RIE. The cross-pattern on the zone plate results from the interaction of the polarized light from the microscope with the periodic structures of the zone plate. The homogeneity of the cross-pattern indicates that periodic structures down to 19 nm linewidth have been generated.

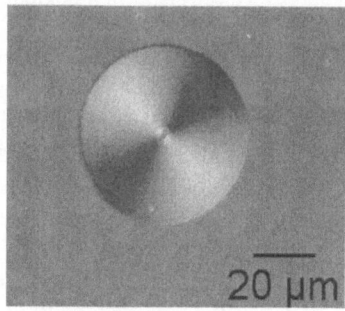

**Fig. 4.** Differential interference contrast image of a zone plate pattern with 19 nm outermost zonewidth after pattern transfer into a copolymer layer by RIE.

## 3 Cross-Linked Styrene Copolymer for Structuring Nickel and Germanium

Electroplating is the favourite method to generate nanostructures of nickel. Earlier investigations have shown that it is possible to use e-beam resist structures (PMMA) as a galvanoform for electrodeposition [4]. However, comparatively thick resist layers have to be used to realize nickel structures of considerable height. The difficulty is to combine small linewidth and high aspect ratio. Our aim was to develop a process in which electron beam lithography generated resist structures with small aspect ratio can be transferred into a high aspect ratio galvanoform by RIE. For this purpose, we developed a copolymer that consists of the monomers styrene and divinylbenzene with a molecule ratio of 3.1:1.

The copolymer has lowest electrical conductivity, which is necessary to ensure that electrodeposition starts exclusively on the plating base between the copolymer structures. The organic copolymer can be structured by RIE with oxygen. High mechanical stability is obtained by the chemical structure of the copolymer, which is highly three dimensionally cross-linked by divinylbenzene. The hydrophobic properties of the copolymer prevent water absorption and changes of the galvanoform during electrodeposition. As opposed to PMMA galvanoforms produced by RIE with $O_2$, the cross-linked copolymer shows no sidewall growth of nickel during electroplating.

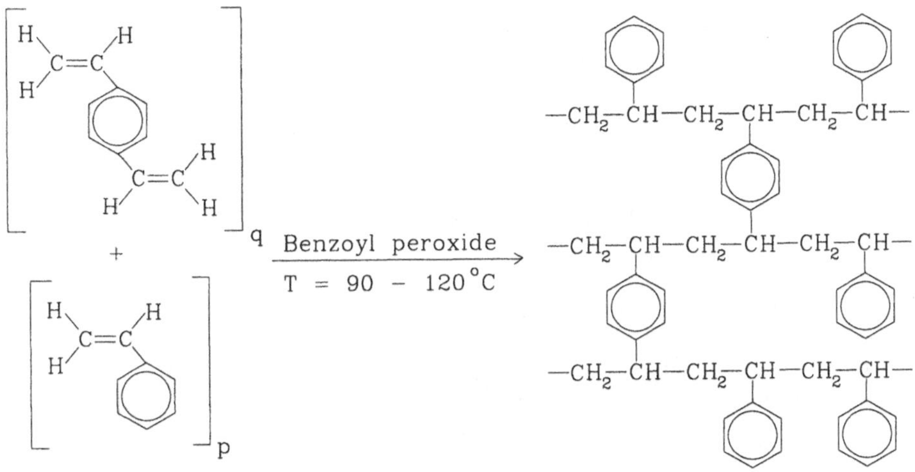

**Fig. 5.** Scheme for the radical reactions of styrene and divinylbenzene initiated with benzoyl peroxide at 90 °C to achieve three-dimensional cross-linked polymer chains.

Figure 5 illustrates the cross-linking reactions between the monomers. In addition, the giant molecular structure containing aromatic compounds make the copolymer well suited to be used as a highly selective etch mask for structuring germanium by RIE with $CBrF_3$.

For the synthesis of the copolymer, 0.08 g dehydrated benzoylperoxide is dissolved in 5 ml styrene and 2 ml divinylbenzene. The solution is heated to 95°C, and the radical starter initiates the polymerization. After about 5 minutes, the co-polymer has the suitable viscosity to build layers of the desired thickness by spin coating, and the polymerization is stopped by cooling. The viscosity of the copolymer can be increased by heating it again to 90°C if the thickness of the spin coated layers is too low. The copolymer layer on the substrate is then highly cross-linked by further radical reactions at 100°C. The temperature is increased to 120°C within 72 hours. The highly cross-linking procedure is still under development in order to enhance the reproducibility of the process.

## 4 Manufacturing Process for Nickel Zone Plates

The zone plate pattern generation was performed with an AKASHI DS 130 C SEM operating at 40 kV, which is adapted to an ELPHY III vector scan lithography device from RAITH Co [5]. A RIE system with asymmetric electrodes working at a frequency of 13.56 MHz was used for the structure transfer.

The layer sequence for the nickel zone plate fabrication is shown in Fig. 6. The layers are deposited by electron beam evaporation and by spin coating on 100-120 nm thick silicon membranes.

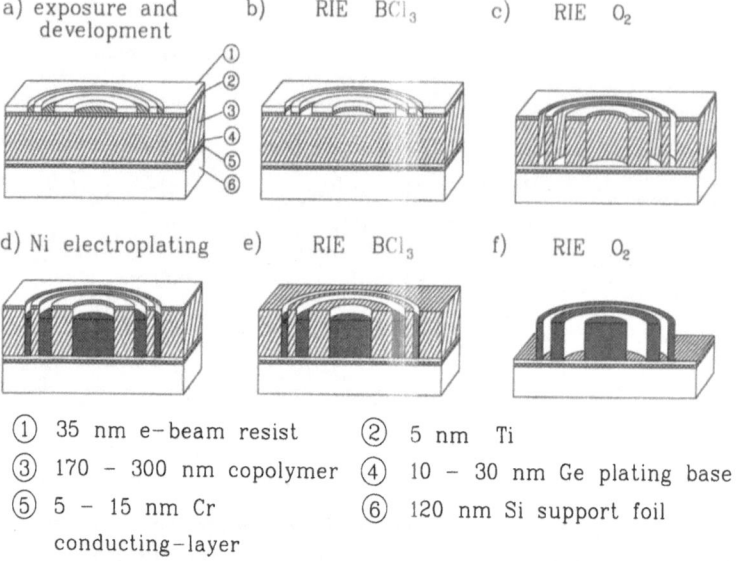

a) exposure and development    b) RIE BCl₃    c) RIE O₂

d) Ni electroplating    e) RIE BCl₃    f) RIE O₂

① 35 nm e-beam resist       ② 5 nm Ti
③ 170 - 300 nm copolymer    ④ 10 - 30 nm Ge plating base
⑤ 5 - 15 nm Cr              ⑥ 120 nm Si support foil
  conducting-layer

**Fig. 6.** Layer sequence and processing steps for the manufacture of nickel zone plates.

After exposure, 30s development in a 1:4 mixture of ethyleneglycolmonoethyl-ether and ethylene-glycol-monobuthylether, and short rinses in propanol and pentane, the zone plate pattern is transferred into a titanium layer by RIE with BCl₃ at 16 mTorr and a bias voltage of 0.4 kV. The etch time was 90 s. Following the co-polymer layer is structured by RIE with O₂ at 10 mTorr and a bias of 0.4 kV. The resulting galvanoform was filled by electrodeposition in a nickel sulphamate bath operating at a temperature of 35°C and a pH-value of 3.5. The electrodeposition starts homogeneous on a germanium layer, and a chromium layer ensures a sufficient electrical conductivity. As opposed to gold, which is often used as a plating base, germanium has a low sputter rate during RIE with O₂ This prevents sidewall growth during electroplating because of plating base material sputtered on the walls of the structures. Finally the titanium and copolymer structures are removed by RIE. Figure 7 shows innermost and outermost structures of a nickel zone plate with 150 nm heigth and indicates the uniform electrodeposition.

**Fig. 7.** SEM micrograph of central and outermost zones of a nickel zone plate.

# 5 Manufacturing Process for Germanium Zone Plates

The generation of high aspect ratio germanium structures by RIE requires a highly selective etching mask. The selectivity of the copolymer during germanium etching in a $CBrF_3$ plasma at 10 mTorr and a bias of 0.4 kV is enhanced by a factor of about 1.6 compared to an AZ 1350 etch mask widely used for structuring germanium [6, 7, 8].

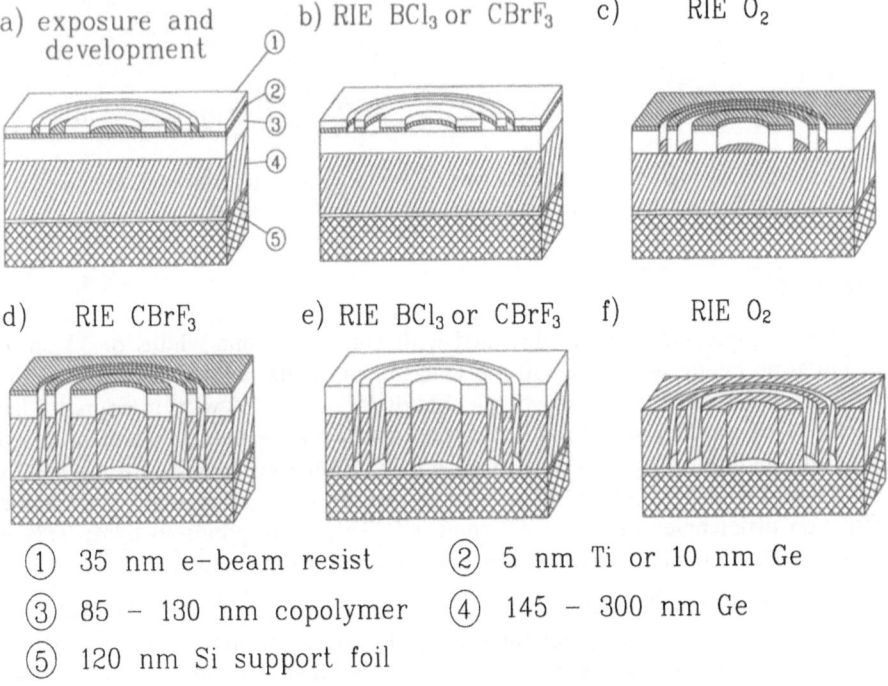

a) exposure and development

b) RIE $BCl_3$ or $CBrF_3$

c) RIE $O_2$

d) RIE $CBrF_3$

e) RIE $BCl_3$ or $CBrF_3$

f) RIE $O_2$

① 35 nm e−beam resist  ② 5 nm Ti or 10 nm Ge
③ 85 − 130 nm copolymer  ④ 145 − 300 nm Ge
⑤ 120 nm Si support foil

**Fig. 8.** Layer sequence and processing steps for the manufacture of germanium zone plates

Figure 8 shows the layer sequence and the processing steps for germanium zone plates. After exposure and development, the resist structures are transferred into an intermediate layer of titanium by RIE with $BCl_3$ at 16 mTorr and a bias of 0.4 kV or germanium by RIE with $CBrF_3$ (15 mTorr, 0.3 kV). We observed that the resist structures adhere much better to titanium, whereas the intermediate germanium layer causes no micromasking during the following process steps. The adhesion of the resist structures becomes especially important for periodic structures smaller than 30 nm linewidth, because the resist structures tend to be displaced on the intermediate layer during the development. The pattern is transferred into the copolymer by RIE with $O_2$ (10 mTorr, 0.4 kV). Afterwards, the copolymer structures serve as an etching mask for structuring germanium (15 mTorr, 0.4 kV). Finally, the remaining mask material is removed by $O_2$-RIE. Figure 9 shows outermost structures of a germanium zone plate.

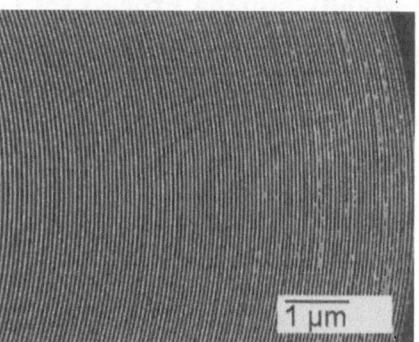

**Fig. 9.** SEM micrograph of outermost structures of a germanium zone plate after pattern transfer into a 220 nm thick germanium layer.

# 6 Results

Nickel zone plates have been fabricated with outermost zone widths of 43 nm and 150 nm zone height as well as 30 nm and 130 nm zone heigth, thereby first order diffraction efficiencies at 2.4 nm wavelenght of 15% and 11% were measured. This is 90 - 80 % of the theoretical value for the achieved zone heigth. Germanium zone plates with outermost zone widths of 40 nm (230 nm zone heigth), 30 nm (200 nm zone heigth) and 19.5 nm (145 nm zone heigth) were fabricated. They achieve diffraction efficiencies of 14%, 10% and 4%. This corresponds to 87%, 77% and 52% of the theoretical efficiency for the corresponding zone height.

The diffraction efficiencies of the germanium and nickel zone plates show the potential of the appropriate nanostructuring processes mentioned above. The decrease of efficiency with decreasing zone width indicates that the pattern transfer process still has to be improved, especially for zone widths below 30 nm and high aspect ratios.

The imaging performances of the zone plates were investigated with a test pattern. Structures of about 24 nm width have been resolved. Fig. 10 shows an X-ray image of a Siemens star taken with the X-ray microscope at BESSY using a nickel zone plate with outermost zone width of 30 nm as objective.

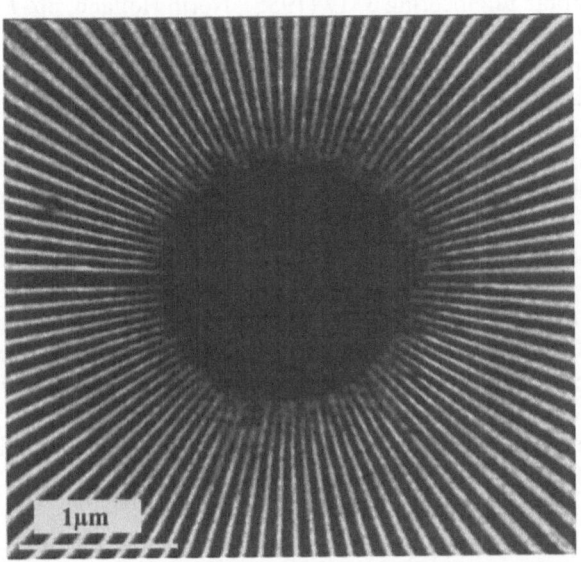

**Fig. 10.** X-ray image of Siemens star taken with a nickel zone plate with 30 nm outermost zone width. Image processing was applied in order to enhance the image contrast of structures near the resolution limit [9]. The inhomogeneity in the inner region of the image is caused by the non perfect test object.

The zone plates have been used in the X-ray microscope at BESSY for investigations on biological, medical, colloidal specimens and in solid state physics [10, 11, 12].

## Acknowledgements

The authors gratefully thank G. Schmahl and D. Rudolph for their encouragement as well as P. Guttmann for the measurement of the efficiencies. Furthermore we thank J. Herbst for assistance in performing the experiments. This work was funded by the German Federal Minister for Education and Research (BMBF) under contract 05 5MGDXB 6.

# References

1    J. Maser, in: X-Ray Microscopy IV, ed. by V.V. Aristov and A.I. Erko, Chernogolovka, Moskow Region, Bogorodski Pechatnik, 1994.

2    F. Emoto K. Gamo, S. Namba, N. Samoto, R.Shimizu and N. Tamura, Microelectronic Engineering 3, 17 (1985), North Holland, pp. 17-24.

3    G. Schneider, T. Schliebe and H. Aschoff, J. Vac. Sci. Technol. B 13, 1995.

4    E.H. Anderson and D. Kern, in: X-Ray Microscopy III, ed. by A.G. Michette, G.R. Morrison and C.J. Buckley Springer Verlag, Berlin, pp. 75-78, 1992.

5    C. David, B. Kaulich, R. Medenwald, M. Hettwer, N. Fay, M. Diehl, J. Thieme and G. Schmahl, J. Vac. Sci. Technol. B 13, 1995.

6    C. David, R. Medenwald, J.Thieme, P. Guttmann, D. Rudolph and G. Schmahl, J. Optics, vol. 23, n° 6, pp. 255-258, 1992.

7    D.M. Tennant E.L. Raab, M.M. Becker, M.L. O'Malley, J.E. Bjorkholm and R.W. Epworth, J. Vac. Sc. Technol. B 8, pp. 1970-1974, 1990.

8    J. Thieme, C. David, N. Fay, B. Kaulich, R. Medenwald, M. Hettwer, P.Guttmann, U. Kögler, J. Maser, G. Schneider, D. Rudolph, and G. Schmahl, in: X-Ray Microscopy IV, ed. by V.V. Aristov and A.I. Erko, Chernogolovka, Moskow Region, Bogorodski Pechatnik, 1994.

9    J.-B. Sibarita, J. Lehr, M. Robert-Nicoud and J.-M. Chassery, this volume.

10   G. Schneider, G. Schmahl, T. Schliebe, M. Peuker and P. Guttmann, this volume.

11   J. Thieme, J. Niemeyer, this volume.

12   P. Fischer, G. Schütz, G. Schmahl, P.Guttmann and D. Raasch, this volume.

# Zone Plates
# for a Scanning Transmission X-Ray Microscope

S. J. Spector[1], C. J. Jacobsen[1], D. M. Tennant[2]

[1] Dept. of Physics, SUNY at Sony Brook, Stony Brook, NY 11794, USA
[2] Lucent Technologies Bell Laboratories, Holmdel, NJ 07733, USA

**Abstract.** We describe the use of a commercial e-beam lithography machine for zone plate fabrication. We have modified the software of a JEOL JBX-6000FS, so as to draw high quality circular figures with a current of up to 500 pA within a 7 nm beam spot. Zone plates fabricated in germanium show good efficiency and resolution for scanning transmission x-ray microscopy applications. Zone plates with diameters larger than a writing field were successfully written by stitching together multiple fields.

## 1 Introduction

Fresnel zone plates are the highest resolution optics available for soft x rays. The transverse image resolution $\delta_t$ is approximately equal to the outer zone width $\delta_{r_N}$ only when all zones are correctly positioned to within about a third of their width [1]. Also, the thickness of the zones should be sufficient to adequately attenuate or phase-shift the transmitted x-ray front. This leads to requirements for patterning accuracy of about $1{:}10^4$, and an aspect ratio of 6:1. Fresnel zone plates therefore offer serious challenges in microfabrication.

Because Fresnel zone plates are key components in many x-ray microscopes, several research groups in the field have developed custom e-beam lithography systems for in-house zone plate fabrication [2, 3, 4]. However, most groups do not have the resouces needed to develop or purchase, and maintain and operate such systems soley for zone plate fabrication. We describe here the fabrication of Fresnel zone plates using a commercially-available electron beam lithography system in a multipurpose microfabrication laboratory. In this standard setup, with no special equipment other than modification of the pattern-generating software, we have fabricated zone plates with outer zone widths as small as $\delta_{r_N} = 30$ nm over diameters of 50–160 $\mu$m, in 180–250 nm of germanium. When used in a scanning transmission x-ray microscope, the zone plates produce the smallest focused spot of electromagnetic radiation of any wavelength.

## 2 Characteristics of Zone Plates for Scanning Microscopy

Zone plates for scanning x-ray microscopes (SXM) have somewhat different requirements than zone plates in transmission x-ray microscopes (TXM). For example, absolute effiency is helpful but not critical for the SXM. If a given number

X-Ray Microscopy and Spectromicroscopy
Eds.: J. Thieme, G. Schmahl, D. Rudolph, E. Umbach
© Springer-Verlag Berlin Heidelberg 1998

of photons are required in the image, reductions in zone plate efficiency will lead to increased radiation dose to the sample in TXM but will simply affect the time required to acquire the image in SXM. In a SXM, a central stop and order sorting aperture are used to block all but the first-diffraction-order, focused x rays. It is convenient to fabricate a thick gold central stop directly in the center of zone plate for this purpose. The placement of the order sorting aperture between the zone plate and the sample sets a minimum on the focal length of usable zone plates. In addition, the scanning photoemission microscope (SPEM) requires even longer focal lengths to allow detection of photo-electrons. Because the focal length is proportional to the diameter of the zone plate, there is motivation for fabricating larger diameter zone plates.

## 3    Electron Beam Lithography

The zone plates were patterned using the JEOL JBX-6000FS electron beam (e-beam) lithography system at Lucent Technologies, Bell Laboratories. The JBX-6000FS system has a thermal field emission source which operates at 50 KeV and can deliver 500 pA of current into a 7 nm spot size. (For all the exposures reported here, a beam current of 100 pA was used). The sample stage position is monitored by an interferometer with $\lambda/128 \simeq 5$ nm precision. The interferometer feeds back to the deflector to correct for any stage movement. The precisely controlled stage movement is also used as a standard for adjusting the gain and rotation of a writing field.

The JBX-6000FS is a rectangular coordinate system which draws figures with a minimum pixel size of 2.5 nm. The hardware is only designed to draw simple figures or primitives such as certain trapezoids, rectangles, and lines. More complicated figures such as circles need to be converted into primitives by software. The software provided by the manufacturer was found to be less than satisfactory for our application, in that it fractures circles into an unnecessarily large number of primitives. Also, if the minimum pixel spacing is used, the intersection between figures is exposed twice. Our solution has been to rewrite the software in a way which achieves a 10x reduction in the number of primitives as shown in Fig. 1. In addition, one edge of intersecting primitives is pulled back to avoid double exposure of the intersection. Because only trapezoids with angles less then 45° are permitted by the JEOL hardware, there are still complications at 45°. These complications have not caused any significant effect on resulting exposures.

## 4    Process Steps

The zone plates are fabricated on $\simeq$ 120nm thick $Si_3N_4$ membranes. The central stop is fabricated first along with alignment crosses which are later used to accurately place the zones around the central stop. The central stop and crosses are both fabricated in 300 nm of Au in a single e-beam lithography step. A lift-off process with a thick bilayer resist is used [5]. Because of the relatively low

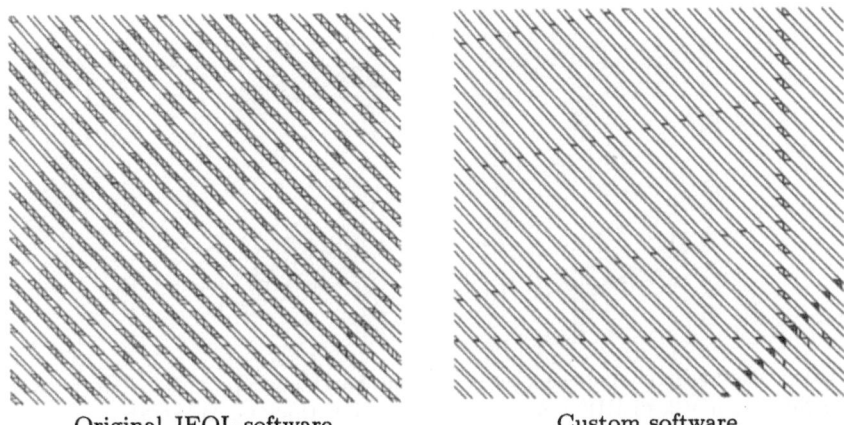

Original JEOL software.                    Custom software.

**Fig. 1.** Conversion of rings into primitives. There are 10x fewer primitives created by the custom software than by the original JEOL software. The subfield boundaries are shown toward the right and bottom of the figure, as well as the complications at 45° shown at the bottom right corner.

resolution necessary, higher currents can be used; over 100 central stops with crosses can be written in an hour.

In order to write the high resolution features in a zone plate, it is necessary to use a thin (40 nm) layer of PMMA as the imaging resist. The PMMA is baked at a temperature of 160° C overnight. Compared to cooler bake temperatures, we have found that a bake temperature of 160° C improves the resolution performance of PMMA, while decreasing its sensitivity. This is consistant with increased cross-linking of the PMMA, which may occur at higher bake temperatures [6]. Exposures are typically done with doses near $800\mu C/cm^2$. This dose is somewhat inflated due to the fact that we expose an $x - 15$ nm line to produce an $x$ nm zone. This is done to correct for a finite beam size as well as proximity effects.

The pattern is transferred from the thin PMMA into the 180 nm Ge layer by a trilayer process shown in Fig. 2 [7]. Although thicker Ge would provide more efficient zone plates, it is very difficult to fabricate such high resolution structures in thicker Ge. The Ge layers are formed by evaporation, and the resist layers are formed by spinning. The AZ resist is hard baked at 190° C for 1 hour. Reactive ion etching with selective gases is used to transfer the pattern through the layers into the germanium. $CF_3Br$ (150V, 10 mtorr) is used to etch the Ge layers, and $O_2$ (300V, 10 mtorr) is used to etch the AZ layer and to clean the zone plate after the final Ge etch.

## 5    Results

Zone plates have been fabricated with outer zone widths of 30 and 40 nm. The zone plates are 80 $\mu$m in diameter and were fabricated in Ge 180 nm thick.

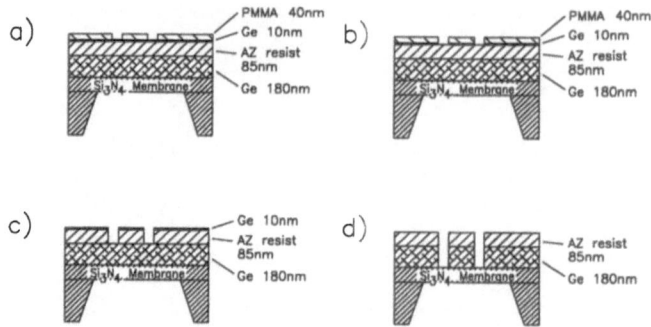

**Fig. 2.** Fabrication with a trilayer resist. a) exposure and development, b) RIE (reactive ion etch) thin Ge mask, c) RIE AZ photoresist d) RIE thick Ge substrate.

Figure 3 shows the outer zones of a 40 nm zone plate, and Figs. 4a&b show a comparison between 30 nm zone plates fabricated before and after the software revision. Notice that the "spokes" or radial line width variations present before the software revision have greatly been reduced. In addition, exposures done after the software revision have indicated greater exposure latitude.

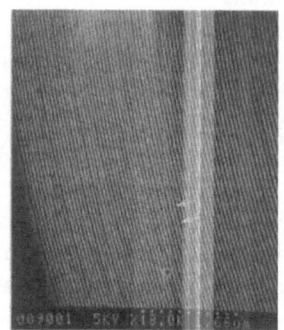

**Fig. 3.** Outer zones of a 40 nm zone plate.

**Fig. 4a.:** 30 nm outer zones before software revision. Arrows point along two adjacent "spokes."

**Fig. 4b.:** 30 nm outer zones after software revision.

The zone plates have been tested in the STXM and their efficiencies have been measured. The diffraction efficiencies of some zone plates are shown in Table 1. The efficiencies are lower than the theoretical value of 10% but more than sufficient for use in the STXM. The diffraction efficiency of a region of a zone plate can be measured by using a pinhole aperture placed in the far field. The relationship between diffraction efficiency and zone width for three zone plates is shown in Fig. 5. As the zone widths become finer, the efficiencies become lower. This drop in efficiency is possibly due to incomplete etching of the Ge and to the less perfect line shape and profiles of the outer zones. The

**Table 1.** Diffraction efficiencies of zone plates at a wavelength of 3.1 nm.

| Outer zone width | Conversion software | Efficiency |
|---|---|---|
| 30 nm | Custom | 5.7% |
| 30 nm | JEOL | 4.3% |
| 40 nm | Custom | 7.5% |
| 40 nm | JEOL | 6.8% |

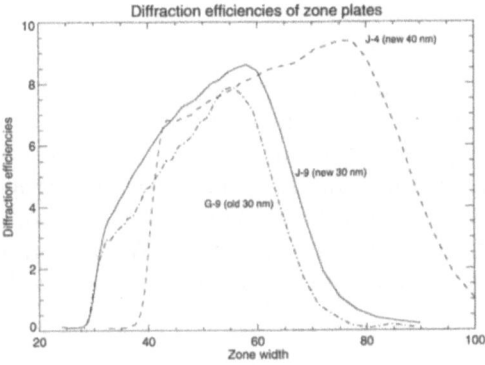

**Fig. 5.** Efficiencies for
regions of three zone plates.

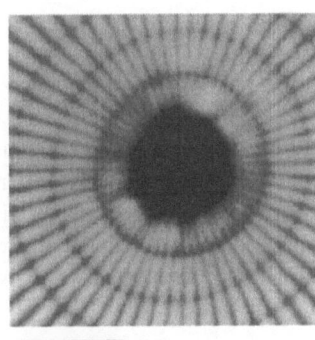

**Fig. 6.** STXM image of test
pattern taken using a zone plate
with 30 nm outer zones.

30 nm zone plate fabricated after the software revision (J-9) has less of a drop
in efficiency than the zone plate fabricated before the software revision (G-9).
This may be due to the improved line shape.

Figure 6 shows an image of a test pattern made using a 30 nm zone plate
in the STXM. At the inner ring of the test pattern features are 40 nm in size.
Features <30 nm in size can be observed in this image. This is an improvement
over the 45 nm zone plates which were the best we previously had available [8].

## 6   Larger Diameter

Increasing the diameter of a zone plate increases the working distance of the
zone plate and also increases the focused flux from the zone plate. (However,
a disadvantage to a larger zone plate is an increase in the necessary spatial
and temporal coherence of the illumination). In high resolution mode the JBX-
6000FS has a maximum field size of 80 $\mu$m. To minimize errors in the zone plate
we typically fabricate zone plates with a maximum diameter of 80 $\mu$m. The
fabrication of larger diameter zone plates can be achieved by stitching together
multiple writing fields using the precise movement of the stage.

Zone plates 160 $\mu$m in diameter with 60 nm outer zone widths were fabricated
by stitching four fields. (The zone plates were used in some of the first Cryo-
STXM experiments at X1A). Nine zone plates were fabricated, and the errors

at the field boundaries varied from very slight to severe. A few zone plates had maximum errors 15 nm or smaller, but a few had errors as large as 60 nm. The source of the errors are yet to be thoroughly investigated, but possibilities include inaccuracies of the stage movement, membrane movement, or beam drift possibly due to electric charging. The nature of the errors indicate beam drift as the most likely source. To allow less time for drift, future exposures will be done at higher current. However, errors due to electric charging may not be improved by this tactic.

# 7   Future

Further improvements in zone plate fabrication may be achieved by fabricating the zone plate in nickel instead of germanium. A nickel zone plate has nearly twice the efficiency of a germanium plate of the same thickness. Because it is difficult to pattern nickel by reactive ion etching, it is necessary to electoplate nickel nanostructures. Other improvements can be made by using an improved imaging resist and also by improving the pattern transfer [9, 10]. We are currently setting up to do electroplating of nickel, and we hope to pursue other methods of improvement.

# 8   Conclusion

Zone plates with outer zone widths as small as 30 nm have been fabricated for use in the STXM at the National Synchrotron Light Source. The fabrication of the zone plates was improved by a software revision which fractured circular patterns in an improved way. The zone plates have good efficiencies and imaging properties. Large diameter zone plates, 160 $\mu$m in diameter, have been fabricated by stitching together multiple writing fields. Zone plates with errors at the field boundaries as small as 15 nm are not uncommon.

# References

1. M. J. Simpson and A. G. Michette. The effects of manufacturing inaccuracies on the imaging properties of fresnel zone plates. *Optica Acta*, 30:1455–1462, 1983. (now Journal of Modern Optics).
2. C. David, B. Kaulich, R. Medenwaldt, M. Hettwer, N. Fay, M. Diehl, J. Thieme, and G. Schmahl. Low-distortion electron-beam lithography for fabrication of high-resolution germanium and tantalum phase zone plates. *Journal of Vacuum Science and Technology*, B 13(6):2762–2766, 1995.
3. E. H. Anderson, V. Boegli, and L. P. Muray. Electron beam lithography digital pattern generator and electronics for generalized curvilinear structures. *Journal of Vacuum Science and Technology*, B 13(6):2525–2534, 1995.
4. P. Charalambous, P. Anastasi, R. E. Burge, and K. Popova. Fabrication of high resolution X-ray diffractive optics at King's College, London. In W. Yun, editor, *X-ray microbeam technology and applications*, volume 2516, pages 2–14, Bellingham, Washington, 1995. Society of Photo-Optical Instrumentation Engineers (SPIE).

5. R. E. Howard, E. L. Hu, L. D. Jackel, P. Grabby, and D. M. Tennant. 400 angstrom line width e-beam lithography on thick silicon substrates. *Applied Physics Letters*, 36:596, 1980.

6. X. Zhang, C. Jacobsen, S. Lindaas, and S. Williams. Exposure strategies for PMMA from *in situ* XANES spectroscopy. *Journal of Vacuum Science and Technology*, B 13(4):1477–1483, 1995.

7. D. M. Tennant, E. L. Raab, M. M. Becker, M. L. O'Malley, J. E. Bjorkholm, and R. W. Epworth. High resolution germanium zone plates and apertures for soft x-ray focalometry. *Journal of Vacuum Science and Technology B*, 8:1970–1974, 1990.

8. C. Jacobsen, S. Williams, E. Anderson, M. T. Browne, C. J. Buckley, D. Kern, J. Kirz, M. Rivers, and X. Zhang. Diffraction-limited imaging in a scanning transmission x-ray microscope. *Optics Communications*, 86:351–364, 1991.

9. G. Schneider, T. Schliebe, and H. Aschoff. Cross-linked polymers for nanofabrication of high resolution zone plates in nickel and germanium. *Journal of Vacuum Science and Technology B*, 13:2809–2812, 1995.

10. J. Fujita, Y. Ohnishi, Y. Ochiai, and S. Matsui. Ultrahigh resolution of calixarene negative resist in electron beam lithography. *Applied Physics Letters*, 68:1297–1299, 1996.

This article was processed using the LaTeX macro package with LLNCS style

# Fabrication of the X-Ray Condenser Zone Plate KZP 7

M. Hettwer and D. Rudolph

Forschungseinrichtung Röntgenphysik, Georg-August-Universität Göttingen,
Geiststraße 11, D-37073 Göttingen

**Abstract.** The condenser zone plate KZP7 is used not only in the Göttingen X-ray microscope at BESSY in Berlin, but also in the X-ray microscopes at ALS in Berkeley, USA, and at ISA in Aarhus, Danmark. The zone plate interference patten of KZP7 is generated with uv laser light using an optical system with two aspheric mirrors. The zone plate pattern is recorded in photoresist. To obtain optimum contrast an optically matched layer system is used. It reduces backreflection to the photoresist layer. The photoresist pattern is transferred into germanium by an trilevel reactive ion etching process.

## 1 Introduction

The condenser zone plate KZP 7 is used since 1989 in the Göttingen X-ray microscope at the BESSY electron storage ring in Berlin as linear monochromator and X-ray condenser. In addition, it is used in the X-ray microscopes of the Centre for X-ray Optics at the ALS storage ring, Berkley, USA, and in the X-ray microscope of the Institute for Storage Ring Facility, ISA, of the university Aarhus, Denmark [1, 2].

In this paper the preparation of the condenser zone plate KZP 7 is presented. The holographic generation of the zone plate pattern and the pattern transfer into the final zone plate material using a tri level etching process is described. With calculations of backreflection of the laser light from the layer system to the photoresist the layer system for this nanostructuring process has been improved by using a different interlayer material.

The condenser zone plate KZP 7 is a zone plate with about 34000 zones. It has an inner radius of 2.0 mm and an outer radius of 4.5 mm. The inner zones have widths of 125 nm and the outermost zones have zone widths of about 54 nm. The condenser is made of germanium and can achieve a maximum theoretical first order diffraction efficiency of 10.8% at an X-ray wavelength of 2.4 nm and with a zone height of 380 nm. The focal length for $\lambda = 2.4$ nm is $f_{2.4nm} = 201$ mm. Figure 1 shows the zone plate exposed in photoresist on a silicon wafer.

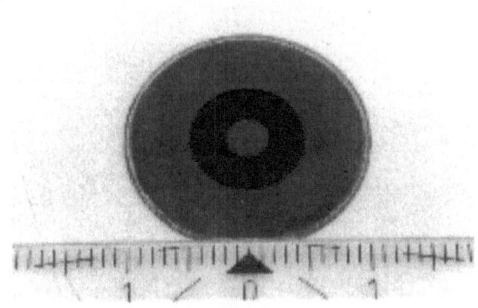

**Fig. 1.** Condenser zone plate KZP 7

**X-Ray Microscopy and Spectromicroscopy**
Eds.: J. Thieme, G. Schmahl, D. Rudolph, E. Umbach
© Springer-Verlag Berlin Heidelberg 1998

## 2 Optical Arrangement

The zone plate pattern of the condenser KZP 7 is holographically generated with an aspherical mirror system which was calculated by D.Rudolph. Up to now it is not possible to generate a zone plate pattern of this size with electron beam lithography techniques.

In Fig. 2a the optical arrangement for the pattern generation is shown. The UV laser light which is used for the exposure is obtained by an argon ion laser with a frequency doubling unit to generate uv radiation with $\lambda = 257$ nm. A quarter wave plate changes the linear polarisation of laser light into circular polarisation. The UV light passes a spatial filter consisting of an objective and an 5 μm pinhole. The optical system is illuminated by an UV objective with a spherical wavefront.

a

**quater wave    objective 1    spatial filter    objective 2    mirror system**
**plate**

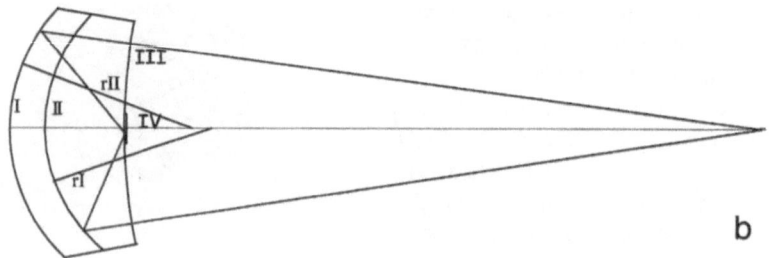

b

**Fig. 2.** Optical arrangement (a) and schematic of the aspherical mirror system (b).
I and II aspheric mirrors, III entrance sphere, IV plane exit surface

This system is made of fused silica. It contains two aspheric mirrors and a a plane-concave lens.

The incoming spherical wavefront passes the spherical entrance surface. Part of the wave is reflected by the semitransparent aspherical surface II. The transmitted light is reflected by surface I (Fig. 2b). The zone plate pattern results from the interference of these two waves. It is recorded in a photoresist layer which is matched to the plane exit surface by the immersion liquid dodecane. The exposure time of the zone plate pattern is about 60-90 s.

## 3 Nanostructuring Process

After exposure, development and postbake of the photoresist, the zone plate pattern is transferred into the final zone plate material germanium using a tri level ion etching process. The layer system for this etching process (see Fig. 3) consists of a 300 nm thick germanium layer as zone plate material followed by a 110 - 130 nm spin coated copolymer layer [3]. On the copolymer layer a 8 nm thick $Ta_2O_5$ intermediate layer is evaporated. On the top of the layer system a 50 nm thick photoresist layer is spin coated.

This layer system is carried by a silicon wafer. Between germanium and the silicon wafer a 6 nm chromium layer as an etch stop is evaporated.

Figure 3 shows the different steps of the nanostructuring process. The first steps in this process are the exposure, development and postbake of the photoresist.

The developed photoresist pattern serves as a mask for the first etching step by reactive ion etching (RIE) with a $CBrF_3$-plasma. With a pressure of 2 pa and a selfbias potential $U_{sb}$ of 400 V tantalum oxide is structured. This interlayer serves as highly selective mask for the etching of the copolymer in an oxygen plasma. The selectivity between tantalum oxide and copolymer is about 1:150. The copolymer is etched with $U_{sb}$ of 400 V and a pressure of 1.3 pa.

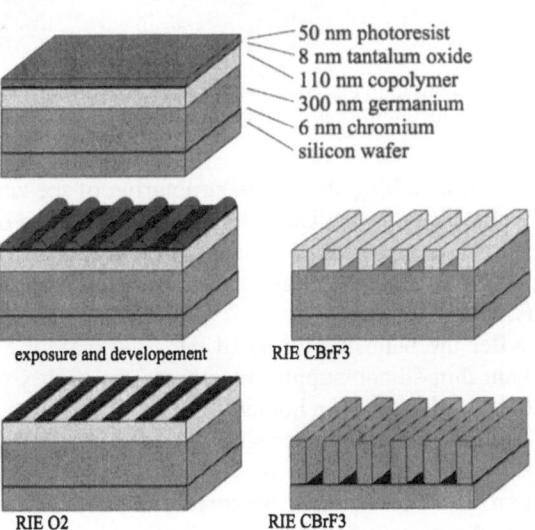

50 nm photoresist
8 nm tantalum oxide
110 nm copolymer
300 nm germanium
6 nm chromium
silicon wafer

exposure and developement

RIE CBrF3

RIE O2

RIE CBrF3

**Fig. 3.** Layer system and nanostructuring process of a KZP 7

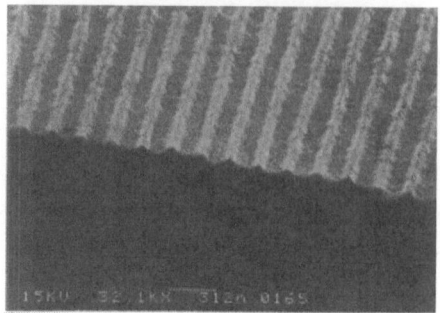

**Fig. 4.** Exposed and developed
photoresist

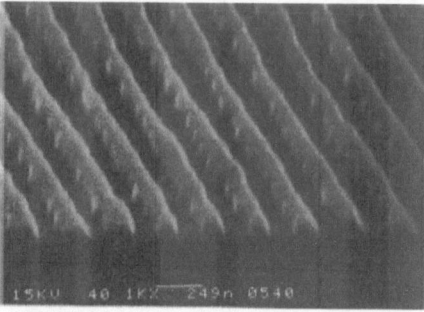

**Fig. 5.** Copolymer mask after oxygen etching; the zone height is about 110 nm

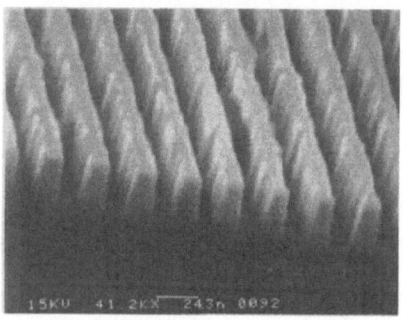

**Fig. 6.** 300 nm high germanium structures;
the zone width is about 120 nm

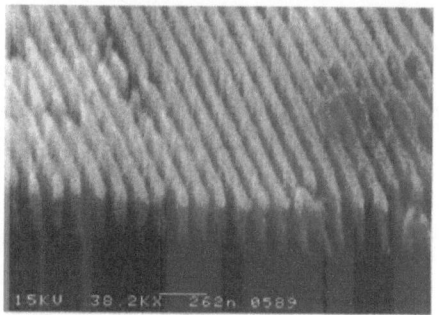

**Fig. 7.** Outermost zones of a KZP 7;
the zone width is about 60 nm and
the zone height is 280 nm

The last etching step is the structuring of the zone plate material germanium with
a $CBrF_3$ -plasma with $U_{sb}$ = 400 V and a pressure of 2 pa. The aspect ratio (ratio between height and width of zones) of the outermost zones is 5:1. During the last etching step the chromium layer serves as etch stop to prevent etching into the silicon
wafer.

After the nanostructuring of the zone plate the silicon wafer is etched to a 150-
250 nm thin silicon support membrane for high x-ray transmission. At last, the zone
plate is prepared onto a holder.

Figures 4 -7 show micrographs of the different etching steps. The samples are exposed under an angle of 80°. Fig. 4 shows the exposed and developed zone plate
pattern in a 50 nm thick photoresist layer, the zone width is about 125 nm. Fig. 5
shows the structured copolymer mask. The height of the zones is 110 nm and the
zone width 100 nm. Fig. 6 and Fig. 7 show the final zones in germanium. The zone
height is about 280 nm. In Fig. 6 the zone width is 120 nm, in Fig. 7 the zone width is
about 60 nm. The inhomogenities of the zones are due to imperfections of the interference patern.

## 4 Optical Matching of the Layer System

The interference pattern of KZP 7 up to now shows some imperfections which cannot be explained by imperfections of the $TEM_{oo}$-mode of the laser beam. Therefore, the first step to increase the efficiency of the zone plate should be to improve the contrast of the interference pattern. Backreflection of the laser light from the layer system is one of the main effects that reduces the contrast of the interference pattern This can be avoided by using an optically matched layer system [4]. Calculations show that backreflection to the photoresist is reduced if the dielectric material tantalum oxide is used instead of the metal titanium that has been used before [3]. Fig.8 shows plots of the backreflection from the layer system to the photoresist at the boundary photoresist – interlayer for tantalum oxide and titanium as interlayer materials. The calculations were done for a height of 110 nm of the mask material and a thickness of 8 nm of the interlayer. The layer system with tantalum oxide as interlayer shows less backreflection to the photoresist over the whole zone plate area then titanium.

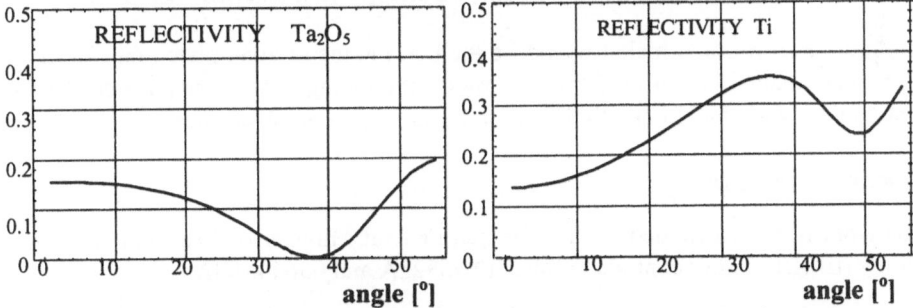

**Fig. 8.** Backreflection of the laser light from the layer system to the photoresist with the two different interlayer materials tantalum oxide and titanium

## 5 Results

With the described reactive ion etching process and the improved layersystem KZP7 condenser zone plates with zone heights of about 330 nm have been built with outermost zone widths down to 54 nm and aspect ratios of 5.5 : 1. These zone plates are made on silicon membranes less than 200 nm thick that corresponds to a transmission of more then 70% at $\lambda = 2.4$ nm. The absolute efficiency is 4.7% which corresponds to a groove efficiency of more than 8% that is about 60% of the theoretical value.

## 6 Further Improvements

A layer system with less then 10% backreflection over the whole zone plate area would consist of a 5 nm tantalum oxide layer and a 180 nm thick copolymer layer. However, this layer system will not work sufficiently as a mask for the described etch process, as the selectivity i too low. Considering a higher germanium layer to achieve maximum efficiency (see Fig. 9) the selectivity has to be improved even more.

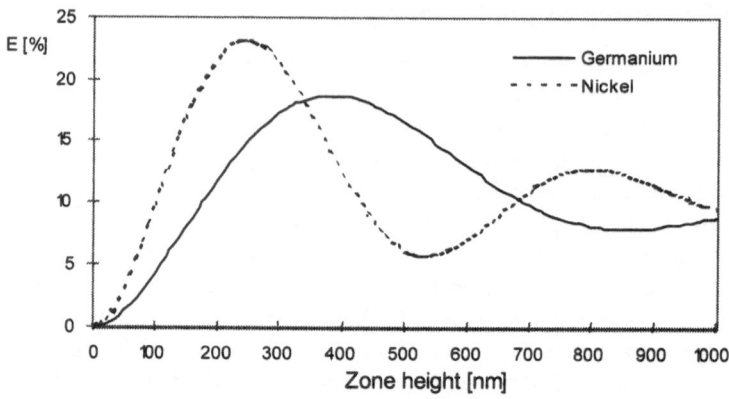

**Fig. 9.** Diffraction efficiency E (groove efficiency) for X-radiation ($\lambda$ = 2.4 nm) as function of the zone height for germanium and nickel as zone plate material

Figure 9 shows that nickel would be the best choice of zone plate material, however, zone plates have to be fabricated by electroplating. This needs a much higher copolymer layer which, in addition, has to be structured much cleaner.

## Acknowledgement

This work has been funded by the German Federal Minister of Research and Technology (BMBF) under contract number 13N5328A and 05644MGA8.

## References

1    W. Meyer-Ilse, H. Medecki, J. T. Brown, J. M. Heck, E. H. Anderson, A. Stead, T. Ford, R. Balhorn, C. Petersen, C. Magowan, D. T. Attwood: X-ray Microscopy in Berkeley, this volume.

2    Joanna Abraham, Robin Medenwaldt, Erik Uggerhøj, P. Guttmann, T. Hjort, J. Jensenius, T. Vorup-Jensen, F.Vollrath, E. Søgaard, J. Tyge Møller: X-ray Microscopy in Aarhus, this volume.

3    M.L. Schattenburg, R.J. Aucoin and R.C. Fleming: Optically matched tri-level resist process for nanostructure fabrication. J. Vac. Sci. Technol. B 13(6), 1995, pp. 3007–3011.

4    G. Schneider, T. Schliebe and H. Aschoff: Cross linked polymers for nanofabrication of high-resolution zone plates in nickel and germanium. J.Vac. Sci. Technol. B 13(6), 1995, pp. 2809–2812.

# Bragg–Fresnel Optics

V. V. Aristov

Institute of Microelectronics Technology, and High Purity Materials RAS
Chernogolovka, Moscow Region, 142432 Russia
E-mail: aristo@ipmt-hpm.ac.ru

## 1 Introduction

Bragg-Fresnel optics (BFO) has been developed in IMT RAS for more than 10 years already [1]. The operation of BFO is based on the combination of two kinds of diffraction: Bragg diffraction of hard X-ray radiation on a perfect single crystal or a multilayer interference mirror and Fresnel diffraction on an artificially created microstructure. At present, BFO is the only instrument of X-radiation focusing in a wide range (up to 100 keV) with a space resolution to 0.1 micron. BFO is used in a number of techniques of X-ray diagnostics and as the basis for the development of new instruments. Bragg-Fresnel lenses successfully tested with use any of synchrotron radiation sources in Russia (Novosibirsk, VEPP-2M), Germany (Hamburg, HASILAB; Berlin, DESY, DORIS), Japan (Tsukuba, Photon Factory), and France (Grenoble, ESRF; Paris, LURE).

**Table 1.** Working parameters of Bragg-Fresnel lenses

| Characteristics | Linear lens | Circular lens crystals | Elliptical lens (multilayers) |
|---|---|---|---|
| Possibility of using various energy values | Yes | Yes | Yes |
| Range | continuous energy 2–100 keV | discrete energy 2–100 keV | continuous energy 0.1–10 keV |
| Spectral band $\Delta E/E$ | $10^{-4}$ | $10^{-5}$– $10^{-6}$ | $10^{-3}$ |
| Minimum resolution | 0.2 μm | 0.2 μm | 0.2 μm |
| Diffraction efficiency | to 40% | to 40% | to 40% |
| Energy gain in focusing from a 100μm aperture to a spot of: | | | |
| 0.3 μm | 300 | $10^5$ | $10^5$ |
| 0.1 μm | 1000 | $10^6$ | $10^6$ |

**X-Ray Microscopy and Spectromicroscopy**
Eds.: J. Thieme, G. Schmahl, D. Rudolph, E. Umbach
© Springer-Verlag Berlin Heidelberg 1998

## 2 Fields of Application (Under Developing Today)

1. X-ray microprobe in the energy range up to 100 keV with a submicron space resolution.
2. Local double and triple crystal diffractometry.
3. Low-angle scattering for structure analysis.
4. X-ray microscopy.
5. X-ray holography.
6. Phase-contrast microscopy and microtomography.
7. X-ray scanning microscopy.
8. X-ray lithography.

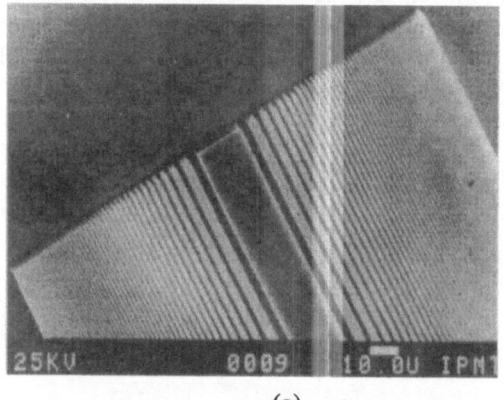

(a)

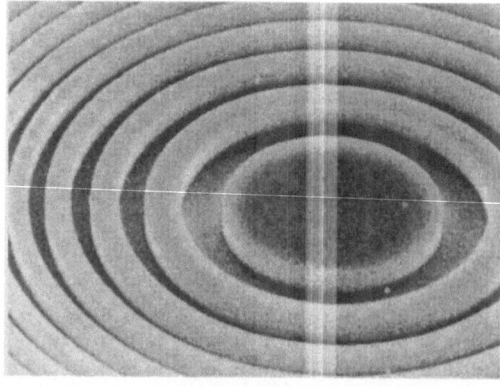

(b)

**Fig. 1.** Examples of Bragg-Fresnel lenses on silicon
(a) Linear lens, (b) Circular lens

# 3 X-Ray Scanning Microscopy on the Base of BFL[1]

Space scanning is due to the X-ray diffraction on an ultrasonic superlattice. Changes in the wavelength of the superlattice give rise to the changes in the space position of X-ray diffraction sattelites. Two-dimensional scanning is achieved by using two ultrasonic superlattices that scan on X-ray beam in space in two mutually perpendicular directions.

Specifications of today X-ray scanning microscope: the range of scanning with an electric switching is 300x200 $\mu m^2$, mechanical scanning 1x1 mm$^2$.

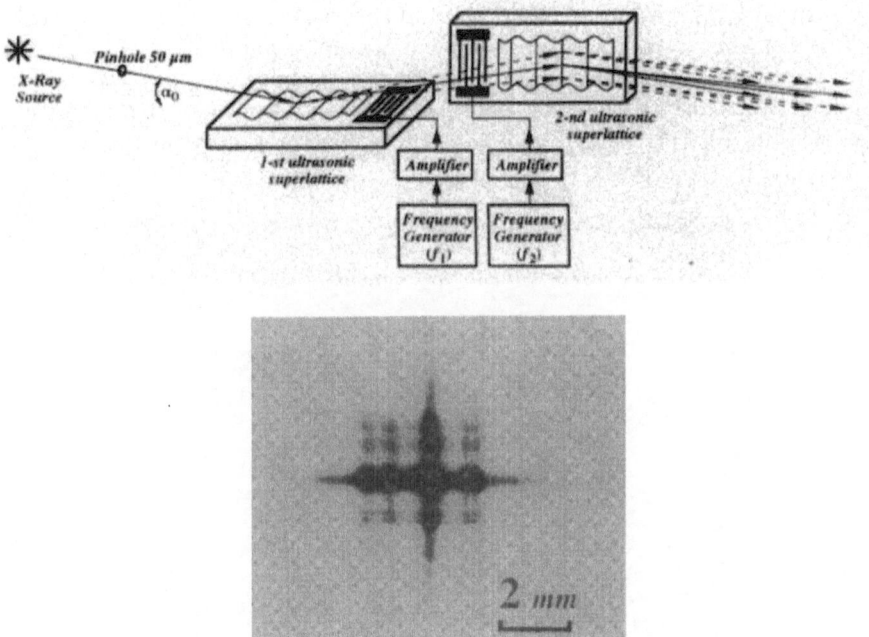

**Fig. 2.** Schematic drawing of an X-ray scanning microscope (two dimensional scanning is achieved by using two ultrasound supergratings which scan an X-ray beam in space on two mutually perpendicular plate) [2]

# 4 Future of BFO

In the period of more than 10 years, lenses have been developed, fabricated and are now operating. What can be the directions of further development of Bragg-Fresnel optics and what improvements should be made? We think that today approaches have been marked and technologies are being developed which allow us to increase the resolution to 0.1–0.05 nm and to decrease the band pass by an order of magnitude and

---

[1] The work is done in collaboration with Laboratoire de Cristallographie at CNRS (Grenoble).

to increase the diffraction efficiency by 1.5–2 times. It also seems expedient to use composite lenses with 1, 2, 3, ... orders of diffraction, to enlarge the set of crystals and reflections (for crystal lenses). Moreover, no investigations have yet been made on the possibility of using asymmetric diffraction geometry, transmittance (Laue) diffraction, and some other interesting problems. Let us now consider the ideas and technological possibilities which would permit achieving the parameters mentioned above.

# 5  Electron-Beam Lithography

A technology has been developed which allows a three-dimensional correction of the proximity effect, with allowance for the resist development. This technology provides the possibilities for drawing and fabrication of complicated structures where large size and fine elements are alternated (Fig. 3.) and possibilities to fabricate kinoform zone plates as well. Moreover, for lens reproducing a special technology was developed including electroplating of metals of resist relief and sequential imprinting in soft materials (Fig. 4.). This technology is of special importance for X-ray optics.

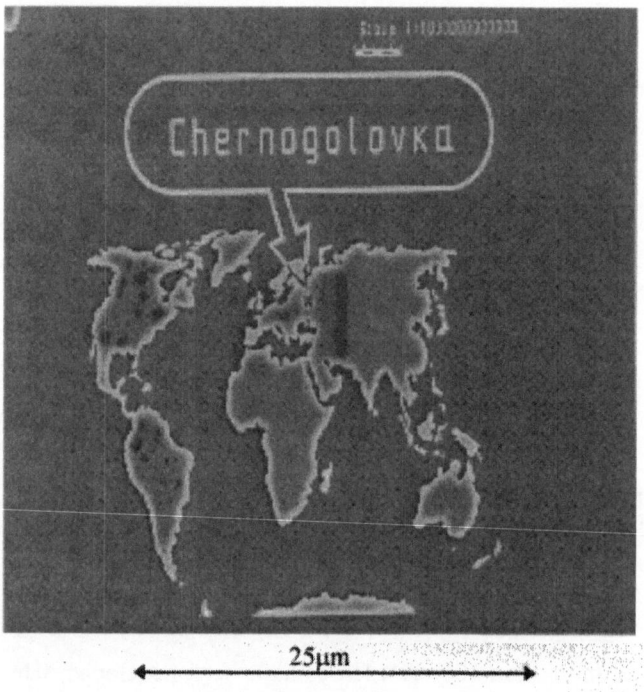

**Fig. 3.** Map of the world made by e-beam lithography (Al on GaAs)

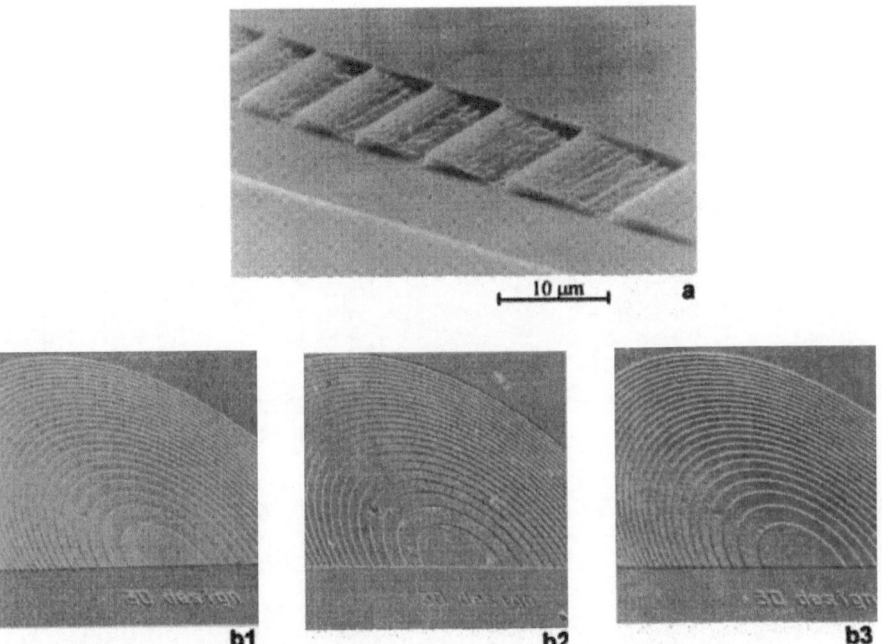

**Fig. 4. (a)** Kinoform structure in Si,
**(b)** Zone plates (diameter 500 μm) in resist **(b1)**,
Cu metal replica from resist **(b2)**,
printing by Cu-replica in polymer **(b3)**

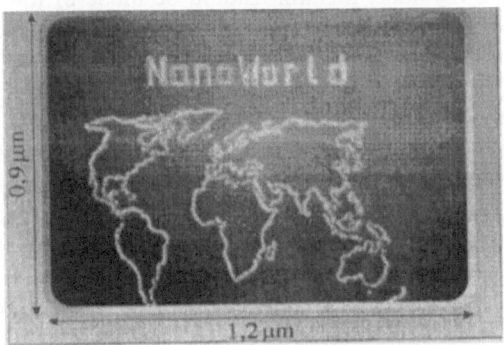

**Fig. 5.** Counter map of the world. Line size near to 5 nm [4]

## 6  Plasmochemical Etching

For Bragg-Fresnel lenses on crystals, this technique is of crucial importance because structures with linear parameters from several microns to 0.1 μm are to be etched to the same depth 2 - 3 μm on one and the same substrate. We have coped with this problem (Fig. 6.), and an appropriate procedure is now being tested [5].

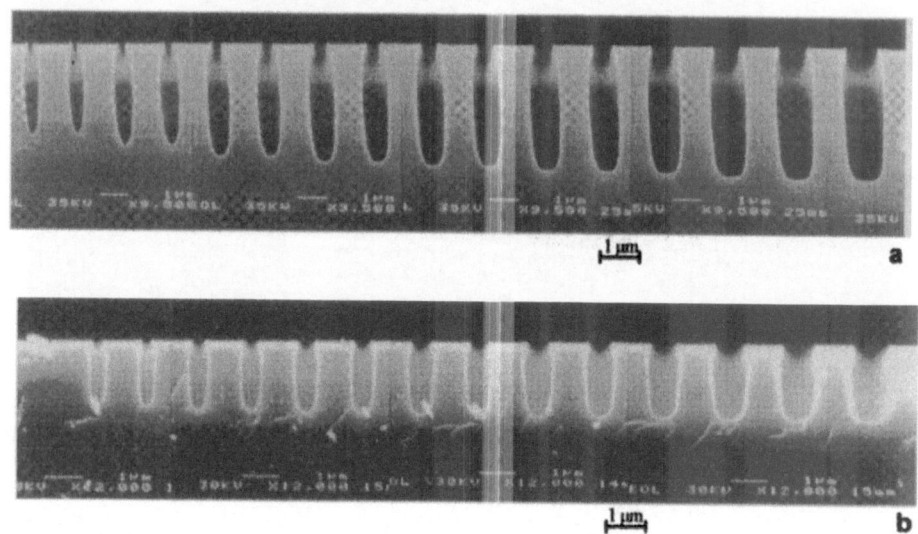

**Fig. 6.** Results of etching in Si crystal depth of profile equal to **(a)** - 4 μm, **(b)** -2 μm

## 7  Composite Lenses

Of other technological methods, mention should be made of ion implantation, MBE for growing gradient layers, and creation of lenses on super mirrors. However, all these methods raise a challenging problem of increasing the band pass. In the case of the diffraction of a plane wave on a gradient grating, the extension of the spectral range is an experimental fact. In the case of a spherical wave, a gradient grating operates as a one-dimensional Fresnel zone plate with its own focus distance [1].

Composite lenses afford an improvement of the lens reflection ability. This is a well known fact, but we consider the possibility of using the second and fourth orders of diffraction from the viewpoint of specific features of the technology of lens fabrication. This possibility can be realized if the zone sizes are 1:3 instead of 1:1. At the same sizes of a zone plate, the intensity gain is 1.5-2 times at a decreasing background [6].

## 8  Conclusion

I would like to emphasize that lenses on crystals and on multilayer mirrors are, at present, only the beginning of the way. Bragg-Fresnel diffraction has great potentialities related to the possibility of varying different parameters. But the discussion of these potentialities is beyond the scope of this paper.

In conclusion, I would like to thank my colleagues from the institute for their efficient work in this hard time for Russian sciences. My thanks are due to the Organizing Committee of XRM-96 for the financial support which made the participation in the work of the conference possible for workers of IMT RAS.

## References

1    V.V. Aristov and A.I. Erko, X-ray Microscopy IV, Proceedings of the 4-th International Conference Chernogolovka, Russia, September 20-24, 1993, Institute of Microelectronics Technology, Chernogolovka, Russia (1994).

2    See papers in this book.

3    S.V. Dubonos, B.N. Gaifullin, V.N. Matveev, H.F. Raith, A.A. Svintsov, and S.I. Zaitsev, J. Vac. Sci Technol. **B 13** (6) Nov/Dec (1995) 25-26.

4    V.V. Aristov, S.V. Dubonos, R.Ya. Dyachenko, B.N. Gaifullin, H.F. Raith, A.A. Svintsov, and S.I. Zaitsev, Microelectron. Eng. **27** (1995) 195.

5    Nano World Picture, Microscopy and analysis **34** (1995) 51.

6    V.A. Yunkin, D. Fischer, and E. Voges,. Reactive ion etching of silicon submicron-sized trenches in $SF_6/Cl_2F_3$ plasma, Microelectronic Engineering **27** (1995) 463.

7    E.V. Shulakov, V.V. Aristov, Surface (Russian) **N 3-4** (1966) 53-59.

8    E.V. Shulakov, V.V. Aristov, Surface (Russian) **N 3-4** (1996) 60-68.

# Bragg–Fresnel Optics
# for High-Energy X-Ray Microscopy Techniques
# at the ESRF

I. Snigireva, A. Souvorov, A. Snigirev

ESRF, B. P. 220, F-38043 Grenoble Cedex, France

**Abstract**. Combination of Bragg-Fresnel optics with high brilliance X-ray beams provided by third generation synchrotron radiation sources like ESRF demonstrated unprecedented and very promising performance in terms of efficiency, resolution, energy tunability and stability. Low emittance and quite small angular source size of the X-ray beam at the ESRF allow along with microprobe techniques like microdiffraction and microfluorescence, new fields of applications such as high resolution diffraction, holography, interferometry and phase contrast imaging based on Bragg-Fresnel optics to be realized. New types of Bragg-Fresnel optical systems have been designed and tested: linear curved Bragg Fresnel lens (BFL); linear BFL composed of 1st and 3d order of diffraction; circular BFL with ultrasound modulation.

## 1 Introduction

Among the focusing elements for hard X-rays (> 6 keV) Bragg-Fresnel Optics (BFO) shows excellent compatibility with the third generation synchrotron radiation sources such as ESRF. In addition to a beam of extremely high brilliance, these X-ray sources are characterized by very small source size. A typical source size at the ESRF is 30-80 µm, that at the source-to-sample distance of 50 m gives an angular source aperture of about 1-2 µrad. The coherence preservation is precisely, an essential feature which is required of the focusing optics. Since the Bragg-Fresnel optics (BFO) is a combina-

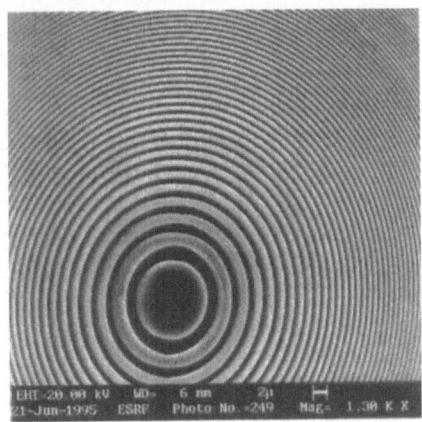

**Fig. 1.** SEM image of the linear and circular BFLs

**X-Ray Microscopy and Spectromicroscopy**
Eds.: J. Thieme, G. Schmahl, D. Rudolph, E. Umbach
© Springer-Verlag Berlin Heidelberg 1998

tion of Bragg reflection from the crystal and diffraction by Fresnel structure grooved into a crystal, it is evident from creating principles, that the BFO is coherent optics and this is the only focusing optics that is able to preserve the coherence of the incoming beam [1-3]. As a coherent focusing element the Bragg-Fresnel optics allows to realize along with standard microprobe the new applications such as high resolution diffraction, ultra small angle scattering, holography, interferometry and phase contrast imaging. At present two types of Bragg-Fresnel lenses (Fig. 1) are mainly used: linear BFL in sagittal geometry and circular BFL in backscattering geometry [2-3].

## 2 BFO Principles and Performance

Bragg-Fresnel crystal optics is based on a superposition of Bragg diffraction by a crystal and dispersion by a Fresnel structure, which is patterned on the surface or grooved into the crystal. The wave reflected by the lower surface of the BFL zone structure gains an additional phase shift $\pi$, as compared to that reflected by the upper surface. The phase shift linearly depends on two parameters, namely, the relief height and angle of incidence. The profile height over a wide range of x-ray energies is measured in micrometers, much larger than the wavelength of the incident radiation. The second distinguishing feature of a BFL is that, for a given reflection and a relief height, the phase delay does not depend on the X-ray energy [2]. Diffraction efficiency of the BFL is very closed to the theoretical performance and is about 40%. The limiting spatial resolution that can be obtained in BFO is given by the width of the outermost zone of the zone structure. Present technologies permit to achieve fractions of a micron.

### 2.1 Linear BFL

A linear BFL in sagittal geometry (Fig. 2) on a flat substrate produces one dimensional focusing of X-rays [2, 4-7]. Phase properties of a linear BFL structure do not depend on the energy. Therefore the same lens can be designed for wide energy range determined only by the Bragg's law. Tests of linear BFL were done at the undulator source and wiggler sources. It was shown that linear BFL is capable of focusing X-ray radiation in the range from 2 to 100 keV[10]. The focal spot of 0.8 μm FWHM with intensity $10^8$-$10^9$ photons/sec for linear BFL was measured at 8 keV and was limited by the source size according to

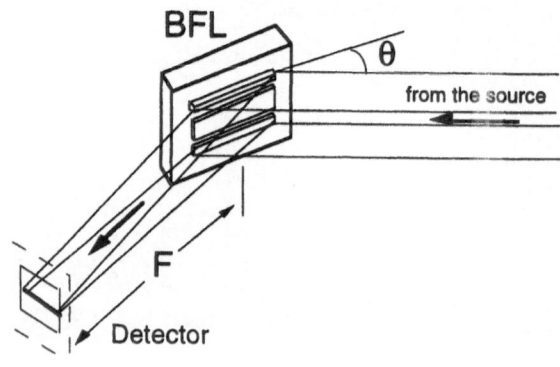

**Fig. 2.** Sagittal focusing by linear BFL

the demagnification ratio [6].

Cylindrical bending of the BFL [8-9] was applied to produce two-dimensional focusing. The experiment was carried out at the Optics beamline (BM 5) at the ESRF. The focusing properties of the curved BFL were tested at energies 18 keV and 28 keV.

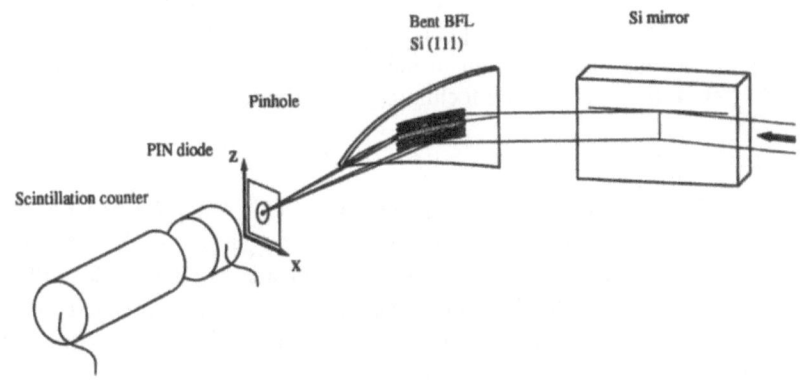

**Fig. 3.** Experimental setup for the curved BFL at energies 18 and 30 keV

The intensity in the focal plane of the BFL was measured by means of 2D mapping with 1 μm pinhole paired with scintillation detector or Si PIN diode. In accordance with demagnification factor and X-ray source size the focal spots of $2*4.5$ μm$^2$ at 18 keV and of $3*6.5$ μm$^2$ at 30 keV were measured. The comparison of the X-ray integrated intensity in the focus spot by the flat and curved linear BFL with same parameters was carried out. A gain by the factor of up to 100 in the focal flux was obtained. To improve the resolution and to increase the focal flux the BFL may be completed with the third order structure which, increases the total aperture of the BFL by a factor of 3 respectively [10]. To perform the experiment a linear BFL on Si (111) substrate with first and third order

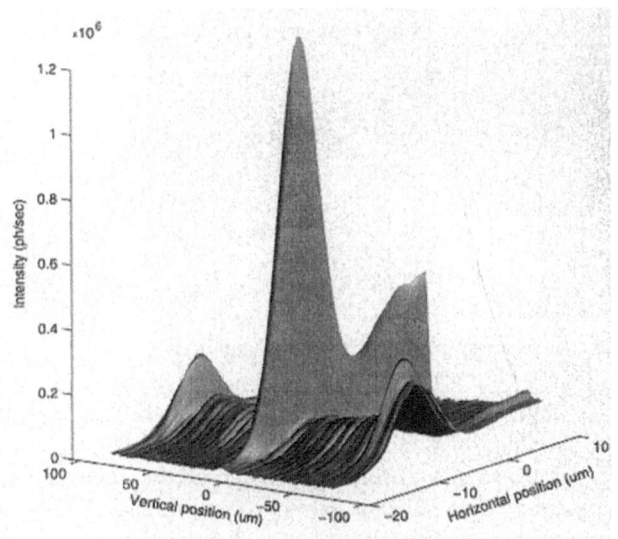

**Fig. 4.** 2D intensity mapping of the focal spot of bent BFL at the white 30 keV X-ray beam

structures was fabricated with the following parameters: total aperture A = 380 μm, lens length L = 8 mm, outermost zone width $\Delta r_n$ = 0.3 μm, focal distance F = 0.25 m at E = 8 keV. The measured focal spot size was in a good agreement with the limitations given by the source size.

Another way to increase the absolute intensity in the focal spot of a BFL is to change the crystal reflectivity, i.e. to modify the crystal lattice. Ion implanted BFL was studied and an enhancement of the focal flux of up to 15% due to crystal lattice deformations was demonstrated [11]. However, due to the complicated scattering of the implanted ions inside micron or submicron size crystal features which make up the BFL relief, the implantation technology was found not very effective.

## 2.2 Circular BFL

A circular BFL in backscattering geometry (Fig. 5) produces two-dimensional focusing at fixed X-ray energies determined by Bragg's law for the different reflection orders [3, 12-13]. It was shown that the focusing can be made with significant diffraction efficiency of the BFL in the energy range of about 2 keV-18 keV when the reflections were varied from Si-111 to Si-999. The focal spot of 0.7 μm FWHM with intensity $10^7$ - $10^8$ photons/sec for circular BFL was obtained at 10 keV [13].

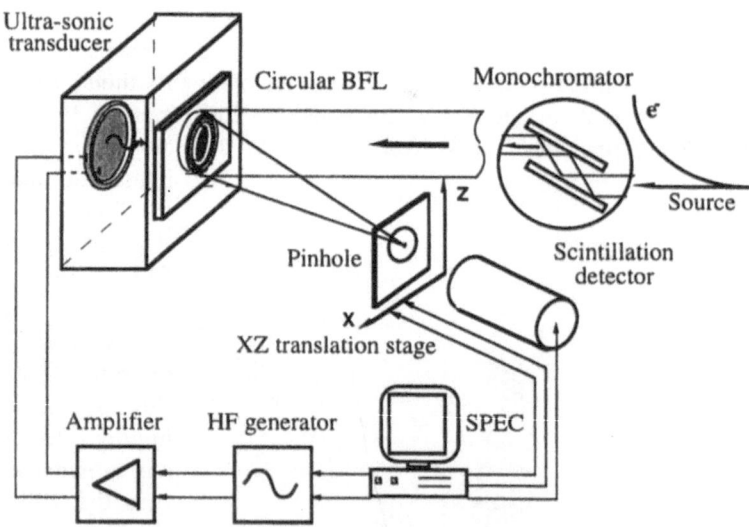

**Fig. 5.** Experimental setup for BFL acoustic modulation.

The phase BFL gives a high efficiency focusing of monochromatic X-rays however no more than a small fraction of the total intensity of a white source is gathered into the focus. To increase the absolute intensity in the focal spot the ultrasonic modulation of crystals which BFL is based on was applied [14]. The influence of an ultra sonic modulation of the Bragg-Fresnel lens as the radiation flux at the focal spot was studied (Fig. 5). The reflectivity and integrated intensity of the crystal substrate depend on the frequency and amplitude of the excited ultra sonic wave. The intensity in the focal plane of the BFL was measured by means of 2D mapping with 10 μm pinhole paired with scintillation detector or Si PIN diode. A gain by the factor of up to 3 in the focal flux was obtained (Fig. 6).

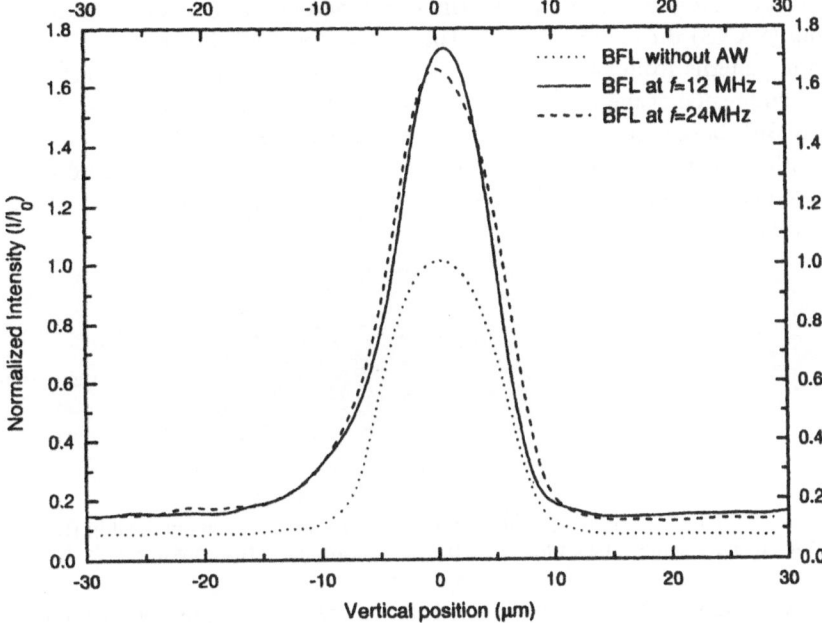

**Fig. 6.** Vertical cross sections of the BFL focal spot at different ultrasonic frequencies

## 3 BFO Applications

It is evident that such intense microfocalized X-ray beams open up new capabilities to develop hard x-ray microimaging and microprobe techniques. Some examples and recent results are described in the following.

### 3.1 BFO Based Microprobe

A varied program of experiments using X-ray microbeam produced by BFLs for fluorescence imaging and elemental distribution was carried out. The experiments have been performed at undulator and bending magnet beamlines at the ESRF. Applications of the developed fluorescence microprobes for elemental distributions in volcanic rocks, Antarctica micrometeorites, bone specimens and human hair slices were demonstrated [13, 15–16].

The X-ray microbeam is very desirable for high pressure experiments, especially when diamond cells has to be transparent for X-rays and very little amount of sample is involved in the measurement [17]. Because of the X-ray absorption in the several millimeter thickness of diamonds, X-rays of greater than 18 keV energy are generally employed. A linear Bragg-Fresnel lens working simultaneously as a monochromator and focusing element was applied for high pressure experiments. A range of the sample s were studied, but most important results were obtained on oxygen. A phase transition in the solid oxygen was observed around 96 GPa [18].

Very promising application for BFO microprobe is for small angle X-ray scattering (SAXS) diffraction studies. The BFL-based SAXS camera for measuring diffraction patterns from samples in specific regions of micron dimensions has been designed and tested at the ESRF Microfocus beamline [19]. The attractive feature of proposed BFL-based camera is the fact that one can measure down to the smallest angles without using a complicated collimation system: defining and guard apertures (slits). For native turkey leg tendon collagen intermediate areas between calcified and non-calcified

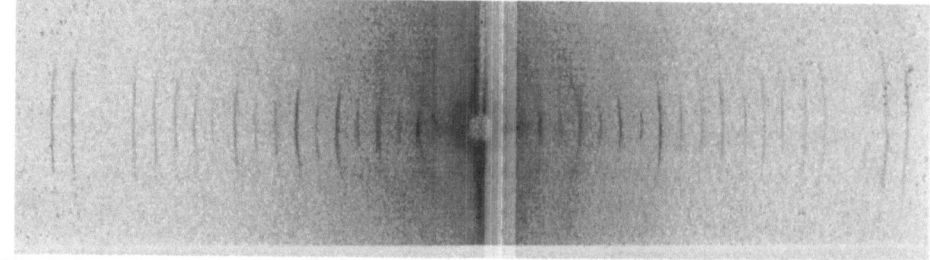

**Fig. 7.** Low angle diffraction pattern of the turkey leg tendon collagen obtained in BFL based SAXS camera using image plate. The first 25 meridional reflections were resolved.

regions were analyzed (Fig. 7). Thus, small-angle scattering is possible from a sample on the scale of a few μm and can be extended to the subμm range. These open the possibility for new applications of SAXS, in particular in the area of surfaces and interfaces.

## 3.2 High Resolution Diffraction

Linear crystal Bragg-Fresnel lens is acting as a focusing monochromator producing cylindrical wave front. This means a sagittally focused beam remains very parallel in meridional plane. This focusing monochromator can be applied in high resolution diffraction technique: double- and triple-crystal diffractometry for a detailed study of nearly perfect semiconductor crystals with topological surface structure. Two examples of double crystal diffractometry with high spatial resolution: Stress analysis of the turbine blades and rocking curve scans across the lateral transition zones in III-V heterostructures are presented.

A high resolution diffractometry technique was applied for peak profiles studies of a turbine blade of the nickel-base super alloy which was subjected to service in an accelerated mission test for several hundred hours. In served turbine blades the hot region near the leading and trailing edges are subjected to temperatures up to 1000°C, whereas the regions near the cooling channels are subjected to temperatures of about 800°C. These temperature gradients cause strong inhomogeneities in the local thermal and mechanical loads. The local lattice parameters were determined from the locally measured line profiles of the Bragg reflections (Fig. 8). The analysis of the data shows that local lattice-parameter changes in the surface regions of a monocrystalline turbine blade. This results can be explained by the superposition of thermally induced and deformation induced long-range internal stresses with local changes of the chemical composition in these region of the turbine blade [20].

III-V heterostructures grown by different selective area epitaxy techniques (planar and embedded selective area MOVPE and MOMBE) were investigated by double crystal diffractometry set-up at the ESRF BM 5 beamline. To obtain optimum spatial resolution for rocking curve measurements, the sample was placed in the image plane of the BFL (Fig. 8). The X-ray beam diffracted from the sample was recorded by a Na(I) scintillation counter. The samples were test structures of InGaAs and InGaAsP layers grown on an InP (001) substrate that was partially masked with $SiO_2$ fields and laser/waveguide devices laterally integrated on an InP (001) wafer. In order to determine the lattice mismatch close to the boundaries of the layer / oxide and laser / waveguide boundary, rocking curve scans with micrometer step width were performed. The lattice distortions of the III-V-heterostructures show changes at the boundaries in the range of 5 µm to 100 µm depending on the selected process [21].

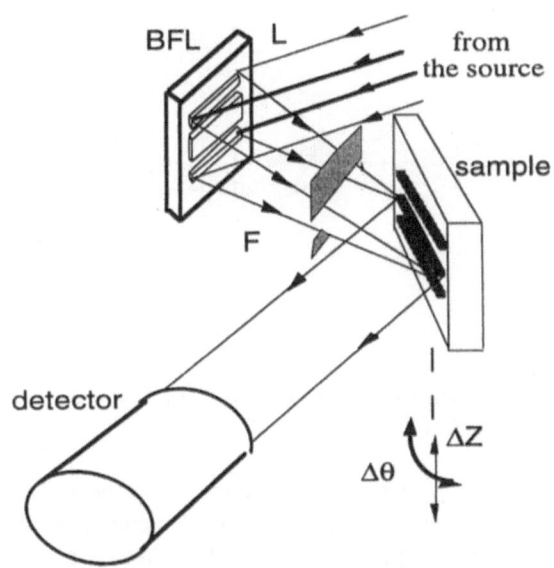

**Fig. 8.** Double crystal diffractometry setup based on BFL

## 3.3 BFO Imaging Applications

Usual principle of the microscopy based on amplitude contrast which arises through differences in the absorption length from material to material. From this concept x-ray imaging microscopy is practically impossible for hard X-rays. It is well known that the contrast of the sample to be imaged can be enhanced considerably by using phase contrast. Zernicke type phase contrast microscopy was realized in soft X-ray domain [22]. In principle the same approach can be apply for hard X-rays using BFO or zone plates. But we have suggested to use another way. High level of collimation and coherency of the X-ray beam provided by the undulators at the ESRF make it possible to develop phase sensitive technique [23-25]. The transmission X-ray

microscope was tested at ID 22 beamline. (Fig. 9). A test object - a fine gold greed-
was attached to a 70 μm pinhole. A torroidal mirror was used as a condenser. The fine
0.5 μm gold grid was clearly resolved at 9.5 keV and a contrast of more than 20%
was measured [26]. It was proved the necessity of applying so called defocusing for
phase contrast imaging of weakly absorbing materials. Moreover, depth of the image
field was experimentally measured to be almost infinite, this confirms a coherent
illumination optical setup.

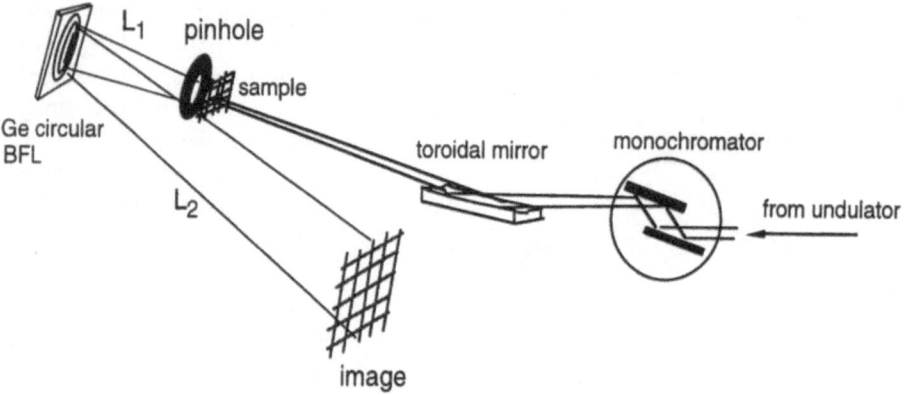

**Fig. 9.** X-ray microscopy setup based on circular BFL, where $L_1$ was 15 cm, $L_2$ was varied
from 50 to 60 cm.

The circular BFL was applied for imaging of the undulator source at the ESRF
[27-28]. To measure the value of emittance in situ a second optical element as an
image expander has been applied. Two optical geometry's have been tested at the
energy 8 keV. In a first set-up the image formed by the long focus (F = 1.25 m) BFL
as an objective lens has been vertically enlarged by asymmetrically reflected Si-422
crystal with magnification factor 15. The enlarged image was recorded by X-ray CCD
camera having a resolution of 30 μm FWHM. In a second set-up classical telescope
geometry was applied where two BFLs were used in tandem. The first objective
forms a real inverted image, which was examined using the second short focus BFL
(F = 0.25 m), the eyepiece. The second focal plane of the objective almost coincided
with the first focal plane of the eyepiece and a 100 μm pinhole was installed in this
plane in order to spurs the zero diffraction order for better image contrast. The image
was recorded by X-ray CCD placed at 1.5m distance from the second lens. The
computer recorded images for both optical set-ups were treated and the deduced
values of the emittance were in a good agreement with other estimates.

## 4 Future BFO Applications

We have presented last results on BFO developments and recent applications at
the ESRF. We plan to further pursue our efforts to improve the resolution of our
optics and our technical development of increasing the photon flux in the focal spot.

In parallel with already demonstrated and applied BFO based applications we want to develop new ones like multicrystal high resolution diffraction and holographic techniques.

BFO is a perfect crystal optics and it is unique candidate for studding fine coherent process by means of double and triple crystal diffractometry with high spatial resolution. High angular resolution can be easily achieved using an asymmetrically cut BFL crystal, so standing wave technique with lateral resolution of about 1 µm is also feasible.

Coherent imaging technique based on Gabor in-line holography setup such as phase contrast imaging, holography, interferometry and tomography are successfully progressing now at third generation synchrotron radiation sources [23-25, 29]. The resolution of these techniques is determined by the resolution of detector system: at present this is 2-10 µm for high resolution X-ray cameras and 1 µm for high resolutions films. The use of additional optics like a Bragg-Fresnel lens installed after the object should lead to a 0.1 µm resolution.

BFL based Fourier transform holography seems to be a promising alternative as well. We propose to utilize a Bragg-Fresnel lens to obtain a point source for the spherical reference wave. The resolution will then be limited by the size of the focal spot, thus by the outermost zone width of the Bragg-Fresnel zone plate.

## Acknowledgments

We thank V. Yunkin and N. Gornakova from the Institute of Microelectronics Technology RAS for BFO fabrication. We are very grateful Prof. V. Aristov for his continuous encouragement. We express gratitude to the ESRF staff for their help in performing experiments. The Bragg-Fresnel Optics project is carried out with the Institute of Microelectronics Technology RAS under collaboration contract No. CL 0048.

## References

1    V. V. Aristov, A. A. Snigirev, Yu. A. Basov, A. Yu. Nikulin, AIP Conf. Proc., 147, 253, (1986).

2    V. V. Aristov, Yu. A. Basov, T. E. Goureev, A. A. Snigirev, T. Ischikawa, K. Izumi, S. Kikuta, Jpn. J. Appl. Phys., vol.31, 2616, (1992).

3    Yu. A. Basov, T. L. Pravdivtseva, A. A. Snigirev, M. Belakhovsky, P. Dhez, A. Freund, , Nucl. Instrum.& Methods A308, 363, (1991).

4    Snigirev, Rev. Sci. Instr, 66(2), 2053, (1995).

5    Snigirev, V. Kohn, "Bragg-Fresnel Optics at the ESRF: microdiffraction and microimaging applications" in X-ray Microbeam Technology and Applications, W. Yun, Editor, Proc. SPIE, vol. 2516, 27, 1995.

6    S. M. Kuznetsov, I. I. Snigireva, A. A. Snigirev, P. Engström, C. Riekel, Appl. Phys. Lett. 65 No 7, 827, ( 1994).

7    U. Bonse, C. Riekel, A. A. Snigirev, Rev. Sci. Instrum., 63, 622, (1992).

8    Ya. Hartman, A. Freund, I. Snigireva, A. Souvorov, A. Snigirev, accepted in NIM.

9    Souvorov, I. Snigireva, A. Snigirev, V. Yunkin, to be published.

10   M. J. Simpson, A. G. Michette, Opt. Acta, **31**, 403 (1984).

11   Suvorov, A. Snigirev, I. Snigireva, E. Aristova, Rev. Sci. Instrum., 67(5), 1733, (1996).

12   E. Tarasona, P. Elleaume, J. Chavanne, Ya. M. Hartman, A. A. Snigirev, I. I. Snigireva, Rev. Sci. Instr., 65(6), 1959, (1994).

13   Snigirev, I. Snigireva, P. Engström, S. Lequien, A. Suvorov, Ya. Hartmann, P. Chevallier, F. Legrand, G. Soullie, M. Idir, , Rev. Sci. Instr., 66(2), 1461, (1995).

14   Souvorov, I. Snigireva, A. Snigirev, E. Aristova, Ya. Hartman, accepted in Appl. Phys. Letters.

15   P. Chevallier, P. Dhez, F. Legrand, M. Idir, G. Soullie, A. Mirone, A. Erko, A. Snigirev, I. Snigireva, A. Souvorov, A. Freund, P. Engström, J. Als Nielson, G. Grübel, NIM A354, 584 (1995).

16   P. Chevallier, P. Dhez, F. Legrand, A. Erko, Yu. Agafonov, L. Panchenko, A. Yakshin, J. Trace and Microprobe Techniques, 14(3), 517, (1996).

17   M. Hanfland, D. Hausermann, A. Snigirev, I. Snigireva, Y. Ahahama, M. Mcmahon, ESRF Newsletters, No. 22, 8, (1994).

18   Y. Akahama, H. Kawamura, D. Hausermann, M. Hanfland, O. Shimomura, Phys. Ref. Letters 74, 23, 4690 (1995).

19   A. Snigirev, I. I. Snigireva, C. Riekel, A. Miller, L. Wess, T. Wess, Journal de Physique IV, Vol. 3, C8, Suppl.JP III, N12, 443, (1993).

20   H. Biermann, B. V. Grossmann, S. Mechsner, H. Mughrabi, T. Ungar, A. Snigirev, I. Snigireva, A. Souvorov, M. Kocsis and C. Raven, to be published in Physika Scripta.

21   Iberl, M. Schuster, H. Göbel, B. Baur, R. Matz, A. Snigirev, I. Snigireva, A. Freund, B. Lengeler, H. Heinecke, J. Phys. D: Appl. Phys., 28, A200, (1995).

22   G. Schmahl, P. Gutmann, G. Schneider, B. Niemann, C. David, T. Wilhein, J. Thieme, and D. Rudolph, in X-ray Microscopy IV, edited by V. V. Aristov and A. I. Erko (Bogorodski Pechatnik, Chernogolovka, Moscow region, 1994), pp. 196-206.

23   Snigirev, I. Snigireva, A. Suvorov, M. Kocsis, V. Kohn, ESRF Newsletters, No 24, 23, (1995).

24   Snigirev, I. Snigireva, V. Kohn, S. Kuznetsov, I. Schelokov, Rev. Sci. Instrum., 66(12), 5486, (1995).

25   Snigirev, I. Snigireva, V. Kohn, S. Kuznetsov, Nucl. Instrum. & Methods, A370, 634, (1996).

26   Snigirev, I. Snigireva, P. Bösecke, S. Lequien, accepted in Optics Commun.

27   Ya. Hartman, E. Tarazona, P. Elleaume, I. Snigireva, A. Snigirev, Journal de Physique IV, Colloq.C9, vol. 4, C945, (1994).

28   Ya. Hartman, E. Tarazona, P. Elleaume, I. Snigireva, A. Snigirev, Rev. Sci. Instrum., 66(2), 1978, (1995).

29   Raven, A. Snigirev, I. Snigireva, P. Spanne, A. Suvorov, V. Kohn, Appl. Phys. Lett., 69 (13), 1826, (1996).

# High Numerical-Aperture X-Ray Condensers for Transmission X-Ray Microscopes

B. Niemann

Forschungseinrichtung Röntgenphysik, Georg-August-Universität Göttingen,
Geiststraße 11, D-37073 Göttingen

**Abstract.** In the near future transmission X-ray microscopes (TXM) will be installed at undulator X-ray sources, which generate highly collimated X-ray beams. To match a TXM to such a source X-ray condenser-monochromators of smaller light collecting area and increased numerical aperture (NA) are required. This can be achieved with the concept of dynamical aperture synthesis. Herein an off-axis zone plate with comparatively "coarse" zones is used, which can be processed with high diffraction efficiency with standard e-beam lithography methods. The numerical aperture of any existing X-ray micro objective can be matched for dark and bright field imaging. A monochromaticity of several thousand is possible and the homogeneity of the intensity distribution in the object field is enhanced compared to a zone plate linear monochromator used in TXMs at the moment.

## 1 Introduction

In the last years X-ray microscopy in the spectral range of 2.4 to 4.5 nm progressed significantly. Scanning transmission and transmission X-ray microscopes were developed, which run routinely and with fast access [1–8] at brilliant X-ray sources. Up to now transmission X-ray microscopes only used X-rays generated by bending magnets of electron storage rings. At the BESSY electron storage ring the condenser-monochromator of the TXM uses the holographically generated condenser zone plate KZP7 [9]. However, the new undulator X-ray sources which are under construction and which generate highly collimated X-ray beams, e.g. BESSY II and the ESRF, require new X-ray condenser-monochromators of increased NA, which better match a TXM to such a source.

## 2 On the Resolution and Numerical Aperture of Zone Plates in a TXM

It is known from the theory of the optical microscope that the $NA_{cond}$ of the condenser of a bright field transmission microscope should always be nearly matched to the $NA_{objective}$ of the objective to get a nearly incoherent object illumination from noncoherent radiating sources. Then a nearly linear relation between the object transmittance and the image intensity results. If $NA_{cond}$ is much smaller than $NA_{objective}$ then partially coherent imaging is given and the linear transformation between the object transmittance and the image intensity is lost for the important high spatial frequencies, which determine the resolution of the microscope. Therefore it is a common practise to use a condenser with $NA_{cond} \approx 1 .. 1.5 \cdot NA_{objective}$.

**X-Ray Microscopy and Spectromicroscopy**
Eds.: J. Thieme, G. Schmahl, D. Rudolph, E. Umbach
© Springer-Verlag Berlin Heidelberg 1998

The resolution $\delta = 1.22 \cdot dr_n$ and the NA $= \lambda / 2 \cdot dr_n$ of a zone plate is determined by the width $dr_n$ of the finest, outermost zone. Consequently, to increase the resolution of a TXM ever finer zones in the micro zone plate objectives are generated; 19 nm smallest zone width were already achieved with the help of e-beam writing-systems. A zone width $dr_n$ as small as 54 nm was achieved in condenser zone plate KZP7. This cannot be improved simply, because it is determined by the optic and by the laser UV light wavelength, which generate the zone plate interference pattern used for lithographic processing. Thus the NA$_{cond}$ of the KZP7 does not match the NA$_{objective}$ of these objectives. Furthermore, large area condenser zone plates with the required small zones cannot be generated with other methods, e.g. with current e-beam writing-systems, as an unrealistic long writing time - in the order of weeks - is required.

## 3 The Optical Set-up of the TXM at BESSY

Figure 1 shows an optical ray diagram of the TXM at BESSY. X-rays from a bending magnet source illuminate the condenser zone plate KZP7 with about 9 mm diameter. The condenser images the source into the object plane. In combination with a mono-chromator pinhole in the object plane the condenser acts as a linear monochromator – due to the wavelength dependence of the focal length of zone plates. The monochromator bandwidth $\delta\lambda/\lambda \cong 2d/D$, D = diameter of condenser zone plate, d = diameter of pinhole [10], is usually matched to the number N of zones - typically some hundred - of the micro zone plate used as an imaging X-ray micro objective. The micro zone plate generates a magnified image of the object in the image field. All TXMs which use a micro zone plate as an micro-objective need a condenser which illuminate the object field from a hollow cone. This is achieved with the help of a central stop on the condenser zone plate. It prevents that the object is illuminated by broad band zero order radiation of the condenser. In addition, increasing the diameter of the central stop on the condenser increases the useful image field region which is free of intense zero order radiation of both the object and the micro zone plate, see Fig. 1. However, it is desirable to increase this object field size and to use a condenser which increases the field diameter at least a factor of 2 and beyond the field limits given by the aberrations of the objective micro zone plate.

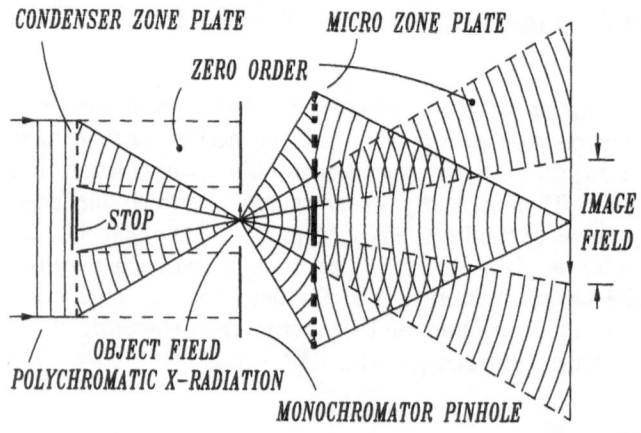

**Fig. 1.** Optical ray diagram of a transmission X-ray microscope. In this example the numerical aperture of the condenser zone plate is not matched to the high resolution micro zone plate and partially coherent imaging is obtained. The size of the image field is comparable small.

At the bending magnet source of BESSY the vertical divergence of the radiation is about 1 mrad FWHM. Therefore at the TXM location at 15 m distance KZP7 is illuminated nearly homogeneously in the vertical direction. It is illuminated homogeneously in the horizontal direction, because the radiating electrons in a bending magnet move on an arc. The FWHM diameter of the source - typically some tenth of a millimetre - which is observed by the condenser zone plate, is determined by the statistically varying electron orbits, whereas the observable projection of each individual electron orbit is much smaller than this source size, typically only a few microns. This means that a condenser of much larger horizontal aperture would increase the flux in the source image significantly, because flux from a much longer arc of the electron orbit is collected. However, the total image size would not be altered but of course the NA in horizontal direction would be increased.

The radiation characteristic of an undulator source is different. Undulator U41 at BESSY II will deliver radiation, which at the microscope location is about 2 x 6 mm² FWHM in diameter. Therefore it is impossible to fully illuminate the large area of

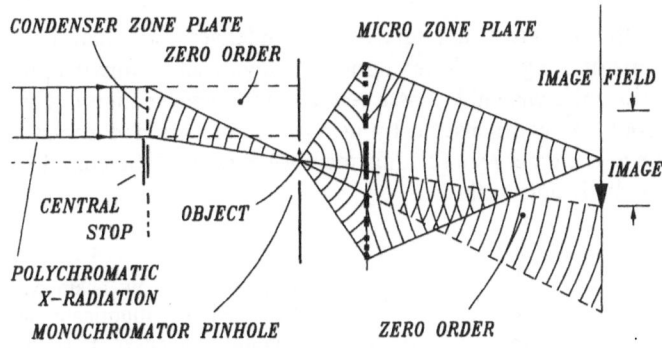

Fig. 2. A well collimated undulator beam illuminates an off axis part of a condenser zone plate. An oblique illumination of low numerical aperture is produced in the object field.

KZP7. Moreover, as the inner area of KZP7 with 4 mm diameter cannot be used: it has got no zones because of the interference-optical design used for its generation - only an eccentric part of KZP7 can be illuminated, see Fig. 2. An asymmetric transfer function of the microscope will result. On the other hand - besides the reason mentioned above - there is another cause to use a condenser zone plate without inner zones, if the TXM uses a zone plate as an objective and thus requires monochromatic object illumination: Each condenser zone plate which is also used to monochromatize the incoming radiation has to have an obstructed centre, because the inner zones - when regarded as a grating - show only low dispersion and would diffract radiation of a large wavelength range into the used object field. This drastically reduces the monochromaticity and therefore is unacceptable.

However, an annular condenser zone plate can be illuminated completely with a narrow beam, if e.g. another additional dispersing zone plate optic in front of the condenser is used to produce a wide beam. However, such a design is not very practical, because much light will be lost at this additional dispersing element. Furthermore the adjustment in such a hypothetical TXM which uses an additional dispersing zone plate to produce a wide beam is much more complicated: In a conventional TXM with *two* dispersing optical elements (one condenser and one objective zone plate) there *always* is a plane between both zone plates which is in focus for one distinct wavelength. Thus moving the object along the optical axis a focused image will be achieved at a special position. This is not the case if the microscope contains three zone plates and focusing will be a long lasting trial and error procedure with the two zone plates or one zone plate and the object to be shifted stepwise one after another along the optical axis.

## 4 On the Zone Plate Concept and Design Consideration of the Optical Set-up of a TXM at an Undulator Source

A better possibility to fully illuminate the condenser zone plate with a well collimated beam can be realised with a pair of parallel plane mirrors, which rotate around the axis of the incident beam and direct the light onto a condenser zone plate, as illustrated in Fig. 3. Thus successively the incident beam is spread over the full annular and a rotational symmetric aperture is synthesised. Nevertheless, even in this case the problem to match the numerical aperture of the condenser to the micro-objective still is not solved.

A possibility to increase the numerical aperture of the condenser zone plate is to combine it with focusing mirrors, e.g. a Wolter optic. However, as a mirror optic shows no chromatic aberration, the ensemble has a reduced monochromaticity. This loss in monochromaticity could be compensated by using a smaller monochromator pinhole. However, this leads to extremely small adjustment tolerances of the complete optical system and is an unrealistic approach.

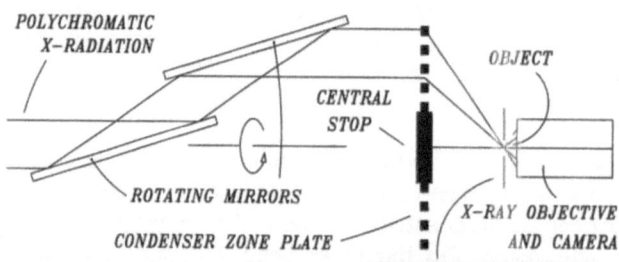

**Fig. 3.** A narrow beam can be used to illuminate an annular condenser, if a pair of rotating plane mirrors is located in front of it. The numerical aperture is given by the condenser zone plate.

Therefore new condenser-monochromator concepts were developed. The new optical systems collect all flux in the central radiation cone, monochromatize it, illuminate the object rotational symmetric, produce an incoherent image recording and especially can be matched to the numerical aperture of any high resolution micro zone plates (with e.g. 17 nm outermost zone width) for bright and dark field imaging. The necessary diffracting optics are of some mm diameter and have line widths of > 100 nm. They can be generated with existing processes using e-beam lithography and reactive ion etching techniques. The electron-beam lithography system LION LV1 (Leica Lithographie Systeme GmbH, Jena) installed in Göttingen can write the required zone patterns on thin substrates with the required accuracy and size. Under realistic conditions diffracting optics with a total line length of about 500 m can be exposed within a week. More than 10 % absolute diffraction efficiency are state of the art for this kind of diffracting X-ray optics.

## 5 A Focusator with Ring Focus and Hollow Cone Mirrors

One of these new systems is shown in Fig. 5 [11] and includes the general concept of focusators [12]. The system consists of a dispersing special focusator followed by of one or two conical mirrors. The radiation originating from the point-like undulator source typically illuminates a region of a few millimetre diameter at some ten meters distance. Here the radiation meets the special focusator, which in its focal plane generates a ring - instead of the spot common for lenses. At an appropriate location between the focusator and the focal plane two hollow-cone mirrors are situated, which reflect the radiation to the optical axis and produce a quasimonochromatic object illumination.

Figure 4 shows the dispersing zones of the required focusator in comparison to a zone plate. It is obvious that the inner zone of the focusator has got the smallest zone width, in opposition to a zone plate. The focusator will generate a "ring image" with a certain diameter D at a distance depending on several design parameters. The focal length of a focusator can be selected independently of the required numerical aperture of the ensemble focusator - hollow cone mirrors. A realistic focusator of 1 m

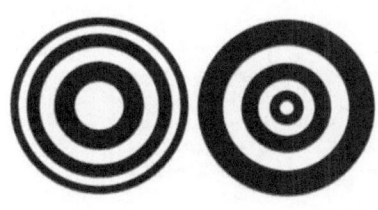

**Fig. 4.** *Left*: Zone plate. *Right*: focusator producing a ring "image" in the focal plane.

focal length at 2.4 nm wavelength and 2 mm diameter has zones of > 100 nm width. The NA of the arrangement is determined by selecting an appropriate reflection angle of the hollow cone mirrors, e.g. 3.8° to match the NA of a micro zone plate with $dr_n$ = 18 nm at 2.4 nm wavelength. S. Oestreich [13] calculated that two (and not one) hollow cone mirrors are required to get a practical system with sufficiently large

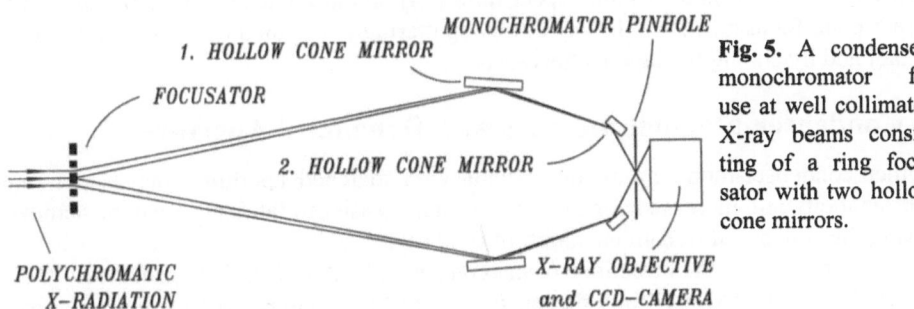

**Fig. 5.** A condenser-monochromator for use at well collimated X-ray beams consisting of a ring focusator with two hollow cone mirrors.

adjustment tolerances. He also calculated that a focusator combined with two hollow cone mirrors will generate a small intense quasimonochromatic spot on the optical axis surrounded by a low intensity halo.

In order to get a uniform object field illumination, the focusator has to be scanned in two dimensions. The cited calculations also show that it is only necessary to scan the focusator and to keep the hollow cone mirrors at a fixed position. Scanning the focusator is easily possible with piezoelectric transducers in the 100 Hz frequency range. Such a scanning method was experienced in the Göttingen TXM to produce a uniform object field illumination with a usual condenser zone plate [8].

The difficulties in realising the focusator concept are to produce conical mirrors with sufficient slope accuracy and of slow surface roughness, which are difficult to achieve at the same time on a curved surface. In practise a slope error of several arc seconds - a standard value achievable by the manufactures - limits the monochromaticity to several hundred [13].

In Fig. 6 another concept is shown. It uses two focusators, each generate a ring focus. The first focusator merely used as a light collector and has coarse zones; the second focusator focuses the convergent beam in a small spot and also monochromatizes it.

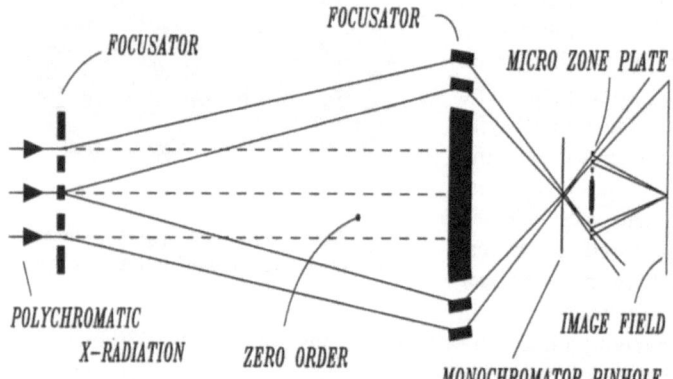

**Fig. 6.** A condenser-monochromator containing two focusators. It collects a narrow beam, monochromatizes and concentrates it with the required high numerical aperture onto the object.

The concept of high efficient, high order diffraction with zone structures of low line to space ratio and of high aspect ratio [14], could advantageously be applied to the second focusator. It could be built by sputtering zones on a conical cylinder or on a ball and by slicing the layers afterwards.

## 6 Condenser-Monochromators with Dynamical Aperture

Other condenser-monochromator systems with matched aperture, increased mono-chromaticity and increased zone width – which is easier to fabricate – can be achieved using the concept of dynamical aperture synthesis.

Figure 7 shows the basic idea of the concept [11]. An off-axis transmission (OTZ) is used to produce a spectrally dispersed image of the source, which is deflected back with a plane mirror to the optical axis of the incident beam. If only a small wavelength range is required, an off-axis reflection zone plate (ORZ) [15] can also be used. By rotating the off-axis zone plate and the mirror synchronously around the optical axis of the microcope, the oblique illumination bundle describes a complete radiation cone and synthesises an annular pupil. A symmetric transfer function with the full spatial frequency response of an incoherent system results, even if coherent radiation is used

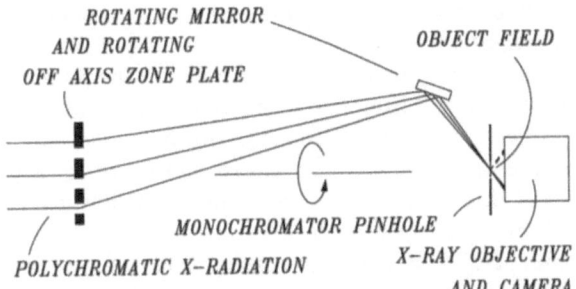

**Fig. 7.** Scheme of the dynamical aperture syn-thesis. An off-axis zone plate collects a narrow beam and monochromatizes it. The required numerical aperture is achieved by deflecting the converging beam with a plane mirror. Off-axis zone plate and mirror are rotating.

[16]. This property is comparable to the image recording in a scanning transmission microscope, where coherent radiation is used to form a diffraction limited scanning spot. An incoherent image is obtained, because the intensity transmitted in different object point is recorded one after another and cannot interfere.

The focal length F of the off-axis zone plate is selected to deliver a source image of about 20–30 μm diameter. This corresponds to the size of the object field, which can be imaged aberration free in an X-ray microscope. At an undulator source F typically is some meters. At a bending magnet with a larger source a higher demagnification is necessary and therefore focal lengths of 0.5–1 m are useful.

At shorter X-ray wavelengths a crystal monochromator and a focusing optic - a mirror or a zone plate - can be used to generate a focused monochromatic source image [17]. The image is reflected into the object region with two rotating plane mirrors, which synthesise the dynamical aperture.

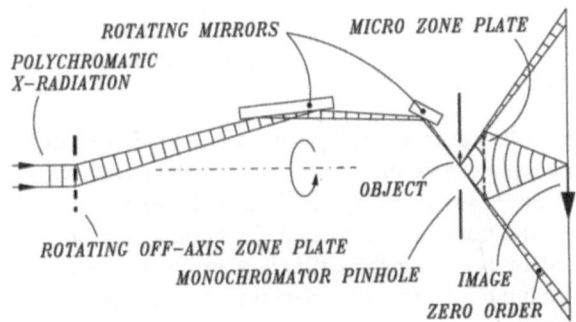

**Fig. 8.** Dynamical aperture syn-thesis with an off-axis zone plate and two plane mirrors. All three optical elements rotate synchronously. The direction of dispersion is rotating, whereas the source image is fixed. In this two respects the set-up is equivalent of the zone plate monochromator shown in Fig. 1.

With the layout of Fig. 8 dynamical aperture synthesis of a condenser-monochromator for bright field and dark field imaging can be achieved.

The orientation of rays having passed a system with two reflection is conserved. Therefore the arrangement of Fig. 8 delivers a fixed source image, only the direction of dispersion rotates. In this two respects the arrangement of Fig. 8 is the equivalent of the zone plate monochromator shown in Fig. 1 (whereas the arrangement presented in Fig. 7 will produce a *rotating* source image). The arrangement of Fig. 8 is much less sensitive to tilts of the rotating mirror holder, which can be produced by backlash in the ball bearings, because these tilts influence both mirrors and the corresponding changes in the reflections nearly compensate.

However, any change of the wavelength requires a new adjustment of the rotating mirrors and the rotating zone plate: the first mirror has to be corrected in tilt and position along the optical axis to collect the photons from the off-axis zone plate, which also has to be shifted along the optical axis. With the second mirror a new aperture angle can be adjusted. As all three optical elements rotate, a sophisticated mechanical design is required to perform such adjustments.

In Fig. 9 a more versatile arrangement is shown, which can easily be used for a wide wavelength range like a zone plate linear monochromator and offers several advantages. The off-axis zone plate is not rotating; therefore the direction of dispersion will be fixed. In opposition to all condenser-monochromator arrangements described above the spectrum will advantageously be located in a stripe, whose width at a focused wavelength is determined by the dimensions of the source image. In general  the source area of an undulator is elliptical. If the direction of dispersion is

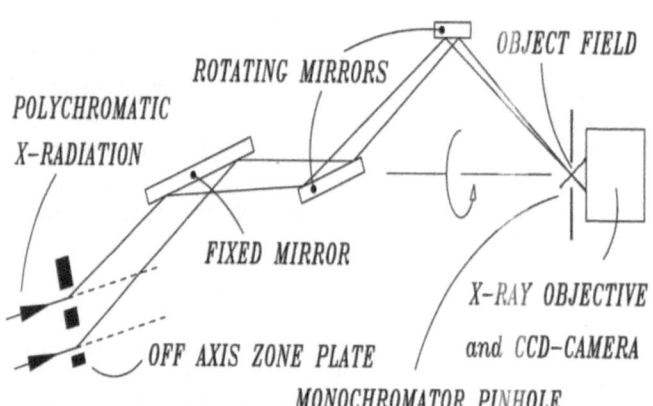

ROTATING MIRRORS

OBJECT FIELD

POLYCHROMATIC
X-RADIATION

FIXED MIRROR

X-RAY OBJECTIVE
and CCD-CAMERA

OFF AXIS ZONE PLATE

MONOCHROMATOR PINHOLE

**Fig. 9.** Dynamical aperture synthesis of a high numerical aperture condenser - linear monochromator. It includes a fixed off-axis zone plate and an adjustable plane mirror followed by a two rotating small plane mirrors. Source image and direction of dispersion are fixed.

orientated parallel to the small axis of this ellipse, the best monochromaticity can be achieved with a given off-axis zone plate and the stripe has got it largest width, resulting in a spectrally pure and homogenous intensity distribution in the focal plane along the direction of dispersion. In the transverse direction the homogeneity of the object illumination can be improved by scanning the off-axis plate in this direction with a piezoelectric transducer. Behind the off-axis zone plate a first mirror is located, which can be tilted. It directs the spectrally dispersed radiation onto a pair of rotating mirrors. As *two* mirrors rotate, a *fixed* source image will be obtained.

The length of the first mirror can be reduced, if the direction of the incoming beam and the axis of rotation are inclined to each other, see Fig. 9. When the working wavelength has to be changed, the off-axis zone plate has to be refocused and the inclination angle of the first mirror has to be matched. However, for spectroscopic applications with the TXM such as XANES the wavelength in the object area has to be changed only by a small fraction. This can be done computer controlled by *only* changing the reflection angle of the first mirror slightly. The two rotating mirrors have not to be adjusted, if the numerical aperture is kept constant.

It has to be noted that the dimensions of the plane mirrors needed in the water window wavelength range are about 10 cm or less. Such plane mirrors can be manufactured with very small roughness and slope error at the same time, as the polishing process of plane substrates will automatically increase the surface flatness.

As already mentioned above and sketched in Fig. 1 and Fig. 2, the size of the usable image field in the TXM is directly linked to the numerical aperture of the illuminating condenser. The concept of dynamical aperture synthesis allows to match the apertures which will increase the usable image field significantly. Thus the size of the usable image field will be given by the spherical aberrations of the zone plate objective to be used; general view images of increased size will be possible.

## 7 Operating Characteristics

The BESSY II undulator U41 at low beta section has got a source area of $x \cdot y = 193 \cdot 71$ $\mu m^2$ FWHM and is emitting in an solid angle of $x' \cdot y' = 183 \cdot 80$ $\mu rad^2$ FWHM [18] at 10% coupling (this coupling is an upper limit; at lower coupling a better performance condenser-monochromator can be achieved.) The spectral brilliance is $SB = 3 \cdot 10^{18}$ photons/(sec, mm², mrad², 0.1% BW, 100 mA). An off-axis zone plate segment of about 3 mm diameter located at about 37 m distance will deliver an image spot of $6 \cdot 18$ $\mu m^2$. It works as a focusing linear monochromator with spectral resolution $r = 30mm / 18 \mu m \approx 1700$, if an object field of 18 $\mu m$ diameter is regarded. Nevertheless, the spectral resolution within this region is roughly 2-3 times higher, depending on the surface slope error of the reflecting mirrors. This high monochromaticity may be especially interesting for microscopy of magnetic layers [19] if the condenser and the microscope are located at a helical undulator source. Multiplying SB with $x \cdot x' \cdot y \cdot y'$, an assumed overall mirror reflectivity of 50%, an off axis zone plate segment of about 3 mm in diameter 10% diffraction efficiency, which is realistic number for such "coarse" zone structures, we get at least $2 \cdot 10^9$ photons/(sec, $\mu m^2$, 0.1% BW, 100 mA). This is roughly 20 x higher in flux and 10 x better in monochromaticity than the current conditions at BESSY and a matched NA will be achieved. Furthermore the high monochromaticity will allow the use of micro zone plate objectives with high zone numbers which have an increased focal length $f = N \cdot dr_n^2 \cdot 16/\lambda$. This is especially important for applications where enough space is needed to rotate the object in front of the micro zone plate objective, e.g. to record pairs of stereo images or series of images for tomography.

On the other hand, for microscopical investigations which require a reduced monochromaticity and a higher photon density in the object area, the coarser, inner zones of the given off-axis zone plate can be used, as this region shows a lower dispersion.

## 8 Conclusion

Significantly improved X-ray microscopes for bright field and dark field imaging can be realised at third generation undulator sources, if the concept of dynamical aperture synthesis is applied to build the necessary X-ray condensers. At these undulators the condensers can realize aperture matching, significantly improve the monochromaticity, efficiency, size of the usable object field and homogeneity of the illumination.

## Acknowledgement

This work has been funded by the German Federal Minister of Education and Research (BMBF) under contract number 13 N 64 91.

## References

1  J. Kirz, C. Jacobsen and M. Howells *Quarterly Reviews of Biophysics* Vol. 28, No. 1 (1995), 33 – 130.

2  W. Meyer-Ilse, H. Medecki, J.T. Brown, J. Heck, E. Anderson, C. Magowan, A. Stead, T. Ford, R. Balhorn, D. Arndt-Jovin, T. Jovin, C. Petersen and D. Attwood "X-Ray Microscopy in Berkeley", this volume.

3  W. Meyer-Ilse, L. Jochum, E. Anderson, D. Attwood, C. Magowan, R. Balhorn, M. Moronne, D. Rudolph and G. Schmahl, *Synchr. Rad. News*, VOL.8, Nr.3, (1995), .29-33.

4  T.P.M. Beelen, W.D. Shi, G.R. Morrison, H.F. van Garderen, M.T. Browne, R.A. van Santen and E. Pantos *J. Coll. Interf. Sci.* 185 (1997) 217-227

5  J. Abraham, R. Medenwaldt, E. Uggerhoj, P.Guttmann, T.Hjort, J.Jensenius, T. Vorup-Jensen, F.Vollrath, E.Søgaard and T. Møller, "X-Ray Microscopy in Aarhus", this volume.

6  N. Watanabe, A. Hirai, K. Takemoto, Y. Shimanuki, M. Taniguchi, E. Anderson, D. Attwood, D. Kern, S. Shimizu, H. Nagata, K. Kawasaki, S. Aoki, Y. Nakayama and H. Kihara  "Imaging soft X-ray microscopy with zone plates, in parallel use of optical microscope for wet biospecimens in air at UVSOR", this volume.

7  G. Schmahl, D. Rudolph, B. Niemann, P. Guttmann, J. Thieme, G.  Schneider, C. David, M. Diehl and T. Wilhein *Optik*, No.3 (1993), 95-102.

8  B. Niemann,G. Schneider, P. Guttmann, D. Rudolph and G. Schmahl "The new Göttingen X-ray microscope with object holder in air for wet specimens", in: *X-Ray Microscopy IV*, (1993) 66-75, Aristov, V.V. and Erko, A.I. (eds.), Chernogolovka.

9  M. Hettwer and D. Rudolph "Fabrication of a condenser zone plate for X-ray microscopes", this volume.

10 B. Niemann, D. Rudolph and G. Schmahl, G. *Opt. Com.*, Vol.12, No. 2, (1974), 160 –163.

11 B. Niemann, patent applications DE 196 330 47.5, DE 196 33 047.5 (1996).

12 V. A. Danilov, I.N. Sisakjan and V.A. Soifer "Focusators - new elements of diffractive optics", in: *X-Ray Microscopy IV*, (1993) 518-522, Aristov, V.V. and Erko, A.I. (eds.), Chernogolovka.

13  S. Oestreich "Untersuchungen über einen Kondensor mit hoher Apertur für ein Röntgenmikroskop an einer Synchrotronstrahlungsquelle", Diploma thesis, Göttingen (1995).

14  G. Schneider and J. Maser "Zone plates as imaging optics in high diffraction orders described by coupled wave theory", this volume.

15  B. Niemann, patent application DE 195 42 679 A1, (1995).

16  D.J. Cronin and A.E. Smith *Opt. Eng.* Vol. 12, no. 2, (1973), 50 – 55.

17  S. Oestreich and B. Niemann "Design of a Condenser for an X-Ray Microscope on a low-ß section Undulator Source at the ESRF", this volume.

18  P.Guttmann, G.Schmahl, B.Niemann, D.Rudolph, G.Schneider and L.Bahrdt "The X-ray microscopy project at BESSY II", this volume.

19  P. Fischer, G. Schütz , G. Schmahl, P. Guttmann and D. Raasch *Z. Physik B*, 101 (1996), 313 – 316.

Part V

**X-Ray Sources**

# Debris-Free Liquid-Target Laser-Plasma X-Ray Sources for Microscopy and Lithography

H. M. Hertz, L. Rymell, M. Berglund, L. Malmqvist

Dept. of Physics, Lund Inst. of Technol., P.O. Box 118, S-221 00 Lund, Sweden
E-mail: Hans.Hertz@fysik.lth.se

**Abstract.** We review the development of compact laser-plasma soft X-ray sources based on microscopic liquid drops or jets as target. It is shown that such sources provides debris-free, high-brightness, narrow-bandwidth operation at water-window wavelengths, making them suitable as compact sources for soft X-ray microscopy. Application of the method for proximity and projection lithography is also reviewed.

## 1 Introduction

High-brightness soft X-ray sources have applications in many fields, e.g., microscopy, lithography or surface science. Large facilities such as synchrotron sources provide high average power. However, many applications would benefit from table-top sources having high peak power and reasonable repetition rate. This is particularly true for X-ray microscopy, where the development of a compact instrument would greatly increase the accessibility to this technology and therefore can be foreseen to have a significant impact on the development of X-ray microscopy applications. Such compact sources include laser plasmas and pinch plasmas [1].

Laser plasmas are attractive table-top soft X-ray sources due to their small size, high brightness, high spatial stability and potentially high repetition rate. This source has been developed for microscopy [1, 2, 3] and lithography [4]. With conventional metal targets, conversion efficiencies of several tens of per cent may be reached with laser intensities of ~$10^{14}$ W/cm$^2$ [5]. However, the conventional LPP solid target produces debris which may destroy or coat sensitive X-ray components, such as masks, multilayer optics or zone plates, positioned close to the plasma. Unfortunately, increasing the distance or introducing filters in order to protect the components result in a significantly reduced X-ray flux. In this paper we review the use of microscopic liquid droplets or jets as target for table-top laser-plasma X-ray generation. This target reduces debris production several orders of magnitude compared to conventional targets, thus increasing the effective photon flux a few orders of magnitude since smaller source-component distances may be employed. Furthermore, it provides narrow bandwidth radiation making it suitable for zone-plate or multilayer optics imaging, allows nearly 4π steradian geometric access, provides fresh target material for full-day operation without interrupts and allows high-repetition-rate lasers to be used.

**X-Ray Microscopy and Spectromicroscopy**
Eds.: J. Thieme, G. Schmahl, D. Rudolph, E. Umbach
© Springer-Verlag Berlin Heidelberg 1998

## 2 Hydrodynamics of Continuous Liquid Jets

For the work described in this paper it is essential to be able to generate microscopic liquid jets or droplets with high spatial stability. The continuous liquid jet provide such a drop-production method. The hydrodynamics of such jets in vacuum is discussed in Ref. 6 and summarized below.

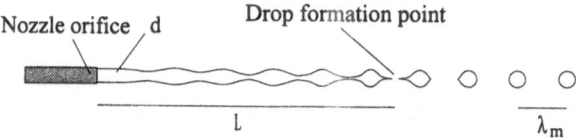

**Fig. 1.** Drop-formation process in continuous liquid jets.

When a liquid is forced through a nozzle, a liquid jet is formed (cf. Fig. 1). The jet eventually spontaneously breaks up in a train of droplets. By applying a piezoelectric vibration approximately at the mechanical resonance frequency of the nozzle, spatially stable drop generation can be achieved. The break-up distance, L, to the drop-formation point is

$$L = 12 \cdot v \left[ \sqrt{\frac{\rho \cdot d^3}{\sigma}} + \frac{3\eta d}{\sigma} \right], \tag{1}$$

where $v$ is the jet velocity, $d$ is the jet diameter, $\rho$ is the density, $\sigma$ is the surface tension and $\eta$ is the viscosity of the liquid.

With simple liquids like ethanol or water in our nozzles, Eq. (1) results in break-up distances of a few mm. However, as will be evident below, in many cases liquids with more complicated hydrodynamic properties must be used in order to achieve the necessary elemental composition for a desired spectral emission. Thus, stable droplet formation in liquids, including solutions, having different hydrodynamic properties must be achieved. In particular, the surface tension must be sufficiently large to allow spontaneous break-up of the liquid jet into a train of droplets according to Eq. (1). Break-up distances longer than a few centimeters are not acceptable since the spatial stability of the jet and the droplets further away from the nozzle exit tends to decrease, resulting in unstable plasma formation. For such liquids, the liquid-jet method discussed in Sect. 6.2. is useful.

In addition to the surface tension, the viscosity of the liquid must be sufficiently large to form a stable jet. For our liquids, this corresponds to a maximum Reynolds number $R_e$ of ~1000. Here

$$R_e = \frac{\rho v_0 d}{\eta}, \tag{2}$$

where $v_0$ is the jet exit velocity.

## 3 Droplet-Target Laser-Plasma X-Ray Source

The principal experimental arrangement for the droplet-target laser-plasma soft X-ray source is shown in Fig. 2. It is described in detail in several publications [7]. For the demonstration of basic source characteristics, ethanol is used as target liquid. 10–15 μm droplets are produced inside an ~$10^{-4}$ mbar pressure vacuum tank by an ~1 MHz vibrating capillary glass nozzle. In the first arrangement, the beam from a frequency-doubled, modelocked, 70 mJ/pulse, 100 ps, 10 Hz Nd:YAG laser was focused on the droplets with a FWHM focal spot diameter of approximately 12 μm. The high spatial stability of the continuous-liquid-jet drop-generation method used here allows each laser pulse to hit a single droplet with high (a few μm) accuracy. This stability is essential for the efficient use of the target material and in order to reduce shot-to-shot fluctuations in the soft X-ray emission. Furthermore, it allows the generation of small droplets (10-15 μm), which is important in order to reduce debris emission [8].

The emission spectrum is characterized with a 1 m grazing incidence monochromator equipped with a CsI coated electron-multiplier detector. The water-window emission is dominated by C V, C VI ($\lambda \approx 2.8-4.0$ nm) O VII, and O VIII ($\lambda \approx 1.5-2.2$ nm) line emission, as shown in Fig. 3. Depending on the laser parameters, the source diameter is typically 10-20 μm, as determined by a pinhole camera to. Flux measurements are discussed in the next section.

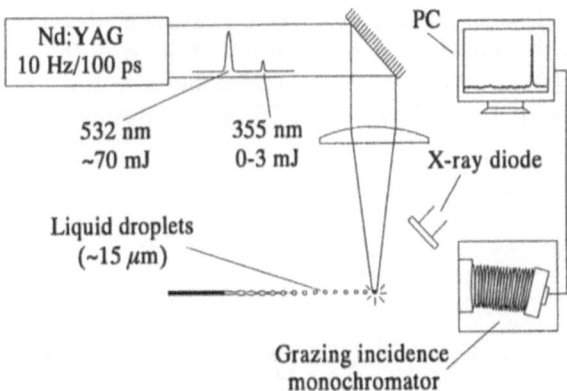

**Fig. 2.** Experimental arrangement for droplet-target laser-plasma X-ray generation.

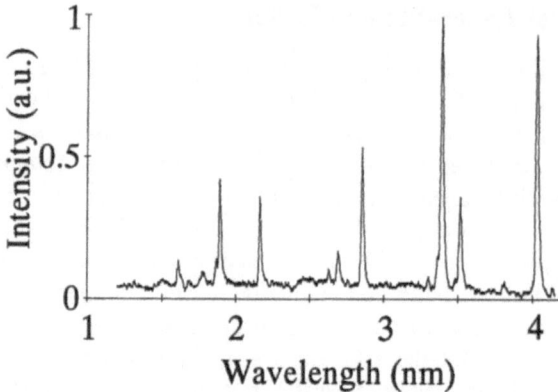

**Fig. 3.** Water-window spectrum from ethanol-droplet target.

## 4 Prepulse Enhancement of X-Ray Emission and Brightness

The emitted X-ray flux is measured with a GaAsP X-ray diode covered by suitable free-standing thin-film metal filters. We have recently improved the emitted X-ray flux as well as the brightness by the use of a small UV prepulse [9]. The UV prepulse is generated by frequency conversion of residual IR laser light and hits the droplet target a few ns before the main visible pulse. Detailed measurements of source size and water-window photon flux as a function of prepulse delay and energy were performed. Due to the prepulse, the brightness is increased approximately a factor 2 and the photon flux approximately a factor 8 compared to when no prepulse is in use. The effect on the photon flux is shown in Fig. 4. With this method the conversion efficiency of a 65 mJ visible green main pulse and 3 mJ UV prepulse is approximately 4% to the 1s-2p C VI line at $\lambda=3.37$ nm, corresponding to $>3\cdot10^{12}$ photons/ (sr·line·pulse).

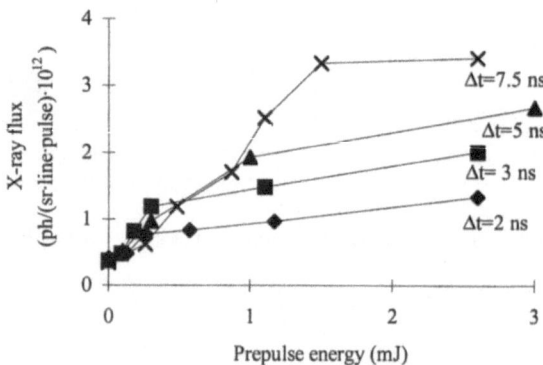

**Fig. 4.** Enhancement of X-ray flux as a function of UV prepulse energy and time delay.

# 5 X-Ray Microscopy

X-ray microscopy allows high-resolution imaging of samples in their natural wet environment with high resolution. Natural contrast for carbon-containing objects is provided in the water window ($\lambda$=2.3–4.4 nm).

## 5.1 Single-Line Source for Microscopy

Due to their lower attenuation in water, the N VII and N VI lines at $\lambda$= 2.5 and 2.9 nm are often better suited for microscopy than the carbon lines at $\lambda$=3.4–4.0 nm. Furthermore, quasi-monochromatic, narrow-bandwidth, single-line emission with low continuum background is important for high-contrast imaging with zone plates due to their chromatic aberration. Figure 5 shows the emission spectrum from such a source using ammonium hydroxide droplets as target and 600 nm Ti filters [10]. The unfiltered flux is ~$1 \cdot 10^{12}$ photons/(sr·line·pulse) and the bandwidth has experimentally been determined to $\lambda/\Delta\lambda \geq 450$ [11]. The debris is reduced more than 2 orders of magnitude compared to the ethanol target, making the source "debris-free". Also in Ref. 10, we show that the droplet target method can be extended to solid substances by dissolving them in a suitable liquid. This extends the range of accessible wavelengths and allows spectral tailoring of the emission.

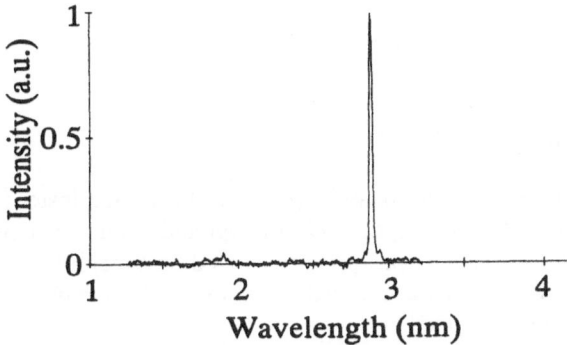

**Fig. 5.** Single-line narrow-bandwidth N VI emission from titanium-filtered ammonium-hydroxide target.

## 5.2 Development Towards a Laser-Plasma Table-Top X-Ray Microscope

In collaboration with Forschungseinrichtung Röntgenphysik, Georg-August Universität, Göttingen, we have attempted employing the single-line source for table-top microscopy. The source is then combined with the microscope originally developed for a pinch plasma source [12]. However, this microscope is based on a elliptical condenser mirror. Due to the strong coma in this mirror, the alignment of our very small source is difficult. To circumvent this problem we will take advantage of the large geometrical access of the droplet-plasma source and combine it with normal-incidence multilayer-coated spherical optics according to Fig. 6. The mirror is manufactured by Osmic [13]. Such mirrors have the advantage that they automatically suppress all but the selected spectral line. Raytracing shows that a 25 μm spot in the object plane can be obtained. Assuming 1.5% reflectivity and that filters absorb 50%, the X-ray intensity in the object plane has been calculated to $1.5 \cdot 10^6$ ph./μm$^2$· pulse·spectral line. Thus, typically a few hundred pulses will be needed for good-quality imaging. This corresponds to an exposure time of a few tens of seconds with the current 10 Hz laser system.

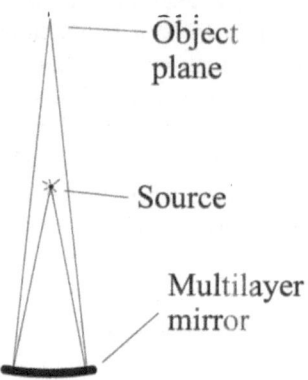

**Fig. 6.** Normal-incidence multilayer condenser mirror for microscopy applications.

# 6 X-Ray Lithography

By using soft X-ray ($\lambda \approx 1$-2 nm) or EUV ($\lambda \approx 13$ nm) radiation instead of visible or UV light for lithography, the packing density of integrated circuits can be significantly increased. Previous development has primarily relied on synchrotron radiation sources. It is of vital interest for the spread as well economy of this development that compact, granular sources can be utilized.

## 6.1 Proximity Lithography with Liquid-Target Source

We have developed a source based on F IX and F VIII ion emission from a liquid fluorocarbon target [14]. The source emits $\sim 2 \cdot 10^{12}$ photons/(sr·line·pulse) into the $\lambda \approx$ 1.2–1.7 nm wavelength window suitable for proximity lithography. Experiments using a chemically enhanced resist produces high-aspect ratio sub-100 nm structures, as shown in Fig. 7 [15]. The exposure time is currently 20 minutes. With higher-repetition-rate lasers, exposure times less than a minute are feasible.

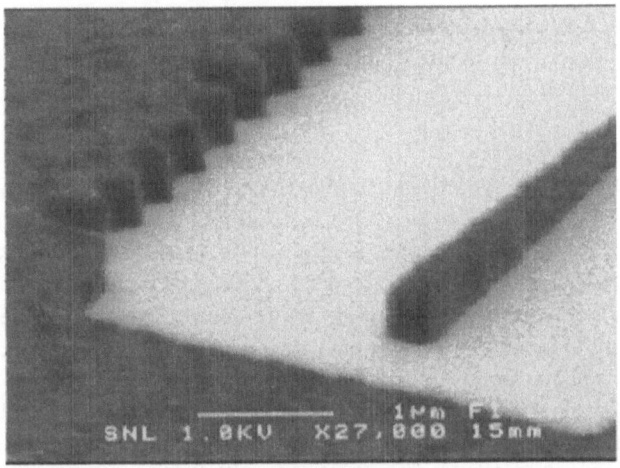

**Fig. 7.** Scanning electron micrograph of high-aspect-ratio structures fabricated by table-top X-ray lithography.

### 6.2 Liquid-Jet Method

For many liquids, stable drop-formation is difficult to achieve. This is particularly true for liquids with low surface tension, which results in drop-formation far from the nozzle orifice (cf. Eq. (1)) making the stability low. The fluorocarbon liquids discussed in Sect. 6.1 do show such problems. For these and similar liquids we have developed the "liquid-jet"-target [16]. Here the laser is focused onto the liquid jet before it breaks up into droplets (cf. Fig. 1). This has several advantages, of which increased X-ray stability and the lack of need for temporal synchronization between the laser and the droplets, are the most important. It is interesting that the debris emission is equally low and the X-ray flux is equally high as with the droplet target.

### 6.3 Source for EUV Projection Lithography

For EUV projection lithography in the $\lambda=10-15$ nm range, the emission from oxygen ions contains several suitable lines. Using an 8 ns frequency-doubled Nd:YAG laser and water/methanol droplets we obtained $\sim 4 \cdot 10^{12}$ photons/(sr·line·pulse) at the 2p-4d O VI line at $\lambda=13.0$ nm [17].

## 7 Summary of Source Characteristics

The continuous-liquid-jet target provides several attractive features for a compact X-ray source:

## 7.1 Granular Source

The X-ray source is truly table-top with a typical foot-print of less than 1×2 m$^2$. The compact size and granular nature makes it suitable for many X-ray applications in the normal-scale research and development laboratory.

## 7.2 Debris

Debris emission from conventional targets may damage and coat fragile soft-X-ray optics positioned close to the plasma. We have shown that the ionic and atomic debris emission from ethanol-droplet target is reduced by more than 3 orders of magnitude compared to low-debris tape targets of plastic with approximately the same elemental contents as the ethanol [8]. The quantitative measurements resulted in a debris emission of 5 pg/sr·pulse. Thin (100 nm) freestanding Al films positioned close (30 mm) from the plasma show no new pinholes after several hours of 10 Hz plasma operation, indicating that larger fragments are not emitted from the droplet plasma. The low debris emission is probably due to that the full droplet is ionized and that there is no target material in the low-intensity radial wings of the laser beam. Using target liquids with solely gaseous compounds, the debris emission is not detectable within the accuracy of our measurement system. It is reduced by more than 2 orders of magnitude compared to the ethanol droplet target to <0.01 pg/sr·pulse, making the source debris-free for practical purposes [10]. Table 1 summarizes our quantitative debris measurements.

**Table 1.** Summary of quantitative debris deposition measurements.

| Target type | Debris emission (pg/sr.·pulse) | Ref |
|---|---|---|
| Thin-film plastic tape | 5000 | 7 |
| Ethanol | 5 | 7, 8 |
| Urea solution | 10 | 10 |
| Ammonium hydroxide | <0.01 | 10 |
| Fluorocarbon | 70 | 12 |

## 7.3 High Repetition Rate

The continuous flow of target material allows high-repetition-rate lasers, or several lasers in parallel, to be used. Conventional target systems do not allow sufficient advance speeds to provide fresh target area at proper rates. Thus, the liquid-jet or droplet target in combination with the high-repetition-rate (up to 1000 Hz) lasers currently being developed opens up the possibilities for a high-average-power X-ray source.

## 7.4 Operating Time

With conventional targets, frequent disruptive interrupts are necessary in order to change the target when the target material comes to an end. This is especially true for high-repetition-rate operation (100-1000 Hz) where even very long tape targets (km) allows only minutes to hours of operation [18]. With the microscopic liquid jet, the operating time is only limited by the volume of the target liquid's container. With typical flow-rates of 10 ml/hour, it is clear the operating time between interrupts can be made very long. Currently we run for full days but indefinite operating time should in principle be possible.

## 7.5 Tailored Spectral Emission

The continuous liquid jet method allows stable droplet generation for a wide range of liquids or solutions. Thus, the target liquid may be chosen to spectrally tailor the emitted laser-plasma X-ray radiation. We have published results on ethanol [7], ammonium hydroxide [10], urea/water [10], fluorocarbon [14] and water/methanol [6,17] and have unpublished work on several other liquids. The liquid-jet method will further extend the range of suitable target liquids [16].

## 7.6 Flux and Conversion Efficiency

The emitted photon flux has been discussed above. With the prepulse arrangement we have obtained 10% conversion efficiency into the water-window [9]. The fluorocarbon source currently has 4–5% conversion efficiency to the $\lambda \approx 1.2$–$1.7$ nm range [14]. In neither case we have tried to maximize the conversion efficiency so higher numbers can probably be obtained.

## 7.7 Plasma Size and Brightness

The size of the X-ray emitting plasma may be tuned from ~10 μm and up by the use of different prepulse parameters or laser pulse widths [9]. This is important for achieving suitable penumbral blur in proximity lithography. In microscopy, the plasma size may be adjusted for maximum useful source brightness, for, e.g., the given condenser optics.

## 7.8 Spatial Stability

We have measured the spatial stability of the operating plasma source to ± a few μm with a pinhole camera. This stability is important in many X-ray applications where even a slight movement of the source results in a significant loss of photon flux, e.g., when optics with high aberrations are used.

## 7.9 Uniformity and Geometric Access

The X-ray emission has been found to be close to uniform. The uniformity in combination with the nearly $4\pi$ sr geometric access allows, e.g., multiple exposure stations for lithography to be used simultaneously. Also, plasma diagnostic equipment can be operated in parallel with the main application.

## 7.10 Target Material Cost

Even for expensive liquids such as the fluorocarbon, the target material cost is estimated to $10^{-7}$/shot [14]. For common liquids such as ethanol, water, ammonium hydroxide etc., this number reduces several orders of magnitude. Since there is no need for preparation of the target (e.g., mechanical forming or polishing), the total target cost is very low.

## Acknowledgements

The authors gratefully acknowledge Terje Rye, Siemens-Elema, for providing us with nozzles, A. Bogdanov and L. Montelius for fruitful collaboration on the lithography project, T. Wilhein for the line-width measurements, and J. Thieme, B. Niemann and T. Wilhein for cooperation on the microscopy project. This work was financed by the Swedish Research Council for Engineering Sciences, Swedish Natural Science Research Council, the Swedish Board for Industrial and Technical Development and the Wallenberg Foundation.

## References

1  See, e.g., W. Neff, D. Rothweiler, K. Eidmann, R. Lebert, F. Richter, and G. Winhart, in *Applications of Laser Plasma Radiation*, Ed. M. C. Richardsson, Society of Photo-Optical Instrumentation Engineers, SPIE Vol. 2015 (Bellingham, Washington, 1994) p. 32.

2  See, e.g., several papers in *X-Ray Microscopy IV*, eds. V. V. Aristov and A. I. Erko, p. 381 (Bogorodskii Pechatnik Publishers, Chernogolovka, Moscow region, 1994).

3  A. G. Michette, Rep. Prog. Phys. **51**, 1525 (1988).

4  See, e.g., F. Bijkerk, E. Louis, M. J. van der Wiel, I. C. E. Turcu, G. Tallents, and D. Batini, J. X-ray Science Technol. **3**, 133 (1992).

5  R. Kodama, K. Okada, N. Ikeda, M. Mineo, K. A. Tanaka, T. Mochizuki, and C. Yamanaka, J. Appl. Phys. **59**, 3050 (1986).

6  H. M. Hertz, L. Rymell, M. Berglund and L. Malmqvist, in *Applications of Laser Plasma Radiation II*, M.C. Richardsson, Ed., SPIE Vol. 2523, pp. 88-93 (Soc. Photo-Optical Instrum. Engineers, Bellingham, Washington, 1995).

7  L. Rymell and H.M. Hertz, Opt. Commun. **103**, 105 (1993).

8  L. Rymell and H. M. Hertz, Rev. Sci. Instrum. **66**, 4916 (1995).

9  M. Berglund, L. Rymell, and H.M. Hertz, Appl. Phys. Lett. **69**, 1683 (1996).

10  L. Rymell, M. Berglund and H.M. Hertz, Appl. Phys. Lett. **66**, 2625 (1995).

11  T. Wilhein, B. Niemann, L. Rymell, M. Berglund, and H.M. Hertz, work in progress.

12  D. Rudolph, G. Schmahl, B. Niemann, M. Diehl, J. Thieme, T. Wilhein, C. David, and K. Michelmann, in *X-ray Microscopy IV*, eds. V. V. Aristov and A. I. Erko, p. 381 (Bogorodskii Pechatnik Publishers, Chernogolovka, Moscow region, 1994).

13  Osmic Inc., Troy, Michigan.

14  L. Malmqvist, L. Rymell, and H. M. Hertz, Appl. Phys. Lett. **68**, 2627 (1996).

15  L. Malmqvist, A. Bogdanov, L. Montelius, and H. M. Hertz, "Nanometer proximity X-ray lithography with liquid target laser-plasma source", Submitted to J. Vac. Sci. Technol.

16  L. Malmqvist, L. Rymell, M. Berglund and H.M. Hertz, "Liquid-jet target for laser-plasma soft X-ray generation", accepted by Rev. Sci. Instrum.

17  L. Malmqvist, L. Rymell, and H. M. Hertz, in *Extreme Ultraviolet Lithography*, OSA Trends in Optics and Photonics Vol. 4, G. D. Kubiak and D. R. Kania, Eds. (Optical Society of America, Washington, DC 1996) pp. 72-75.

18  S. J. Haney, K. W. Berger, G. D. Kubiak, P. D. Rocket, and J. Hunter, Appl. Opt. **32**, 6934 (1993).

# X-Ray Spectromicroscopy
# of 120-fs Laser-Produced Plasma

A. Ya. Faenov[1], T. A. Pikuz[1],
A. A. Firsov[2], L. A. Panchenko[2], Yu. I. Koval[2],
M. Fraenkel[3], A. Zigler[3]

[1] Multicharged Ions Spectra Data Center of VNIIFRTI, Russian Committee of Standards,
Moscow region, 141570, Russia
[2] Institute of Microelectronics Technology and High Purity Materials RAS, Chernogolovka,
Moscow region, Russia
[3] Racah Institute of Physics, The Hebrew University of Jerusalem, Jerusalem 91904, Israel

## 1 Introduction

X-ray spectra of various plasma microsources can be recorded simultaneously with spectral and spatial resolution using spectrograph with flat, convex or concave crystals and a slit placed parallel to the dispersion direction (see, for example, [1, 2]). This approach is simple for realization, but the obtained spatial resolution is limited by the slit size. The slit size is usually not smaller than 20–30 μm, due to the dramatic reduction of the system throughput. High spatial resolution (below 10 μm) combined with relatively high throughput was obtained using a shadow technique [21]. A substantial improvement of spatial resolution (4–10 μm) has been achieved through the use of the crystals bent to a complex surfaces (such as, for example, sphere or tori or elliptically [3–10]). Further improvement of the spatial resolution toward achievement of the micron to the submicron level has been pursued using Fresnel structures such as transmission zone plates. However, the spectral resolution obtained by this approach is very limited.

High spectral resolution of $\lambda/\Delta\lambda \sim 1000$–$10000$ and spatial resolution of micron or even submicron can be obtained simultaneously with the help of so-called Bragg-Fresnel X-ray elements [11–18]. Such elements are composed of a zone plate structure etched into a multilayer mirror or crystal surface. The advantages of the Bragg-Fresnel linear zone plate structure deposited parallel to the crystal dispersion, compared to the traditionally slit placed in the same direction, are connected with both much better spatial resolution and much higher luminosity due to the focusing properties of the Fresnel lens. It is necessary to point out that previously [11, 14–18] various Bragg–Fresnel lenses have been used for focusing synchrotron radiation for the submicron probing. Only recently such lenses have been used for obtaining a high-temperature plasma source spectrum with high spatial resolution. In works [12, 13], linear Bragg–Fresnel lenses formed on the multilayer W/Si mirror and on mica crystal have been used to obtain X-ray spectrally resolved images of z and x-pinches. These devices are bright sources of X-ray radiation and high spatial resolution can be obtained in the single shot. A much more complicated experimental task is to obtain such images of plasmas produced by pico- or femtosecond laser radiation due to their very small flux in the X-ray spectral range.

**X-Ray Microscopy and Spectromicroscopy**
Eds.: J. Thieme, G. Schmahl, D. Rudolph, E. Umbach
© Springer-Verlag Berlin Heidelberg 1998

In the present paper we are demonstrated achievement of high spatial resolution for X-ray spectrum of plasma produced by 20mj, 120 fs laser using a Bragg-Fresnel linear zone plate structure on the mica crystal surface. Very good spectrally (up to $\lambda/\Delta\lambda$ ~10000) and spatially (up to 10 μm) resolved images of such femtosecond laser-produced plasma were also obtained by using FSSR-1D spectrograph [2÷5] with spherically bent mica crystals. The generation of intense, collimated monochromatic X-ray beams ($\lambda$ ~9.5 Å) was presented too.

## 2 Bragg–Fresnel Lens Fabrication

A Linear Bragg–Fresnel lens was designed and fabricated in the Institute of Microelectronics Technology of the  Russian Academy of Sciences. The parameters of the Fresnel zone lens are dictated by the specific requirements of the particular experiment. In principal, the geometry of Fresnel zone plate is determined by

$$r_n = nf\lambda + (n^2\lambda^2)/4, \tag{2.1}$$

where: $n$  is the zone number; $r_n$ - the Fresnel $n$ zone radius; $\lambda$  is the wavelength of radiation; $f$ is the focal distance.

Our experiments were carried out in the spectral range around a central wavelength of 9 Å. As it was indicated by the preliminary experiments this was the shortest spectral range where intensity of X-ray radiation was still enough reasonable. In order to increase the luminosity of experimental set up, the length of focus lens must be chosen as small as possible and in our case was chosen to be $f$ = 5 cm. As it was mentioned above, an improvement  of spatial resolution is strongly connected with the possibility of reduction the size of the last Fresnel zone. In our case the high quality equipment allowed us to have minimum zone width $\Delta z_n$ = 300  nm. Thus, using  equation (2.1) and our experiment parameters: $\lambda$ = 9.16 Å, $f$ = 5 cm, $n$ = 100, $\Delta r_n$ = 300 nm, yields a total width of zone plate of $2r_n$  = 122.6 μm.

For the purposes of plasma diagnostics it is very important to obtain spatially resolved spectra in wide spectral range. The coverage of wide spectral range can be obtained, mainly  by increasing the lengths of Fresnel lens. However, production of large parallel zone plate that contains a last zone of size of 300 nm is a very difficult technological problem. The lens of length 10000 μm was built for this experiment.

The lens used here was made by means of electron beam lithography and ion beam milling processes. It was created on the surface of 100 μm thick mica crystal, over coated by tungsten layer with thickness 1500 Å magnetron sputtering ALCATEL SCM-651 coating system was used for the coating preparation.

The process of writing the primary pattern of the linear zone  plate has been carried out  with ZRM-12 lithography  machine.  The process of zone formation along the long axis has some  technological limitations. The most important of them is the  stability of the lithographic machine operation , since  the  process takes several hours and the precision of matching and  joining of the drawing fields during the movement of the  lithographic table must remain constant and be about 10% of the  minimum zone size. The pattern of linear  zone  plate  has  been created by special program ZON in the format which is  suitable for next step - proximity correction. Tungsten milling  was performed by Ar⁺ ions with energy of 500 eV thought 0.5 mm  resist mask of lens. The lens profile in the layer of the tungsten was created out by milling through the mask resist using  the Kaufmann ion source with Ar as a processing gas.

## 3 Experimental Setup and Results

The experiments were carried out at Racah Institute of Physics of the Hebrew University. High-temperature plasma was generated by heating flat solid target of Magnesium, Lanthanum or Samarium by the 120 fsec laser pulses with energy of about 20 mJ per pulse. Laser  radiation was focused into a spot of about 20 μm. The laser is  based on a Ti: Sapphire oscillator generating 80 fsec pulses at a wavelength of 800 nm. The pulses are amplified in the chirped  pulse amplification (CPA) technique. In this technique, the  pulses, which consist of spectral band of 10 nm around 800 nm are  temporarily stretched to a pulse width of ~1 ns. Then they are amplified in a regenerative amplifier followed by a double-pass amplifier. Finally the pulses are compressed again to a pulse width  of ~120 fs. The repetition rate was 10 pulses per second. It was necessary to shoot up to 1000 pulses in order to obtain a good quality image.

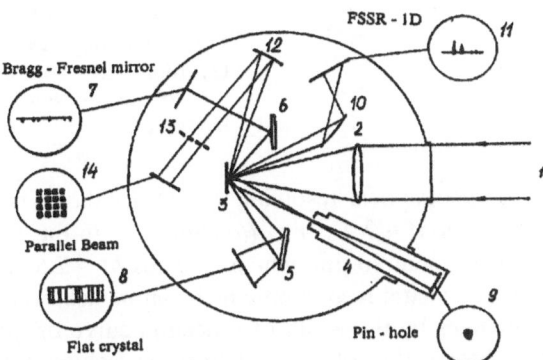

**Fig. 1.** Scheme of experiment (1) Laser radiation; (2) Focusing system; (3) La target; (4) pin-hole; (5) flat spectrograph; (6) Bragg-Fresnel lens; (7–9, 11, 14) films; (10) FSSR-1D; (12) spherical mica crystal with R = 186 mm; (13) test grid.

The experimental set up that includes, X-ray pin-hole  (4), flat spectrograph with RbAP crystal (5), Bragg-Fresnel lens (6), FSSR-1D spectrograph (10), spherically bent mica crystal for obtaining parallel, monocromatic X-ray beam (12) are presented in Fig. 1. The data registered by pin-hole camera allowed us to obtain images of laser-produced plasma without spectral resolution. The flat crystal gave us a general information  about the spectral distribution of the plasma radiation in wide spectral range: λ = 8,5–13 Å without spatial resolution. Bragg-Fresnel lens and FSSR-1D spectrograph allowed us to obtain spectra with spatial resolution simultaneously. Mica spherical crystal gave us opportunity to generate parallel, monochromatic X-ray beam.

## 3.1 Spectrally Resolved Image of Plasma Obtained with Bragg–Fresnel Lens

Figure 2 describes the set up for the Bragg-Fresnel lens experiment. Magnification of M = 1.5* was chosen. Therefore the distance between laser-plasma and Bragg-Fresnel structure was set to $a$ = 83.3 mm. The distance between the Bragg-Fresnel structure to the film (in our case Kodak DEF film was used) was $b$ = 125 mm. We used second order of mica crystal reflection. In this case the middle Bragg angle θ was about 67°. The obtained images of plasma in the spectral range 9.12–9.31 Å are shown on Fig. 3. Simple geometric relation $D_x=(\lambda/a+b)\cdot ctg\theta$ gives us linear dispersion $D_x \sim 0.018$ Å/mm for our experiment.

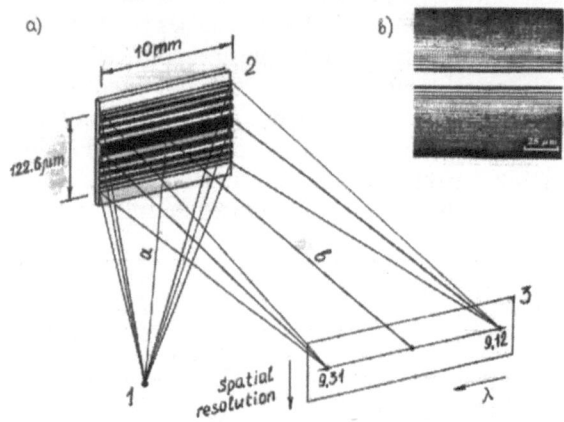

**Fig. 2.** a) Scheme of obtaining spectra with space resolution by Bragg-Fresnel lens. (1) Laser produced plasma X-ray source; (2) Bragg-Fresnel lens on mica crystal; (3) film DEF: b) Foto of part of the Bragg-Fresnel lens surface.

We can also estimate the geometrical resolving power (without including the role of crystal reflection curve) as $M = \lambda/D_x$  $r = ((a+b)/r)\cdot tg\theta$. In our case the plasma source size (see Fig. 3) was about 20 μm, which gives us $M \sim 2.5\cdot10^4$. It is necessary to underline that the best spectral resolution which can be achieved is not more than $\lambda/\Delta\lambda \sim 10000$. It is dictates by the width of reflection curve of the mica crystal [19]. Due to the plasma broadening of spectral lines we achieved experimentally $\lambda/\Delta\lambda \sim 2000$. Some other parameters were also calculated: i) focus depth of Bragg-Fresnel lens ΔF ~200 μm, spatial resolution D ~0.4 μm, efficiency ~ 40%. From Fig. 3 one can see also that the size of the image in the direction perpendicular to the direction of spectral dispersion was not more than 20 μm. The same size of plasma X-ray emission zone was measured by the pin-hole camera.

## 3.2 Spectrally Resolved Image of Plasma Obtained with FSSR-1D Spectrograph

The plasma emission spectra in the X-ray region were measured by the focusing spectrograph with spatial resolution (FSSR-1D). See Fig. 4 and ref.[2÷5] for more details about this scheme. The crystal, film, and source must be placed at special positions. A definite relation between the glancing angle θ and the distance from the source to the top of the spherical crystal $a$ must be satisfied for each wavelength

$$a = -R \sin\theta/\cos2\theta ,$$   (3.1)

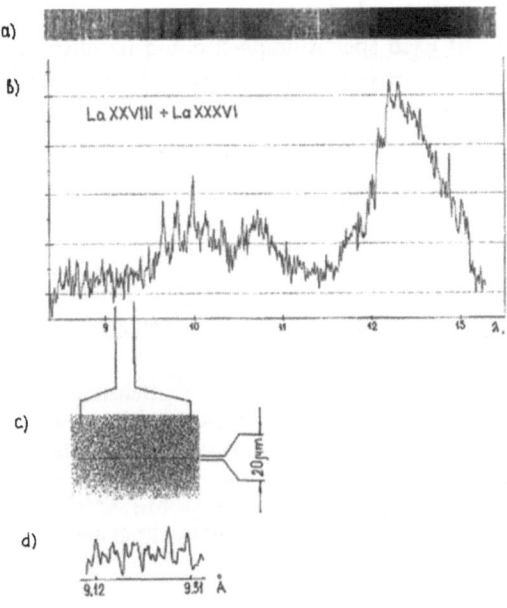

**Fig. 3.** a), b) Spectra and densito-gram of Lanthanum, obtained by flat X-ray spectrograph; c), d) image and densitogram of La plasma in the spectral range: 9.12–9.31 Å.

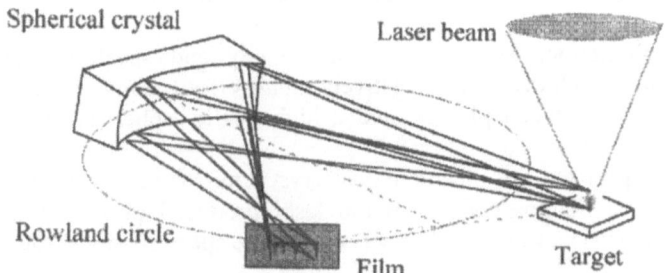

**Fig. 4.** The scheme of Focusing Spectrograph with Spatial Resolution (FSSR-1D).

where $R$ - the radius of crystal curvature. The glancing angle $\theta$ is related to the incident wavelength by Bragg's law $2d\ sin\theta = n\lambda$, where $2d$ is twice the interplanar distance for the crystal, $n$ - is the order of reflection. In the present case, a spherical mica crystal with dimensions 15 x 50 mm and radius of curvature 186 mm was used. The spectra near the resonance line of He-like Mg XI were measured in second order of the mica crystal. The high quality of the spherical mica crystal used produced a spherical resolution better than 10 μm and a spectral resolution given by $\lambda/\Delta\lambda$ up to 10000. The spectrograph was very sensitive and the plasma emission of all above mentioned targets could be recorded during 1 minutes after about 600 shots.

Examples of the plasma emission spectrum for different distances from the target in the region 9.15 ÷ 9.35 Å near He$_\alpha$ - line of the Mg XI is shown in Fig. 5. We can see that the spectra in Fig. 5 contain a big number of resolved spectral lines with significant intensities. Such lines belong to the satellite lines of Li and Be -like ions and can be measured with the accuracy ±0.0005 Å. It is necessary also to underline that due to the high spatial resolution of used spectrograph it could be clear seen how dramatically changed intensities of satellite lines even in the scale of changing distances about 10 μm.

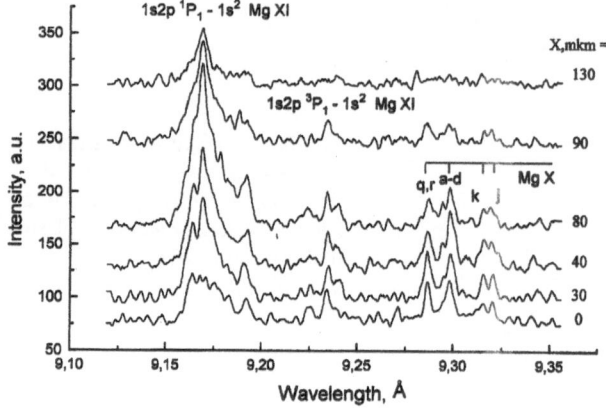

**Fig. 5.** Densitograms of typical spectra near resonance line of He-like Mg XI: q,r.-satellite lines of Li-like Mg X; x, mkm - distance from the target.

## 3.3 Generation of Intense Collimated Monochromatic X-Ray Beam

The high degree of collimation, relative monochromaticity and spectral range selection ability can now be achieved by a relatively small effort, using an X-ray source as X-pinch or laser produced plasma, and a high quality spherical crystal mirror[2÷5].

In this paper we report the use of a spherical crystal mirror made of a bent mica crystal and an intense femtosecond table-top laser produced plasma X-ray source to generate intense pulses of parallel X-ray beams in the spectral range around 9.5 Å.

The parallel beam are formed by using a spherical crystal mirror, in which efficient reflection is possible only for wavelengths that fulfill the Bragg condition. A narrow range of wavelengths, determined by the crystal properties, is reflected by the crystal due to the fact that the X-ray source is a point source and the radiation arrives to the crystal in a range of angles. The spectral range of the reflection is $\Delta\lambda=2d(\sin\theta_{max}-\sin\theta_{min})$, where $\theta_{max}$ and $\theta_{min}$ are the maximal and minimal angles of incidence. In our experiments, we set $\theta_{max}=90°$ so that $\lambda_{max}=2d$. $\Delta\lambda$ is determined by the geometry of the crystal mirror through the relation:

$$\Delta\lambda=(2d/n)[1-cos(D/4F)], \qquad (3.2)$$

where $D$ is the crystal diameter and $F$ is the focal length. Thus, the spectral spread of the reflected beam, $\Delta\lambda/\lambda$ can be easily monitored by changing the crystal parameters.

The plasma formed by 120 fsec laser has near solid density and high temperature and it emits intense short bursts of X-rays. The length of the X-ray pulse in the wavelength range under our consideration is about 1 psec. Thus, a proper selection of target material and spectral range makes it possible to receive intense soft X-ray pulse in the desired wavelength.

The laser pulse is focused on the Samarium target and the X-rays are reflected by the spherical crystal mirror and a flat spectrometer. The X-ray plasma source is placed in the mirror focus. We used a 15 mm by 50 mm mica crystal mirror (the working size was slightly smaller, 12 mm by 46 mm, due to geometrical constrains) with interplanar spacing of $2d$=19.94 Å and radius of curvature of $R_c$=186 mm (focal length $F$=93 mm). In this configuration, the spectral spread of the reflected beam is $\Delta\lambda/\lambda$=7.6·10$^{-3}$. The central wavelength was chosen to be 9.5 Å and the crystal was aligned for second order of reflection. A spectrometer with a flat RbAP crystal was also set to give us a general information about the spectral distribution of the plasma radiation in a wide spectral range: $\lambda$=6,5-11.5 Å. The spectrum observed shows the intense emission around 9.7 Å and from it we can estimate that the total energy produced at the source in the relevant spectral range was 9 mJ. Taking into account the solid angle occupied by the crystal mirror and the mirror's peak reflectivity (which is about 0.15 for second order of reflection in the relevant angles of incidence range), the total energy that was reflected from the crystal mirror was 7 μJ. The energy per shot was 3 nJ, and taking 1psec for the X-ray pulse duration, leads to power of 3 kW per shot. The parallel X-ray beam that was reflected from the crystal mirror passed through a metal grid with wires of 130 μm thick and spacing of 570 μm, placed in various distances from the film. The quality of the generated parallel beam can be monitored by viewing the grid image formed by the beam on the film.

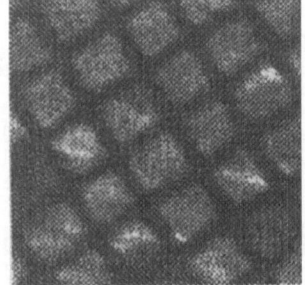 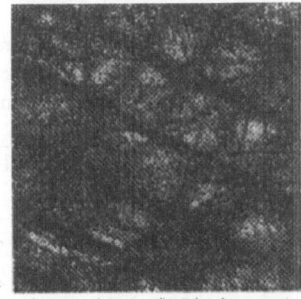

a                              b

**Fig. 6.** Image of the grid formed by the collimated X-ray beam: a) The grid was placed 5 cm from the film; b) the grid was placed 10 cm from the film and a bunch of glass wires was place 6 mm from the film.

Figure 6a shows the formed image when the grid was placed 5 cm from the film. Fig. 6b shows the image when the grid was placed 10 cm from the film and a bunch of glass wires (about 50 μm thick) was placed 6 mm from the film. The quality of the image is conserved even at large distances from the grid and even after few thousands shots. Comparing the images formed at various distances can let us estimate the divergence of the beam to be less then 1 mrad. Estimation of the energy per shot made by direct measurement of the films shown in Fig. 6 leads to similar numbers as above (about 3 nJ, 3 kW per pulse).

The principle presented here for generating intense collimated X-ray beams has many advantages over X-ray laser and over other techniques. The X-ray source can easily be tuned in a wide spectral range dictated by the required application, especially in terms of laser intensity and target material. The high repetition rate of the laser allows a continuous experiment, in contrast to large installations where the repetition rate is much lower (several shots per hour or even per day) [22]. The short X-ray pulse duration generated by the ultrashort laser produced plasma makes it possible to receive high intensity collimated X-ray beams. The quality of the beam was conserved even at large (10 cm) distances, in contrast to previous experiments where the maximum distance was 2 cm [22]. Table 1 summarizes the characteristics of the collimated beams obtained in our experiment, compared to parallel beams obtained by similar means (crystal mirror) using X-pinch and picosecond laser produced plasma, and the shortest wavelength X-ray laser presently available. The table shows that X-ray laser is best apparatus in terms of monochromaticity and energy per pulse, but the monochromatic collimated beams are better in terms of divergence, and wavelength selection ability. The pulse duration of the femtosecond laser is the smallest, creating high intensity beams even with low energy per pulse. The high pulse repetition rate (compared to other plasma sources) makes it possible to collect a lot of energy in short time.

**Table 1.** Comparison between X-ray laser based on the Ta XLVI ion, and monochromatic collimated x-ray beams from X-pinch plasma, picosecond laser produced plasma (from ref. 8) and femtosecond table-top laser produced plasma, described in this paper.

| X-ray source | $\lambda$ (Å) | mono-chroma-ticity ($\Delta\lambda/\lambda$) | beam size (mm) | energy per pulse ($\mu J$) | pulse duration (ns) | Diver-gence ($10^{-4}$ rad) | power (kW) | repeti-tion rate, pulses per hour |
|---|---|---|---|---|---|---|---|---|
| X-pinch | 9.87 - 9.94 | $4\cdot10^{-3}$ | 10x45 | 3.2 | 5-10 | 5x5 | 0.32-0.64 | 1-2 |
| TaXLVI X-ray laser | 45 | $10^{-4}$ | 0.075 x0.05 | 10 | 0.2 | 100x 200 | 50 | 1-2 |
| pico-second Nd:YAG laser | 9.22 | $3\cdot10^{-3}$ | 10x30 | 0.3 | 0.01-0.02 | 6x6 | 15-30 | 3-6 |
| femto-second laser | 9.43 - 9.57 | $7.6\cdot10^{-3}$ | 12x46 | 0.003 | 0.001 | >10x 10 | 3 | 36000 |

# 4 Conclusion

In the present paper  the possibility to obtain  high resolution spectrally resolved image of high Z plasma, produced by the short pulse of 120 fs laser, was demonstrated for  the first time . For high Z elements (such as Lanthanum), the size of the plasma zone that emits X-ray spectral lines of La XXVIII - La XXXVI [20] was found to be 20 μm. This size is equal to the laser spot size. The spatial resolution in the  direction perpendicular to the target was limited due to position deviation caused by the target movement during the several hundreds shots that were required to collect in order to produce a high quality image. Therefore the obtained spatial resolution estimate is just an upper limit of the proposed approach.

The generation of intense, collimated monochromatic X-ray beams was presented. The X-ray source was an ultrashort laser produced plasma and the X-ray pulse duration was ~1 psec. The X-ray pulse was reflected from a spherical crystal mirror and the resulting beam divergence was better then 1 mrad. We should point out that the only tool for generation of collimated X-ray beams is X-ray laser. Nevertheless, for many applications it is useful to use the intense collimated X-ray beams described in this paper.

## Acknowledgment

Part of this work was supported by Russian Fundamental Science Foundation grant N96-02-16111.

## References

1    V.A. Boiko, A.V. Vinogradov, S.A. Pikuz, and A.Ya. Faenov.  J.Sov.Laser.Res., **6**, 851 (1985).

2    I.Yu. Skobelev, A.Ya. Faenov, B.A. Bryunetkin et al. Zhournal Experimental and Theoretical Physics, **108**, 1266 (1995).

3    A.Ya. Faenov, S.A. Pikuz, A.I. Erko et al. Physica Scripta, **50**, 333 (1994).

4    T.A.Pikuz, A.Ya.Faenov, S.A.Pikuz et el. Journal of X-ray   Science and Technology, **5**, 323 (1995).

5    A.Ya.Faenov et al. Phys.Rev.A. **51**, 3529 (1995).

6    Chukhovskii, W.Z. Chang, and E. Forster. J.Appl.Phys. **77**, 1843  (1995), 1849 (1995).

7    Disksmoller, O. Rancu, I. Uschmann, P. Renaudi, C. Chenais-Popovics, J.C. Gauthier, and E. Forster. Optics Comm. **118**, 379  (1995).

8    Forster, K. Gubel, and I. Uschmann. Laser and Particle Beams  **9**, 135 (1991).

9    Forster, E.E. Fill, K. Gabel, H. He, Th. Missalla, O.  Rennner, I. Uschmann, and J. Wark, JQSRT, **51**, 101 (1994).

10   Hockaday, M.D. Wilke, R.L. Blake, J. Vaninetti, and N.T. Gray. Rev. Sci. Instrum. **59** (8), 1822 (1988).

11   Aristov and A.I. Erko. X-ray Optics (Nauka Publishing,  Moscow, 1991).

12  Yu.A. Agafonov, A.I. Erko, B.A. Bryunetkin, A.Ya. Faenov, A.R. Mingaleev, S.A. Pikuz, V.M. Romanova, T.A. Shelkovenko, and I.Yu. Skobelev. J. Sov. Tech. Phys. Lett. **18**, 16 (1992).

13  A.I. Erko, L.A. Panchenko, S.A. Pikuz, A.R. Mingaleev, V.M. Romanova, T.A. Shelkovenko, A.Ya. Faenov, B.A. Bryunetkin, T.A. Pikuz, and I.Yu. Skobelev. Rev.Sci.Instrum., **66** (2), 1047 (1995).

14  W.Z. Chang, E. Forster. J.Appl.Phys. 78 (6) (1995).

15  A.I. Erko. Journal de Physique IV, Colloque C 9, 4, C9-245 1994.

16  Erko, Yu. Agafonov, L.A. Panchenko, A. Yakshin, P. Chevallier, P. Dhez, and F. Legrand. Optics Comm., **106**, 146 (1994).

17  Snigirev, I. Snigireva, P. Engstrom, S. Lequien, A. Suvorov, Ya. Hartman, P. Chevallier, M. Idir, F. Legrand, G. Soullie, and S. Engrand. Rev.Sci.instrum. **66** (2), 1461 (1995).

18  Ya. Hartman, E. Tarazona, P. Elleanme, I. Snigireva, and A. Snigirev. Rev. Sci. Instrum., **66** (2), 1978 (1995).

19  T.A.Pikuz, A.Ya.Faenov, E.Foerster et al "Measurements and calculations of flat and spherically bent mica crystals reflectivity and using then for different applications in the spectral range 1-19 Å. Proceedings of SPIE-95, v.**2512**, p.468-486.

20  A.Zigler, P.Mandelbaum, J.L.Schwob, and D.Mitnik. Physica Scripta, **50**, 61 (1994).

21  A. Zigler, Y.Komet , and H. Zmora Phys. Lett. **60A,** 319 (1977)

22  A.Ya. Faenov et al. Kvan. Elektron., **20**, 457 (1993).

# Investigations on Laser-Generated Plasma Sources

T. Wilhein

Georg-August-Universität Göttingen, Forschungseinrichtung Röntgenphysik,
Geiststraße 11, D-37073 Göttingen, Germany

**Abstract.** In order to measure the spectral brilliance of laser plasma X-ray sources, a spectrograph has been developed which allows simultaneous recording of the wavelength depending source diameter and the source spectrum. The optical system is a new single element X-ray optic [7], which produces a series of lateral displaced enlarged images of the X-ray source at different wavelength on the detector, a cooled slow scan CCD camera with a thinned, back illuminated CCD. First brilliance measurements of different laser plasma sources in the wavelength range $\lambda=1.5...5$nm have been performed, showing that laser plasma sources can serve as bright laboratory X-ray sources for applications in the soft X-ray region.

## 1 Introduction

During the last years, the fields of research using soft X-rays have experienced a remarkable growth. Applications like, e.g., X-ray microscopy, X-ray photoelectron spectroscopy or X-ray lithography have prooved to be powerful scientific tools [1]. Experiments utilizing these technics are mostly performed at synchrotron radiation facilities, which provide quasi-continuous highly collimated X-radiation over a wide spectral range [2]. To extend the applications of these methods, intense laboratory X-ray sources have to be developed. Laser generated plasmas promise to become bright X-ray sources, supplying pulsed radiation at wavelength down to around 1 nm using commercially available laser systems and even much shorter, if terrawatt laser pulses are employed [3, 4]. In order to decide wether a laser plasma is well suited to serve as an X-ray source for a particular application, its emission properties have to be quantified. The main figure of merit used to characterize X-ray sources is the spectral brilliance, which is defined as the number of photons emitted per unit time, source area, solid angle and relative bandwidth [2]. However, for pulsed sources it is more suitable to define the pulse brilliance $b_{pulse}$ as

$$b_{pulse} = \frac{\text{number of photons}}{\text{pulse, source area, solid angle, rel. bandwidth}}$$

Units for the pulse brilliance are photons/(pulse$\times\mu$m$^2\times$sr$\times0.1\%$ BW). The exposure time required to carry out a certain experiment can be estimated by integrating the pulse brilliance over the corresponding experimental properties, e.g. the angle of acceptance, and taking into account the repition rate of the source. To be able to measure the pulse brilliance of laser generated plasma sources in the soft X-ray range, a spectrograph has been developed which allows simultaneous measurement of the spectral distribution and the source size by means of an *off-axis reflection zone plate* (ORZ) [7] in combination with CCD image detection.

**X-Ray Microscopy and Spectromicroscopy**
Eds.: J. Thieme, G. Schmahl, D. Rudolph, E. Umbach
© Springer-Verlag Berlin Heidelberg 1998

## 2 Off-Axis Reflection Zone Plates for Spectral Imaging

A universal idea of point-to-point image formation is shown in Fig. 1 (cp.[5]). A spherical wave emitted by a point source S will converge into image point I if the surface of equal phase describing the wave deflection is given by an ellipsoid with focal points at S and I. Because phase differences being integer multiples of $2 \cdot \pi$ will result in constructive interference, a set of ellipsoidal shells corresponding to optical path differences $2 \cdot \pi \cdot \lambda$ for every two adjacent shells characterize the imaging process. Any single element non-refractive optic can be described using this scheme. For thin, plane diffractive optical elements the pattern providing the appropriate wave deflection is given by the figure which results from cutting the system of ellipsoidal shells with the substrate plane. The general shape of this figures are non-concentric ellipses, which may in special cases lead to, e.g., concentric circles, describing on-axis zone

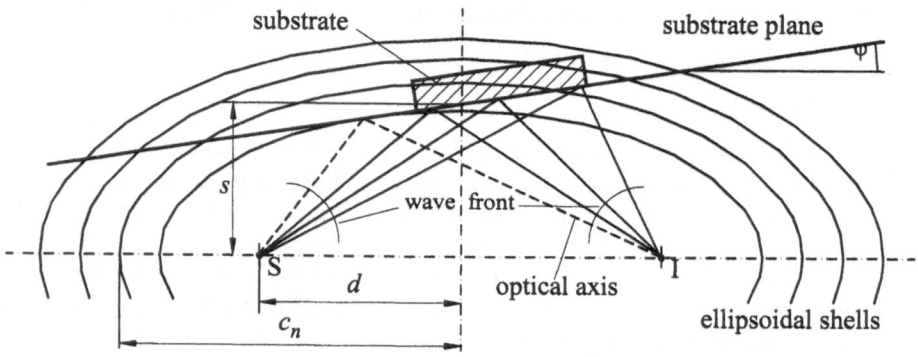

**Fig. 1.** Origin of the off-axis reflection zone plate pattern (see text)

plates. If the straight line between S and I cuts the substrate plane, the optic is operating in transmission, whereas the opposite case leads to a reflective element. For the off-axis reflection zone plate, the pattern is given (see Fig. 1 for descriptions) by

$$\frac{q^2}{a_n{}^2 - s^2 + w^2/v} + \frac{\left(p + w \cdot \sqrt{1 + m^2}/v\right)^2}{\left(1 + m^2\right)/v \cdot \left(a_n{}^2 - s^2 + w^2/v\right)} = 1$$

$p, q =$ coords in substr. plane, $m = \tan(\varphi)$, $c_n = n \cdot \lambda/2 + const$, $n = 0, 1, 2... =$ zone number

$$a_n{}^2 = c_n{}^2 - d^2, \quad v := m^2 + \frac{a_n{}^2}{c_n{}^2}, \quad w := m \cdot s, \quad \varepsilon = \frac{d}{c_n} \cdot \sqrt{\frac{1}{1 + m^2}} = \text{normalized excentricity}$$

Off-axis reflection zone plates (ORZs) [7] are manufactured by e-beam writing of the desired part of this pattern into PMMA and subsequent microstructuring steps. The ORZs used in the below described experiments consists of about 7000 line pairs, written with a LION LV1 e-beam writer (LEICA, Jena, Germany) and structured in ≈20 nm Ge on a glass substrate, in some cases coated with a thin Ni layer for enhanced reflectivity.

**Fig. 2.** Scheme of the first ORZ

**Table 1.** Design parameters of the first ORZ

| size: 2 stripes, each 1×8mm², 1mm spacing, ≈7000 line pairs, grating constant along optic axis $g$ =1131-1138nm, +1. diffr. order | | |
|---|---|---|
| wavelength | $\lambda_0$ | 2.4nm |
| object distance | $a_0$ | 1500mm |
| image distance | $b_0$ | 3000mm |
| Focal length | $f_0$ | 1000mm |
| Magnification | $M_0$ | 2 |
| Input angle (gr.inc.) | $\alpha_0$ | 1.5° |
| Output angle (gr.inc.) | $\beta_0$ | 4.0° |
| Angular disp. at $\lambda_0$ | $d\beta/d\lambda$ | 13mrad/nm |
| Linear disp. at $\lambda_0$ | $dx/d\lambda$ | 38mm/nm |
| Collected solid angle | $\Omega_0$ | $2\times10^{-7}$sr |
| Spatial diffr. limit | $\delta y$ | 3µm |
| Spectral diffr. limit | $\lambda/\Delta\lambda$ | > 1000 |
| (dep. on source size) | | |

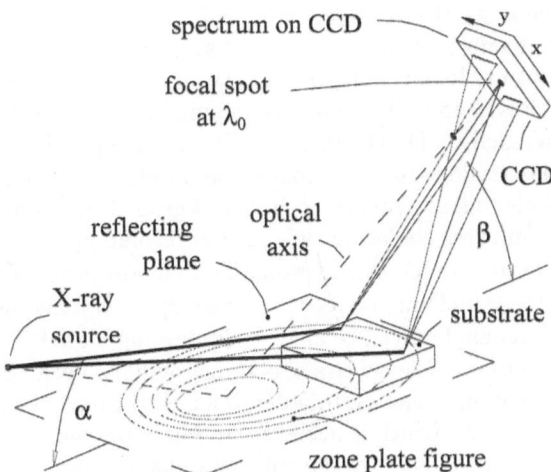

**Fig. 3.** Image formation with an off-axis reflection zone plate [7]

Detailed reports on the e-beam lithography and microfabrication processes are presented in ref. [6–8]. A scheme of the first built ORZ is given in Fig. 2, its paramters listed beside it. In a first order approximation, the optic can be treated as a diffraction grating with slightly variing grating constant $g$ on the zone plate axis and curved lines. Figure 3 explains the appearance of spectrally displacedimages: for its design paramters, that is geometry and wavelength, the ORZ forms a diffaction limited image of a point source at the location of the image point. Because the focal length of zone plates vary as $f \propto \lambda^{-1}$, longer / shorter wavelength will have their focus in front of / behind the detector plane, leading to defocused and, due to the off-axis character of the ORZ, laterally displaced images on the detector. Thus, the image consists of a spectral and a spatial direction, the latter allowing to measure the source size at the focused wavelength. The diffraction limits given in the table beside Fig. 2 for both spatial resolution and spectral resolving power takes into account the effective aperture, the spectral resolving power in addition being influenced by the average angular dispersion of the optic. The off-axis nature of the ORZ in combination with the grazing incidence properties give rise to remarkable aberrations when used in geometries different from the design geometry. A ray-tracing program has been developed to calculate the spectral and spatial behaviour of ORZs under arbitrary conditions, e.g., extended or displaced sources, changes in the input angle etc. For example, it has been found that the aberrations arising from setting the focus of the ORZ to a wavelength $\lambda$ different from the design wavelength $\lambda_0$ can be minimized by keeping input angle and source distant constant, and placing the detector in image distance corresponding to the focal length $f_\lambda = f_0 \cdot \lambda_0 / \lambda$, where $f_0$, $\lambda_0$ are design parame-ters, and with respect to the change in the diffraction angle.

## 3 The ORZ-Based Imaging Spectrograph

The spectrograph consists of three vacuum chambers, connected with vacuum tubes (Fig. 4). The first chamber contains the target system, the second the ORZ and a metal filter (Al or Ti) to block visible, IR and UV light, the third chamber carries the detector, a cooled clow scan CCD camera (Photometrics AT 200L), equipped with a thinned, back illuminated CCD (TK1024AB). The camera provides cooling down to $-40°C$, 16 bit/40 kHz readout and a readout noise of $\approx 10e^-$. The high dynamic range allows to make precise measurements of line/background intensities in the spectral images. In the present setup the achievable spatial resolution is limited by the pixel size ($24 \times 24 \ \mu m^2$) of the CCD and approx. $2 \times$ magnification to $\approx 20 \mu m$ and the spectral resolution to $\approx 0.002$ nm. To record a wider spectral range than that fitting to the CCD size, the input angle $\alpha$ can be choosen by turning the ORZ, keeping the total deflection angle $\gamma$ constant. The spectra shown below have been obtained by taking images at 3...6 different input angles. In order to adjust the optical axis of the ORZ in a way that it hits the source, which is necessary to minimize aberrations, the ORZ can be rotated precisely perpendicular to the substrate plane using a stepper motor with 80000 steps/360°.

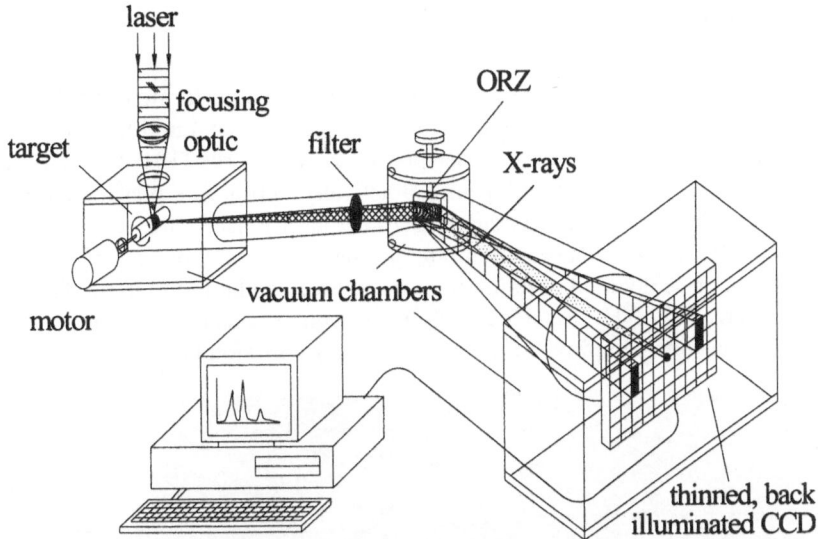

**Fig. 4.** Scheme of the imaging spectrograph [15]

Measuring absolute brilliances means determination of absolute photon numbers. To be able to do so, all components of the spectrograph have to be absolutely calibrated. The filter transmittance and the diffraction efficiency of the ORZ have been measured using the X-ray test chamber of the FE Röntgenphysik at BESSY for wavelength $\lambda=1...5$nm. The quantum efficiency of the thinned, back illuminated CCD was determined at the radiometry beamline of the Physikalisch Technische Bundesanstalt (PTB) at BESSY [9] in the wavelength region $\lambda=1.5...20$nm. ORZ diffraction efficiencies up to 6% have been obtained (at $\lambda=2.2$nm) [10], and the quantum efficiency of the CCD was found to be $\geq 0.6$.

## 4 First Experiments with the Imaging Spectrograph

Three experiments with the new imaging spectrograph are presented below, carried out in collaboration with three different groups using different laser and target systems. In each case the spectrograph was adapted to the respectively existing target chambers. For applications like X-ray microscopy or X-ray photoelectron spectroscopy strong emission in well separated lines is preferable, thus target elements are choosen to have lines from H-like and He-like ions in the desired spectral range. In the water window, $\lambda\approx2.3...4.3$nm, which is of particular interest for investigations on biological samples, nitrogen and carbon fullfill this condition and therefore targets containing these elements were choosen for the experiments. Due to the calibration procedure of the ORZ and the CCD the error in the given photon numbers can be expected to be less then 50%.

## 4.1 Experiment with a Droplet-Target Laser Plasma Source at the Lund Institute of Technology

The first experiment with the ORZ spectrograph has been performed in collaboration with H. Hertz, L. Rymell, M. Berglund at the Dep. of Physics, Lund Institute of Technology, applying the droplet-target laser-plasma X-ray source [10]. This source uses ≈10 μm droplets as target, resulting in a practically debris-free plasma source. The frequency doubled Nd:YAG laser supplies 70 mJ laser pulses with 100–120 ps pulse width at $\lambda$=532 nm and 10Hz repition rate, giving an intensity of ≈$5\cdot10^{14}$ W/cm$^2$ in a 12 μm focal spot. A detailed description of the source can be found in ref. [11, 12]. In this experiment, 30% $NH_3$ dissolved in $H_2O$ has been used as target liquid, and a 3 mJ UV prepulse has been employed for enhanced X-ray emission [13].

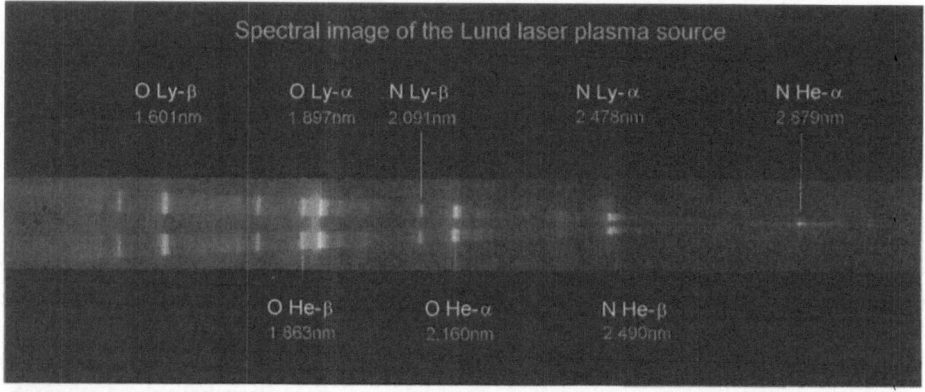

**Fig. 5.** Spectral image of the Lund laser plasma source taken with the imaging spectrograph. Focus set to 2.88 nm. Target: droplets, 30% $NH_3$ dissolved in $H_2O$, laser pulse energy 70 mJ, pulse width ≈120 ps, intensity on target ≈$5\cdot10^{14}$W/cm$^2$ [12], no prepulse

Figure 5 shows a spectral image of the Lund laser plasma source. The focus of the spectrograph was set to the nitrogen He-α line (N VI 1s$^2$-1s2p) at $\lambda$=2.879 nm, because this line is of particular interest for X-ray microscopy. Due to the target composition (only oxygen and nitrogen contribute to the X-ray emission) the spectrum below 5 nm ends with the focused line, thus the spectral image shows only defocused lines in the short wavelength direction. The picture is composed of three different exposures with variing input angle α.

A part of the spectrum taken with the imaging spectrograph is represented in Fig. 6. The expected nitrogen and oxygen lines can easily be identified. The pulse brilliances under the given experimental conditions was found to be ≈$9\times10^7$photons/(pulse×μm$^2$ ×sr×0.1%BW) in the N Ly-α line ($\lambda$=2.478nm), and ≈$5\times10^7$photons/ (pulse×μm$^2$ ×sr×0.1%BW) in the N He-α line ($\lambda$=2.879nm), the relative linewidth were measured to $\lambda/\Delta\lambda$≥350 and 450 (FWHM), respectively.

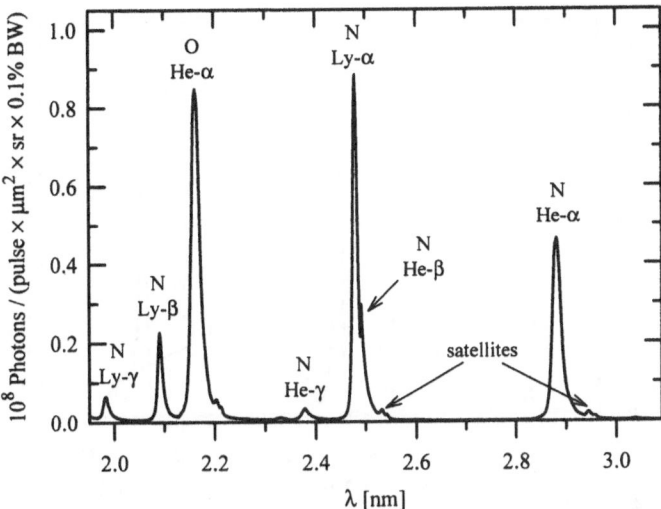

**Fig. 6.** Part of the spectrum of the Lund laser plasma source, 30% $NH_3$ dissolved in $H_2O$, laser pulse: E=70mJ, $\tau \approx$ 120ps, $\lambda$=532nm, 3mJ UV prepulse [10]

The measured linewidth is probablylimited by the effect of the extended source on the spectrograph performance. The source size of $\approx$30µm is in good agreement with other measurements and mainly influenced by the expansion of the plasma during the time delay between pre- and main pulse [13]. Integrating over the entire linewidth and the source size gives for the Ly-$\alpha$ line $4.4 \times 10^{11}$photons/(pulse×sr) and $3.2 \times 10^{11}$photons/(pulse×sr) for the He-$\alpha$ line [10].

### 4.2 Experiment with 8 ns IR Laser Pulses
### at the Fraunhofer Institut für Lasertechnik in Aachen

An experiment using $\approx$8ns IR laser pulses from a Nd:YAG laser has been carried out in collaboration with G. Schriever, S. Mager, K. Gäbel and R. Lebert at the Fraunhofer Institut für Lasertechnik in Aachen [16]. The $\approx$1000 mJ laser pulses at $\lambda$=1064 nm supplied a target intensity of $\approx 10^{13}$W/cm$^2$. Spectra have been taken from cryogenic (frozen nitrogen) and solid (boron nitride) targets, the target refresh limiting the repition rate to 0.03–1 Hz in the given experiment [17]. Figures 7 and 8 represent the measured pulse brilliances. The relatively large measured source size of $\approx$80µm corresponds to the comparatively long laser pulses and may influence the appearance of the spectral line width, as discussed above. With the frozen nitrogen target, relative linewidth in the Ly-$\alpha$ line at $\lambda$=2.478 nm and the He-$\alpha$ line at $\lambda$=2.879 nm have been found to be $\lambda/\Delta\lambda \geq$330 and 200, integration over line width and source size give $1.0 \times 10^{12}$ photons/(pulse×sr) and $2.4 \times 10^{12}$ photons/(pulse×sr), respectively.

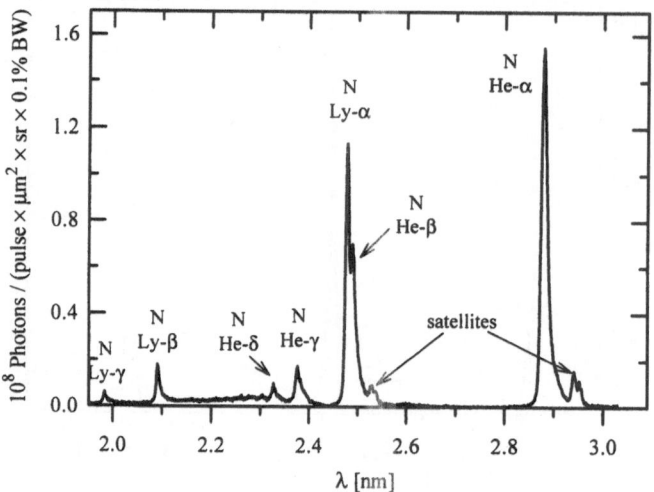

**Fig. 7.** Spectral brilliance of the Aachen laser plasma source, target: cryogenic, frozen nitrogen
laser pulse: E≈1000mJ, τ≈8ns, λ=1064nm [16]

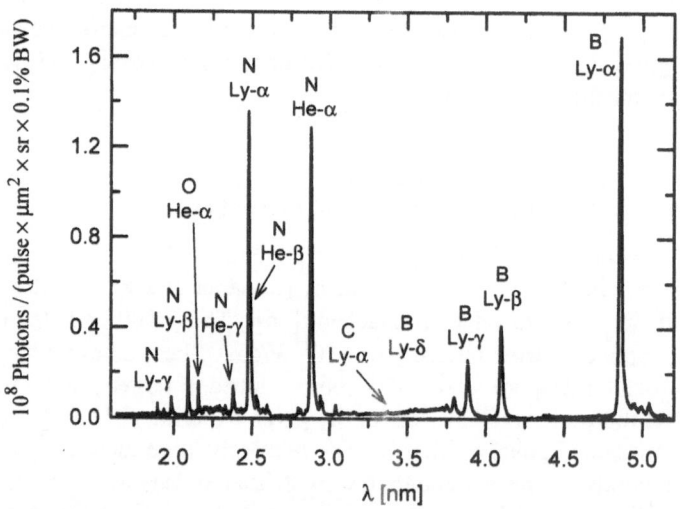

**Fig. 8.** Spectrum of the Aachen laser plasma source, target: solid, boron nitride
laser pulse: E≈1000mJ, τ≈8ns, λ=1064nm [16]

### 4.3 Experiment with High Intensity UV-Laser Pulses at the University of Jena

In collaboration with the U. Teubner, D. Altenbernd, and E. Förster, Max-Planck-Arbeitsgruppe "Röntgenoptik", W. Theobald, R. Haeßner, and R. Sauerbrey, Institut für Optik und Quantenelektronik, University of Jena, investigations on a laser plasma created with a KrF*-Laser [14] have been performed [15]. The $\approx$20mJ UV laser pulses, $\lambda$=248nm, pulse width $\approx$700fs, produced an intensity of $\approx 10^{16}$W/cm$^2$ onto the solid target (used targets: boron nitride and carbon) with a repition rate of 0.5-1Hz. Employing a carbon rod as target, strong emission in the C Ly-$\alpha$ ($\lambda$=3.373nm) and the C He-$\alpha$ ($\lambda$=4.027nm) line could be observed, the photon numbers found in these lines being $1.3 \times 10^{11}$photons/(pulse$\times$sr) and $9 \times 10^{10}$photons/(pulse$\times$sr) [15], the relative line width $\lambda/\Delta\lambda \geq$100 and 320, respectively. The relatively broad appearance of the Ly-$\alpha$ line may indicate that the high laser intensity lead to line broadening due to the Stark effect. The source size was found to be $\approx$25µm, thus, close to the limit given by spectrograph (s. above).

### 4.4 Discussion of the Experimental Results

The data from the described experiments show that it is possible to achieve pulse brilliances in the range of $10^8$photons/(pulse$\times$µm$^2\times$sr$\times$0.1%BW) in the water window with the laser plasma sources under investigation, the Lund droplet-target laser plasma source due to the small source size being the most effective in converting laser pulse energy into X-ray pulse brilliance. The absolute photon numbers calculated by integrating over line width and source size divided by the laser pulse energy gives approximately the same values for the Aachen frozen nitrogen and the Lund $NH_3/H_2O$ droplet target, where in addition to the nitrogen emission strong oxygen lines are observed. The Lund laser plasma source already operating very stable at a repition rate of 10Hz seems to be well suited to act as a bright laboratory X-ray source for applications like, e.g., X-ray microscopy, because it provides high average photon numbers in narrow bandwidth lines. Also the Aachen laser plasma source promise to become an intense X-ray source, if the repition rate can be increased. Laser pulses with very high intensities create comparatively broad line emission in the soft X-ray region, thus a monochromator may be needed if the plasma should work as a monochromatic X-ray source.

## 5 Summary

A spectrograph using off-axis reflection zone plates has been developed and tested with laser produced plasma sources in the spectral range $\lambda$=1.5...5nm. The spatial resolution has been found to be better than 25µm, the spectral resolving power $\lambda\Delta\lambda \geq$450 at 2.88nm. A ray tracing program allows to control the design of ORZs for a particular application. With the calibrated spectrograph absolute measurements of the pulse brilliance of laser generated plasma sources have been performed, the results prooving that pulse brilliances $b_{pulse} > 10^8$photons/(pulse$\times$µm$^2\times$sr$\times$0.1%BW) and integrated photon numbers $N > 10^{12}$ photons/(pulse×sr) in a single are achievable in narrow bandwidth line emission.

## Acknowledgements

The author gratefully acknowledges the excellent collaborations with all the researchers in the labs in Lund, Jena and Aachen. Thanks to B. Niemann, T. Schliebe, G. Schneider, P. Guttmann, J. Herbst, P. Nieschalk from the FE Röntgenphysik, Göttingen, R. Plontke and his team from LEICA, Jena, and G. Schmahl and D. Rudolph for their continous support of this work.
This project has been funded by the German Federal Minister for Education and Research (BMBF) under contract number 13N6491.

## References

1    See, e.g., G. Schmahl, D. Rudolph, B. Niemann, P. Guttmann, J. Thieme, G. Schneider, Naturwissenschaften **83** (1996), 61.
F. J. Wuilleumier, Y. Petroff, and I. Nenner, eds., *Vacuum Ultraviolet Radiation Physics*, Proc. of the 10th VUV conference (World Scientifc, Singapore, 1993).

2    E.-E. Koch, D. E. Eastman, and Y. Farge, *Handbook on Synchrotron Radiation* (North Holland, Amsterdam, 1983).

3    M. C. Richardson, G. A. Kyrala, eds. *Applications of Laser Plasma Radiation II*, Proc. SPIE Vol. 2523 (1995).

4    C. Tillmann, A. Persson, C.-G. Wahlström, S. Svanberg, K. Herrlin, Appl. Phys. B **61**, 333 (1995).

5    A. G. Michette and C. J. Buckley, eds. *X-Ray Science and Technology*, (Institute of Physics, Bristol, 1993).

6    T. Schliebe, *Diffraktive Kondensoroptiken für die Röntgenmikroskopie, Elektronenstrahllithographie und Nanostrukturierung*, PhD-thesis, (Göttingen, in preparation).

7    B. Niemann, T.Wilhein, T. Schliebe, R. Plontke, O. Fortagne, I. Stolberg, M. Zierbock, Microelectr. Eng. **30**, 49 (1996).

8    B. Niemann, *New X-ray optical elements generated by the electron beam lithography system LION LV1*, this volume.

9    T. Wilhein, D. Rothweiler, A. Tusche, F. Scholze, and W. Meyer-Ilse, in *X-ray Microscopy IV*, V.V. Aristov and A.I. Erko, eds. (Bogorodskii Pechatnik Publishers, Chernogolovka, Moscow region, 1994) p. 470.

10    T. Wilhein, D. Hambach, B. Niemann, M. Berglund, L. Rymell, H. M. Hertz, Appl. Phys. Lett. **71** (28), (1997).

11    L. Rymell and H. M. Hertz, Opt. Commun. **103**, 105 (1993).

12    L. Rymell, M. Berglund and H. M. Hertz, Appl. Phys. Lett. **66**, 2625 (1995).

13    M. Berglund, L. Rymell, and H. M. Hertz, Appl. Phys. Lett. **69**, 1683 (1996).

14   U. Teubner, C. Wülker, W. Theobald, E. Förster, Phys. Plasmas **2**, 972 (1995);
      U. Teubner, W. Theobald, C. Wülker, J. Phys. B **29** (1996) 4333.
      S. Szatmari, F. P. Schäfer, Opt. Comm. **68**, 196 (1988).
15   T. Wilhein, R. Häßner, D. Altenbernd, U. Teubner, W. Theobald, E. Förster,
      R. Sauerbrey, *X-ray brilliance measurements of a subpicosecond laser-plasma
      using an elliptical off-axis reflection zone plate,* subm. to Opt. Lett.
16   T. Wilhein, G. Schriever, S. Mager, K. Gäbel, R. Lebert, to be published.
17   R. Lebert, G. Schriever, S. Mager, A. Naweed, O. Treichel, K. Bergmann,
      W. Neff, *Laser Produced and Pinch Plasmas: Narrowband X-Ray sources for
      Applications.* X-Tech 96 Workshop, Berlin, Sept. 29–Oct. 2, 1996.